Remediation *of* Firing Range Impact Berms

Remediation *of* Firing Range Impact Berms

Edited by

Donald F. Lowe
Rice University, Houston, TX

Karen L. Duston
Rice University, Houston, TX

Carroll L. Oubre
Rice University, Houston, TX

C. Herb Ward
Rice University, Houston, TX

Authors
Douglas A. Hlousek
Thomas A. Phillips

CRC Press
Taylor & Francis Group
Boca Raton London New York

CRC Press is an imprint of the
Taylor & Francis Group, an **informa** business

First published 2000 by Lewis Publishers

Published 2019 by CRC Press
Taylor & Francis Group
6000 Broken Sound Parkway NW, Suite 300
Boca Raton, FL 33487-2742

ISBN 13: 978-0-367-45551-4 (pbk)
ISBN 13: 978-1-56670-462-5 (hbk)

Visit the Taylor & Francis Web site at
http://www.taylorandfrancis.com

and the CRC Press Web site at
http://www.crcpress.com

Library of Congress Cataloging-in-Publication Data
Hlousek, Douglas A.
Remediation of firing range impact berms / edited by Donald F. Lowe ... [et al.]; authors, Douglas A. Hlousek, Thomas A. Phillips.
p.   cm. — (AATDF monographs)
Includes bibliographical references and index.
ISBN 1-56670-462-6 (alk. paper)
1. Soil remediation — Technological innovations —United States. 2. Rifle ranges — Environmental aspects — United States. 3. Soils — Lead content — Environmental aspects — United States. 4. United States — Armed forces —Environmental aspects. I. Phillips, Thomas A. II. Lowe, Donald F. III. Title. IV. Series.
TD878.2 .H58 2000
628.5'5—dc21   99-086802
CIP

Library of Congress Card Number 99-086802

Although the information described herein has been funded wholly or in part by the United States Department of Defense (DOD) under Grant No. DACA39-93-0001 to Rice University for the Advanced Applied Technol-ogy Demonstration Facility for Environmental Technology Program (AATDF), it may not necessarily reflect the views of the DOD or Rice University, and no official endorsement should be inferred.

Foreword

To accomplish its mission, the Advanced Applied Technology Demonstration Facility (AATDF) at Rice University selected 11 projects that involved the quantitative demonstration of innovative remediation technologies. Remediation of Lead Contaminated Soils at Small Arms Firing Ranges, Using Mining and Beneficiation Technologies (Lead Remediation) was one of the funded projects.

The primary objective of the Lead-Remediation project demonstration was to provide reliable, detailed performance data to evaluate the feasibility and cost of implementation of a full-scale system. An important secondary objective was to demonstrate the effectiveness of the technology in producing a "clean" soil based on analytical results. The project included the following steps:

- Bench scale treatability studies of prospective soils
- Selection of a demonstration site
- Pilot plant equipment design, installation, and shakedown
- Execution of the demonstration at the selected DOD site
- Preparation of the Final Technical Report (FTR) and the Technology Evaluation Report (TER)

The original Principal Investigators for the Lead Remediation project were:

- Mr. W. Richard McDonald at the Salt Lake Research Center (SLRC) of the U.S. Bureau of Mines (BOM) in Salt Lake City, Utah
- Ms. Barbara M. Nelson at the Naval Facilities Engineering Service Center (NFESC) in Port Hueneme, California

However, the pilot plant was fabricated and operated by former BOM personnel assigned to the U.S. Bureau of Reclamation (BOR) at the Salt Lake Research Center. The Principal Investigators assigned by the BOR to complete the project were:

- Mr. Thomas Phillips
- Dr. Jerry Montgomery

Ms. Nelson remained as an advisor to AATDF. Throughout the project, Mr. Mark Bricka at the U.S. Army Engineer Waterways Experiment Station (WES) provided technical advice to AATDF.

Metcalf & Eddy, Inc. (Metcalf & Eddy) acted as the engineering consultant for all phases of this project. The treatability studies associated with this project were conducted by the Colorado Minerals Research Institute (CMRI) under the direction of Mr. John E. Litz. The analytical data for the samples collected during the demonstration were completed by Curtis & Thomkins, Ltd.

The project team carried out the Lead Remediation demonstration at Naval Air Station Miramar (NAS Miramar) in San Diego, California. Ms. Martha F. Gandy was the Director of the Environmental Protection Division at NAS Miramar and Mr. Kyle Nguyen was NAS Miramar's project contact for the AATDF project team. The NAS's Safety Officer was Mr. Carlos Vigo.

The fieldwork at NAS Miramar began on August 18, 1996 and ended on September 25, 1996. The pilot plant was operated at the demonstration site by BOR personnel. Soil preparation, demonstration assistance, and health and safety were provided by Metcalf & Eddy. The team that carried out the demonstration at NAS Miramar consisted of the following personnel:

- Mr. Thomas A. Phillips, Dr. Jerry Montgomery, and Mr. Christian R. Ferguson
- Mr. Thomas P. Young of Metcalf & Eddy
- Mr. Michael F. Warminsky of Metcalf & Eddy and Brice Environmental Systems

AATDF's Assistant Program Manager was Dr. Donald F. Lowe, who provided invaluable technical assistance to the BOR and Metcalf & Eddy personnel.

The field data, calculated results, mass and energy balances, conclusions, and recommendations from the Lead Remediation demonstration were initially reported in the FTR by the team from the BOR and Metcalf & Eddy. Metcalf & Eddy prepared the hypothetical full-scale plant design and project costs in the TER. The FTR, TER, and this monograph were co-authored for the AATDF program by Mr. Douglas A. Hlousek of Metcalf & Eddy and Mr. Thomas A. Phillips of the BOR. Besides the authors, the major contributors to the completion of the FTR were as follows:

- Mr. Thomas P. Young
- Dr. Jerry Montgomery
- Mr. Christian R. Ferguson

The authors worked closely with the AATDF staff and a team of Project Shepherds who participated in both design and review of the FTR and TER. The Project Shepherds were:

- Mr. Michael F. Warminsky
- Ms. Barbara M. Nelson
- Dr. Irene A. Legiec, DuPont Corporation

Special recognition goes to Ms. Leslie Karr, NFESC, and Mr. Phillip D. Walling, DuPont Corporation.

The AATDF Technology Advisory Board (TAB) and other Advisory Committees provided valuable assistance at the start and during the project in such areas as technology selection, directional changes in projects, technology transfer, and reporting. Mr. Richard Conway played a pivotal role in facilitating the publication process.

The AATDF Program was funded by the United States Department of Defense under Grant No. DACA39-93-1-0001 to Rice University. The Program was given oversight by the U.S. Army Engineer Waterways Experiment Station in Vicksburg, Mississippi.

AATDF Monographs

This monograph is one of a ten-volume series that records the results of the AATDF Program:

- Surfactants and Cosolvents for NAPL Remediation: A Technology Practices Manual
- Sequenced Reactive Barriers for Groundwater Remediation
- Modular Remediation Testing System
- Phytoremediation of Hydrocarbon-Contaminated Soil
- Steam and Electroheating Remediation of Tight Soils
- Soil Vapor Extraction Using Radio Frequency Heating: Resource Manual and Technology Demonstration
- Subsurface Contamination Monitoring Using Laser Fluorescence
- Remediation of Firing Range Impact Berms
- Reuse of Surfactants and Cosolvents for NAPL Remediation
- NAPL Removal: Surfactants, Foams, and Microemulsions

Advanced Applied Technology Demonstration Facility (AATDF)
Energy and Environmental Systems Institute MS-316
Rice University
6100 Main Street
Houston, TX 77005-1892

U.S. Army Engineer
Waterways Experiment Station
3909 Halls Ferry Road
Vicksburg, MS 39180-6199

Rice University
6100 Main Street
Houston, TX 77005-1892

Preface

Following a national competition, the Department of Defense (DOD) awarded a $19.3 million grant to a university consortium of environmental research centers led by Rice University and directed by Dr. C. Herb Ward, Foyt Family Chair of Engineering. The DOD Advanced Applied Technology Demonstration Facility (AATDF) Program for Environmental Remediation Technologies was established on May 1, 1993 to enhance the development of innovative remediation technologies for the DOD by facilitating the process from academic research to full-scale utilization. The AATDF's focus is to select, test, and document performance of innovative environmental technologies for the remediation of DOD sites.

Participating universities include Stanford University, The University of Texas at Austin, Rice University, Lamar University, University of Waterloo, and Louisiana State University. The directors of the environmental research centers at these universities serve as the Technology Advisory Board (TAB). The U.S. Army Engineer Waterways Experiment Station manages the AATDF Grant for the DOD. Dr. John Keeley is the Technical Grant Officer. The DOD/AATDF is supported by five leading consulting engineering firms: Remediation Technologies, Inc., Battelle Memorial Institute, GeoTrans, Inc., Arcadis Geraghty and Miller, Inc., and Groundwater Services, Inc., along with advisory groups from the DOD, industry, and commercialization interests.

Starting with 170 preproposals that were submitted in response to a broadly disseminated announcement, 12 projects were chosen by a peer-review process for field demonstrations. The technologies chosen were targeted at the DOD's most serious problems of soil and ground water contamination. The primary objective was to provide more cost-effective solutions, preferably using *in situ* treatment. Eight projects were led by university researchers, two projects were managed by government agencies, and two others were conducted by engineering companies. Engineering partners were paired with the academic teams to provide field demonstration experience. Technology experts helped guide each project.

DOD sites were evaluated for their potential to support quantitative technology demonstrations. More than 75 sites were evaluated in order to match test sites to technologies. Following the development of detailed work plans, carefully monitored field tests were conducted and the performance and economics of each technology were evaluated.

One AATDF project designed and developed two portable Experimental Controlled Release Systems (ECRS) for testing and field simulations of emerging remediation concepts and technologies. The ECRS is modular and portable and allows researchers, at their sites, to safely simulate contaminant releases and study remediation techniques without contaminant loss to the environment. The completely contained system allows for accurate material and energy balances.

The results of the DOD/AATDF Program provide the DOD and others with detailed performance and cost data for a number of emerging, field-tested technologies. The program also provides information on the niches and limitations of the technologies to allow for more informed selection of remedial solutions for environmental cleanup.

The AATDF Program can be contacted at: Energy and Environmental Systems Institute, MS-316, Rice University, 6100 Main, Houston, TX, 77005, phone 713-527-4700; fax 713-285-5948; or e-mail <eesi@rice.edu>.

The DOD/AATDF Program staff include:

Director:
  Dr. C. Herb Ward
Program Manager:
  Dr. Carroll L. Oubre
Assistant Program Manager:
  Dr. Kathy Balshaw-Biddle
Assistant Program Manager:
  Dr. Stephanie Fiorenza

Assistant Program Manager:
  Dr. Donald F. Lowe
Financial/Data Manager:
  Mr. Robert M. Dawson
Publications Coordinator/Graphic Designer:
  Ms. Mary Cormier
Meeting Coordinator:
  Ms. Susie Spicer

This volume, *Remediation of Firing Range Impact Berms*, is one of a ten-monograph series that records the results of the DOD/AATDF environmental technology demonstrations. Many have contributed to the success of the AATDF program and to the knowledge gained. We trust that our efforts to fully disclose and record our findings will contribute valuable lessons learned and help further innovative technology development for environmental cleanup.

Donald F. Lowe
Karen L. Duston
Carroll L. Oubre
C. Herb Ward

AATDF Authors and Editors

Douglas A. Hlousek

Douglas A. Hlousek is a senior project manager with Metcalf & Eddy, Inc. He has over 25 years of experience in construction management, project management, project controls, and estimating. His expertise is in design-build construction, hazardous waste site remediation, underground storage tank remediation, earthwork, site drainage, water distribution, and sewage disposal projects. He has a B.S. in general engineering from the United States Coast Guard Academy, and an A.S. in business management from Fisher College. He received a certification in environmental hazardous waste management at Northeastern University. He is a technical director for Metcalf & Eddy's NoVOCs innovative technology, providing project control support for the design, pricing structure, and scheduling of NoVOCs in-well stripping projects nationwide. Mr. Hlousek was the program manager for the technology evaluation for Remediation of Lead Contaminated Soils at Small-Arms Firing Ranges Using Mining and Beneficiation Technologies. Mr. Hlousek has presented papers and published articles in the areas of remediation and innovative technologies.

Thomas A. Phillips

Thomas A. Phillips is the President of Hydro GeoTech, Inc. where he is a consultant involved in applying technical solutions to environmental problems at industrial plants and military bases. He has a B.S. in chemical engineering from the University of Maryland. Since 1973, Mr. Phillips has worked in a variety of capacities while employed by the U.S. Bureau of Mines (BOM). In his most recent position, he was research supervisor for the environmental program at the Salt Lake Research Center. In this position, Mr. Phillips supervised 35 researchers who investigated mineral-related environmental issues. He also led or was part of interagency teams involved in acid-mine drainage using water treatment technologies, mine closures, military base closures, and BOM realignment and environmental planning. Mr. Phillips' experience includes serving as a staff engineer where he was responsible for several major programs for primary minerals and recycling and for long range planning, reviewing research plans, and assessing their progress. He has also served as a special assistant to the BOM Director of the Solidad Canyon Project. Mr. Phillips has published several articles and many internal documents involving mineral industry and remediation technologies.

Donald F. Lowe

Donald F. Lowe is an Assistant Program Manager with AATDF at Rice University where he managed four projects involving the field demonstration of innovative technologies. Dr. Lowe has a Ph.D. in metallurgy-chemical engineering from the University of Arizona, a M.S. in metallurgical engineering from the University of Wisconsin-Madison, and a B.S. in mining engineering from the University of North Dakota. In his capacity as project manager for AATDF, Dr. Lowe provided the necessary managerial guidance and technical expertise to bring each project through a successful demonstration. He has also been an active participant in the preparation of the reports for each project. Since 1986, Dr. Lowe has been involved as a technical manager, proposal manager, and senior engineer with several environmental firms. His management responsibilities included cost estimation of remediation projects and economic feasibility studies for numerous processes. He also has provided technical guidance for many remediation projects. Prior to 1986, Dr. Lowe was employed for 25 years as a research supervisor or research engineer with four primary mining or metals recycling companies. He has several patents and publications that are related to metals extraction and recycling processes.

Karen L. Duston

Karen L. Duston is a Research Scientist in the Energy and Environmental Systems Institute (EESI) at Rice University. As a technical reviewer and editor, Dr. Duston has been active in the preparation of the reports for the AATDF projects. She has held the position of Coordinator of the Superfund University Training Institute (SUTI), a federally funded program offering environmental assessment and remediation training to state and federal employees, and as manager of the biological remediation laboratory at Rice University. She was responsible for the daily operation of SUTI and coordinated government, academic, and industrial efforts to offer and prepare instructional materials for the courses. Dr. Duston has a Ph.D. in environmental science and engineering from Rice University, a M.S. in microbiology from Texas A&M University, and a B.S. in chemistry and in metallurgical engineering from the University of Texas at El Paso.

Carroll L. Oubre

Carroll L. Oubre is the Program Manager for the DOD/AATDF Program. As Program Manager he is responsible for the day-to-day management of the $19.3 million DOD/AATDF Program. This includes guidance of the AATDF staff, overview of the 12 demonstration projects, assuring project milestones are met within budget, and that complete reporting of the results are timely.

Dr. Oubre has a B.S. in chemical engineering from the University of Southwestern Louisiana, a M.S. in chemical engineering from Ohio State University, and a Ph.D. in chemical engineering from Rice University. He worked for Shell Oil Company for 28 years, with his last job as Manager of Environmental Research and Development for Royal Dutch Shell in England. Prior to that, he was Director of Environmental Research and Development at Shell Development Company in Houston, Texas.

C. H. (Herb) Ward

C. H. (Herb) Ward is the Foyt Family Chair of Engineering in the George R. Brown School of Engineering at Rice University. He is also Professor of Environmental Science and Engineering and Ecology and Evolutionary Biology.

Dr. Ward has undergraduate (B.S.) and graduate (M.S. and Ph.D.) degrees from New Mexico State University and Cornell University, respectively. He also earned the M.P.H. in environmental health from the University of Texas.

Following 22 years as Chair of the Department of Environmental Science & Engineering at Rice University, Dr. Ward is now Director of the Energy and Environmental Systems Institute (EESI), a university-wide program designed to mobilize industry, government, and academia to focus on problems related to energy production and environmental protection.

Dr. Ward is also Director of the Department of Defense Advanced Applied Technology Demonstration Facility (AATDF) Program, a distinguished consortium of university-based environmental research centers supported by consulting environmental engineering firms to guide selection, development, demonstration, and commercialization of advanced applied environmental restoration technologies for the DOD. For the past 18 years, he has directed the activities of the National Center for Ground Water Research (NCGWR), a consortium of universities charged with conducting long-range exploratory research to help anticipate and solve the nation's emerging ground water problems. He is also Co-Director of the EPA-sponsored Hazardous Substances Research Center/South & Southwest (HSRC/S&SW), whose research focus is on contaminated sediments and dredged materials.

Dr. Ward has served as President of both the American Institute of Biological Sciences and the Society for Industrial Microbiology. He is the founding and current Editor-in-Chief of the international journal *Environmental Toxicology and Chemistry*.

AATDF Advisors

Acronyms and Abbreviations

μm or μ	micrometer or micron, 1×10^{-6} m
%	weight percent (unless otherwise specified), grams (solid, liquid, lead) per 100 grams (solid or slurry)
%(wb)	weight percent, grams solid per 100 grams solid
%V	volume percent, volume of solid per 100 volumes of slurry
$	U.S. dollar
AATDF	Advanced Applied Technology Demonstration Facility for Environmental Technology at Rice University
A_{eff}	effective screen area for trommel screen, m^2 (ft^2)
A_{pool}	spiral classifier pool area, m^2 (ft^2)
ARAR	Applicable or Relevant and Appropriate Requirements
BOM	U.S. Bureau of Mines
BOR	U.S. Bureau of Reclamation
C	concentration, mole/L
CAA	Clean Air Act
CERCLA	Comprehensive Environmental Response, Compensation, and Liability Act
cf	cubic feet
cfm	cubic feet per minute
CFR	Code of Federal Regulations
CIH	Certified Industrial Hygienist
COE	U.S. Army Corps of Engineers
CONS	concentrate
CMRI	Colorado Minerals Research Institute
cm	centimeter
CWA	Clean Water Act
D	diameter (m, cm)
DOD	U.S. Department of Defense
EESI	Energy and Environmental Systems Institute, Rice University
ft	foot (feet)
ft^2	square foot (feet)
ft^3	cubic foot (feet)
FTR	Final Technical Report
g	gram
gal	gallon
gpm	gallons per minute
Hp	horsepower
hr, hrs	hour, hours
HSWA	Hazardous and Solid Waste Amendments to RCRA
I	ionic strength
ID	inside diameter, m (ft)
in.	inch
I/O	input-output
IX	ion exchange
kg	kilogram
kg/hr	kilograms per hour
kt	kilotonne or 1,000,000 kg
kt/hr	kilotonne per hour
kW	kilowatt
L	liter
L_B	length of trommel blind section, m (in.)
Ln	length, m or cm (in.)
L_S	length of trommel screen section, m (in.)
L_T	total length of trommel, m (in.)

L/hr	liters per hour
lb	pound
lb/hr	pounds per hour
LDR	Land Disposal Restriction
m	meter
m^2	square meter
m^3	cubic meter
m^3/hr	cubic meters per hour
ma	milliamps
M&E	Metcalf & Eddy, Inc.
mesh	number of screen openings per inch, U.S. Standard Series
mids	middlings
min	minute
mg	milligram
mg/kg	milligrams per kilogram
mg Pb/kg	milligrams lead per kilogram
mg/L	milligrams per liter
mm	millimeter
NA	Not Applicable
NAS	Naval Air Station
NASA	National Aeronautics and Space Administration
NCP	National Contingency Plan
NFESC	Naval Facilities Engineering Service Center
O&M	operations and maintenance
OSHA	Occupational Safety and Health Administration
PCB	polychlorinated biphenyls
PI	Principal Investigator
PPE	personal protective equipment
SARA	Superfund Amendments and Reauthorization Act
SG	specific gravity (kg/L)
SLRC	Salt Lake Research Center
TAILS	tailings
ton	short ton or 2000 lb
tonne	metric unit for 1000 kg
t	tonne or 1000 kg
TCLP	Toxicity Characteristic Leaching Procedure
TER	Technology Evaluation Report
USAE	U.S. Army Engineer
V_T	total volume, m^3 or L (ft^3 or gal)
WBS	Work Breakdown Structure (FRTR, 1995)
WES	U.S. Army Engineer Waterways Experiment Station
yd	yard (3 ft)
yd^3	cubic yard (27 ft^3)
Z	ionic charge

Contents

Appendices

List of Figures

List of Tables

Executive Summary

The Department of Defense (DOD) funded the Advanced Applied Technology Demonstration Facility (AATDF) in 1993 with the mission of enhancing the development of innovative remedial technologies for the DOD by bridging the gap between academic research and proven technologies. To accomplish its mission, the AATDF selected projects that involved the quantitative demonstration of innovative remediation technologies. Field demonstrations were completed for these projects. To further assist with the commercialization and dissemination of information, the AATDF prepared monographs for several of the remediation technologies.

This monograph discusses the results of a soil washing and leaching project, "Remediation of Lead Contaminated Soils at Small Arms Firing Ranges Using Mining and Beneficiation Technologies." The project consisted of performing bench scale treatability studies to evaluate sites and to optimize the design of a combined soil wash and leach pilot plant; conducting the demonstration; designing a full-scale treatment plant; and evaluating the technical and economic merits of the treatment processes.

The primary objective of the pilot test program was to provide reliable, detailed performance data to evaluate the feasibility and cost for implementing full-scale soil wash and leach systems. The secondary project objectives were to:

- Establish process design procedures for particle sizing and gravity separation to remove heavy metal contaminants from firing range berm soils;
- Test the effectiveness of vat leaching to produce a clean soil;
- Test materials handling methods and equipment designed to decrease soil particle sizes while minimizing the amount of slimes feeding into leach systems;
- Provide data and information needed to establish equipment design and selection parameters for a full-scale treatment plant;
- Provide information for developing a comprehensive operating plan for a commercial size plant, including health and safety, and industrial hygiene;
- Demonstrate the effectiveness of the technology based on an assessment of sampling and analytical results;
- Obtain a processed soil that has a total lead concentration <500 mg/kg; and
- Produce a clean soil that has a TCLP below the lead criteria of 5.0 mg/L.

The bench-scale treatability study was completed in May, 1996 by the Colorado Mineral Research Institute (CMRI), Golden, Colorado. Design, fabrication, and shakedown of pilot plant equipment were conducted at the former U.S. Bureau of Mines (BOM), Salt Lake Research Center (SLRC), Salt Lake City, Utah by U.S. Bureau of Reclamation (BOR) personnel. The site selected for the demonstration was Naval Air Station Miramar (NAS Miramar), San Diego, California. The pilot plant demonstration was conducted from August 18 to September 25, 1996.

The pilot plant used in the demonstration at NAS Miramar was designed to process about 907 kg/hr (0.907 t/hr) of firing range soils or 2000 lb/hr (1.0 ton/hr). Over a 7-day period, the pilot plant processed 50 t (55 tons) of firing range soil.

The project demonstrated that a combined gravity separation and leach system could remove lead contamination from firing range soil. The combined system can also decrease the total lead content below 500 mg/kg and the TCLP lead to below 5 mg/L. However, the pilot plant, using available equipment, was not optimally configured. Several pieces of the process equipment were not properly sized, and the unit operations lacked the flexibility to alter the process flowsheet during the demonstration. However, sufficient data and information were collected to:

- Establish procedures for the selection of unit operations and completion of process designs to effectively treat contaminated firing range soils;

- Establish specifications for the design and operation of a commercial soil washing plant; and
- Prepare a comprehensive operating plan that would include health, safety, and industrial hygiene parameters.

BASE SELECTION

Three military installations with firing ranges were proposed as possible demonstration sites. The three bases were U.S. Army Base Fort Dix (Fort Dix), Naval Air Station Miramar, and Marine Corps Air Ground Combat Command Twentynine Palms (Twentynine Palms). Detailed particle size and gravity separation treatability tests were performed on representative samples from each base. Additional leaching studies were also completed on gravity-treatment fractions from NAS Miramar and Twentynine Palms soils, in which the total lead concentrations exceeded 500 mg/kg. After evaluating the treatability test results and site location factors, NAS Miramar was selected for the pilot plant demonstration.

CONCLUSIONS

Gravity separation combined with vat leaching is one technology that can be successfully applied for the remediation of lead contaminated soils at small arms firing ranges. The demonstration at NAS Miramar illustrated that it was possible to design and construct a soil washing pilot plant, which was mobile, required minimum setup time, and required only three to four operators.

The treatability study of NAS Miramar soil samples showed that gravity separation, vat leaching, and agitation leaching easily decreased the lead concentration of the soil to less than 500 mg/kg. The study also established a workable treatment flowsheet that consisted of the following unit operations:

- Soil deagglomeration and separation of gravel-size particles and lead bullets in an attrition scrubber,
- Separation of coarse and sand soil particles in a trommel screen,
- Removal of coarse lead particles from coarse soil particles in a mineral jig,
- Separation of the sand fraction from slimes fractions in two spiral classifiers,
- Removal of fine lead particles from the sand fraction in a two-stage Reichert spiral concentrator,
- Separation of clay-size particles from the slimes fraction in a hydrocyclone to produce an acceptable feed for a Knelson bowl concentrator,
- Removal of lead particles from the slimes fraction in the Knelson concentrator,
- Separation of the slimes particles from water in two roll-off boxes (thickeners), and
- Extraction of residual lead from coarse and sand fractions in a vat leach system.

Using the flowsheet, a mass balance was derived for a pilot plant that could successfully treat 0.9 t/hr (1.0 ton/hr) of soil. The mass balance was used to obtain required design specifications and capacities of the equipment selected for the pilot plant flowsheet.

The gravity separation and the vat leaching systems for the pilot plant were each constructed on a flatbed trailer. The pilot plant required only 2 days for setup and 2 days for startup. These factors illustrated that a soil washing pilot plant can be constructed and placed into operation rapidly, reducing the initial costs.

During the demonstration, the preferred operating rate for the pilot plant was between 0.9 to 1.8 t of soil/hr (1.0 and 2.0 tons/hr). At lower rates, heavy particles in the slurry lines tended to settle out, plugging flow meters and piping. Operation at higher rates excessively overloaded low capacity equipment, especially the trommel screen and the spiral classifiers.

The particle sizing and gravity separation unit operations were effective in removing free lead particles from firing range soils. The gravity separation units included the mineral jig, Reichert spiral concentrators, and a Knelson bowl concentrator. This result was expected, since the high specific gravity of lead indicated that it can be easily separated from lighter soil particles by gravity methods. However, it was quite evident that some of the pilot plant equipment was not suitable for use in the pilot plant demonstration at NAS Miramar. Improperly sized equipment and lack of flowsheet flexibility affected both plant operation and test data. Therefore, the design specifications and capacities were compared to the actual values for the equipment installed in the pilot plant. This comparison showed that the operating capacities for trommel screen and spiral classifiers were below their design capacities to treat 0.9 t/hr (1.0 ton/hr) of NAS Miramar soil. The pilot plant hydrocyclone was oversized, which affected the operation of the Knelson concentrator to produce a low-grade lead tailings and high-quality lead concentrate. The proper operation of the attrition scrubber and Reichert concentrators were also affected with soil treatment rates above 0.9 t/hr (1.0 ton/hr). However, the remaining pilot plant equipment could easily treat up to 1.8 t/hr (2.0 ton/hr) of soil. These results emphasized the need for treatability tests to optimize the process flows and equipment arrangement and capacities before assembling pilot plant equipment.

The pilot data illustrated that the attrition scrubber was not completely effective in attaining the required lead bullet and coarse soil separation and transport. By design, a small fraction of soil within the feed material was too large to be suspended by the agitators. The large soil particles settled to the bottom of the scrubber. An auger removed both the lead bullets and coarse soil from the bottom of the scrubber. Unfortunately, the auger also ground and shredded the bullets into finer particles. The auger discharge material was fed into the trommel or the spiral classifier for recovery of lead by the mineral jig, Reichert spiral concentrators, and Knelson bowl concentrator. However, the additional load of fine lead particles caused some operational problems in the spiral and Knelson bowl concentrators.

The treatability study data pointed out that for soils with similar characteristics, the pilot plant should have achieved particle-size separations similar to those in the treatability study, and then concentrated the sized streams using gravity separation unit operations. Data from the treatability study also indicated that if the pilot plant were able to make size separations similar to those made in the treatability study, three clean soil fractions would have been produced. Generally, the unit operations in the pilot plant produced results similar to the treatability data. In fact, the Knelson bowl concentrator produced cleaner tailings than were indicated by the treatability data.

Vat leaching pilot plant tests were conducted during the pilot demonstration, but the results were inconclusive due to early equipment failure. However, the partial data was sufficient to demonstrate that leaching occurred and then the combined cementation-ion exchange system permitted the recovery of dissolved lead and recycle of the leach solution. Several samples of the leach solution leaving the cementation-ion exchange columns had lead levels below detection limits.

The agitation leach tests on slimes fractions in the treatability study showed that a combination of aggressive leaching conditions and multiple leaching stages were effective in decreasing total lead concentrations to below 500 mg Pb/kg. Since elemental lead is inert, it must be oxidized before leaching. Hydrogen peroxide was used to oxidize the lead, while nitric acid and hydrochloric acid were used to leach the oxidized lead and control pH. Although nitric acid can also oxidize lead, the fine soil fraction required hydrogen peroxide as the oxidant to yield a low lead content in the leach residue.

RECOMMENDATIONS

The goal of the pilot plant demonstration was to test various unit operations and site-specific conditions and to provide process selection, design, and scale-up information. Soil washing plants are not "off-the-shelf" plants; rather, they are a series of individual unit operations arranged in the most effective and efficient manner based on site-specific conditions. The individual unit operations must therefore be modularized to allow maximum flexibility.

The treatability study is the most critical step undertaken for soil washing demonstration studies or full-scale operations. The arrangement of the individual unit operations depends on the results of the detailed treatability study. If the arrangement of unit operations is incorrect, a pilot plant or full-scale operation will not reproduce the treatability results, nor will it accomplish the removal of lead as required.

For successful soil washing at small arms ranges, the following procedures are recommended:

1. Collect sufficient soil samples that represent the area and depth of the small arms firing range to be treated;
2. Conduct a detailed treatability study including particle size and contaminant distribution, gravity separation and analysis, attrition scrubbing, and leaching, if necessary;
3. Evaluate the treatability study results to determine the optimized process flowsheet to produce the maximum volume of clean soil. The objective is to minimize the amount of soil that requires leaching or other non-gravity treatment;
4. Establish the processing rate that can economically treat the soil;
5. Use the established design procedures to prepare a process flowsheet and to select equipment that can treat the soil at the required processing rate;
6. Arrange individual, modularized unit operations based on the optimized process flowsheet;
7. If necessary, perform a soil washing demonstration at the small arms range; and
8. Select and assemble properly sized modularized equipment for full-scale treatment.

The AATDF pilot plant provided valuable data and information and proved the validity of the soil washing technology. Evaluation of the performance of the pilot plant resulted in the following recommendations:

1. Off-the-shelf process equipment should be used wherever possible. Detailed operational data for the equipment are available, and maintenance and repair are much easier to perform.
2. For treated soils, all clean tailings streams should be combined before final sampling.
3. The operation of the pilot plant required more flexibility. In the demonstration pilot plant, the individual unit operations were arranged without any flexibility to make changes in the treatment processes that could produce size separations necessary to generate clean soil fractions by gravity separation unit operations. Flowsheet development, plant design, and equipment configuration must optimize modularization concepts.
4. The attrition scrubber in the pilot plant should have been replaced with alternative size and gravity separation equipment. The first unit operation in the pilot plant should be a 6.3 mm (1/4 in) vibrating wet screen that would have removed the coarse lead and soil particles. At a minimum, two mineral jigs were needed in the demonstration pilot plant. The +6.3 mm fraction would have been treated in the first mineral jig. The second mineral jig would have treated the +1.19 mm (+16 mesh) fraction from the trommel screen.
5. If needed, a commercial attrition scrubber could be used to treat the -6.3 mm (-1/4 in) fraction from the vibrating screen to remove fine lead particles and contamination from the soil particles before entering the trommel.
6. Larger spiral classifiers were needed to handle the -1.19 mm (-16 mesh) fraction from the trommel.
7. Water removal (dewatering) equipment for the -105 μm (-150 mesh) soil fraction was needed. The dewatering system should have contained a clarifier and centrifuge or a plate and frame pressure filter capable of handling the water requirements. Dewatered soil materials should contain about 60% solids so that they can be easily stacked on a daily basis.
8. The process control platform with all ancillary equipment needed a separate circuit breaker from the process equipment.
9. An industrial computer case with a more robust power supply and an air cooling-filtering system should be considered for future systems.
10. All equipment must be sized based on the maximum capacity of the pilot plant, at least 1.8 t/hr (2.0 tons/hr).

PILOT PLANT EQUIPMENT ASSESSMENT

The majority of equipment that was used for the pilot plant worked. Some equipment was very old and did not operate efficiently. Because of age or inadequate capacity, pilot plant equipment that required replacement included the attrition scrubber, spiral classifiers, mineral jig, trommel, water feed system, Reichert spiral feed and recirculation systems, slimes circuit, hydrocyclone, and pumps. Approximately 50% of the existing equipment was in good condition and had the potential capacity of 1.8 t/hr (2.0 tons/hr).

The estimated cost to buy new replacement equipment and to modify the demonstration pilot plant was approximately $350,000 to $500,000. The time to fabricate, assemble, and start up the modified plant was estimated at 12 to 16 weeks.

HYPOTHETICAL DESIGN FOR FULL-SCALE PROJECTS

The treatability study and the demonstration at NAS Miramar showed that a combined gravity separation and leach system removed coarse lead contamination from firing range soil. The combined system decreased the total lead content below 500 mg/kg and TCLP lead below 5 mg/L. The data generated from the NAS Miramar demonstration, as well as data from other demonstration projects were used to prepare a full-scale technology suitability analysis and a full-scale system cost analysis. In the suitability and cost analyses, equipment design and selection parameters for a full-scale plant were established and a cost analysis completed. The hypothetical design and costs for a full-scale project were intended to be used by remedial product managers, contractors, and decision makers in evaluating the soil washing technology for further consideration as an applicable option for removal of heavy metals from soils at small arms firing ranges.

ECONOMIC CONSIDERATIONS

In a full-scale system operating at 13.6 t/hr (15 tons/hr), the gravity separation process would cost about $132/t ($120/ton) when treating about 5.26 kt (5800 tons) of soil. This treatment cost assumed a 3-year capital depreciation and one operating shift per day. The addition of aggressive leaching conditions and multiple leaching stages could increase the total cost by an additional $65 to $110/t ($60 and $100/ton). However, the equipment and operating costs, associated with using larger capacity equipment, do not apply for the smaller volumes of soil processed in field pilot studies.

If the full-scale treatment plant was operated at its maximum capacity of 18.1 t/hr (20 tons/hr) using two shifts per day, the capital equipment depreciated over 10 years, and about 7.0 kt (7,750 tons) of soil treated per project, the cost decreased to $74/t ($67/ton).

CHAPTER 1

Introduction

In 1995, the Department of Defense (DOD) Advanced Applied Technology Demonstration Facility (AATDF) at Rice University funded the U.S. Bureau of Mines (BOM) project "Remediation of Lead Contaminated Soils at Small Arms Firing Ranges Using Mining and Beneficiation Technologies." After the closing of the BOM in 1996, the U.S. Bureau of Reclamation (BOR) completed the project. The engineering contractor for the field demonstration and report preparation was Metcalf & Eddy (M&E).

The technology demonstration involved removal of lead from contaminated soils on a former DOD firing range. Plant design, costs, and performance analysis for potential future use of this technology on firing ranges were determined and are presented in this monograph. The chapters are divided into the following main topics:

- Introduction
- Soil Washing Treatability Studies
- Technology Demonstration
- Design and Operation of Hypothetical System
- Economic Analysis
- Performance and Potential Application

The appendices contain information and step-by-step procedures that can be used to evaluate the operational capabilities of pilot plant equipment and to design a full-scale system for a particular firing range soil.

1.1 PROJECT OVERVIEW

The soil within firing ranges across the U.S. has been heavily contaminated with lead bullets that hit the ground and impact berm. The lead bullets can be found as whole bullets or fragments. Lead smears on soil particles and lead oxide also exist. Consequently, there is a growing need to treat these firing range soils to prevent public exposure to the lead-laden soil and the leaching of the lead into the ground water. The U.S. Army alone has over 3000 small arms firing ranges. In addition to military facilities, there are numerous firing ranges that are used by law enforcement agencies and by private groups and organizations.

The primary objective of the soil washing demonstration was to provide reliable, detailed performance data that would help evaluate the feasibility and cost of implementing full-scale treatment systems. An important secondary objective was to demonstrate the effectiveness of the technology in producing a "clean" soil based on analytical results. The project included the following steps:

- Bench scale treatability studies of prospective firing range soils,
- Selection of the demonstration site,
- Pilot plant flowsheet development and equipment selection and installation,
- Execution of the demonstration at the selected DOD site,
- Data evaluation and report preparation, and
- Design and economic evaluation for a full-scale treatment system at a hypothetical DOD firing range site.

Colorado Mineral Research Institute (CMRI) in Golden, Colorado carried out the bench-scale treatability studies for the lead remediation project. Pilot plant design, fabrication, and shakedown were completed at the former BOM Salt Lake Research Center in Salt Lake City, Utah. The demonstration was conducted on soil from a firing range within Naval Air Station Miramar (NAS Miramar), San Diego, California. The fieldwork began on August 18, 1996, and ended on September 25, 1996.

The pilot plant consisted of gravity and leach systems. The gravity system contained a soil feed system, an attrition scrubber, a trommel, a mineral jig concentrator, two spiral classifiers, two Reichert spiral concentrators, a hydrocyclone, a Knelson bowl concentrator, and two fine soil settling basins. The plant also had a small batch leach system that treated coarse and sand soil fractions containing residual lead contamination.

During the last seven days of operation, about 50 t (55 tons) of soil were processed through the gravity separation system. The equipment was operated at varying throughputs while data were collected to assess process performance. Solid samples were analyzed to determine lead distributions. Due to equipment operating problems, only a limited amount of material was processed through the vat leaching plant during the demonstration period.

Using the summarized demonstration results, a technology suitability analysis was developed. In addition, equipment design and selection parameters for a full-scale plant were established. The plant design was based on a typical DOD firing range containing 5.261 kt (5800 ton) of soil to be processed. The plant had a nominal capacity of 13.6 t/hr (15 ton/hr). The major unit operations included particle sizing, gravity separation, and soil dewatering equipment.

1.2 SITE IDENTIFICATION

Three military installations with small arms firing ranges were proposed as possible demonstration sites for this project by representatives from the Waterways Experiment Station (WES) and the Naval Facilities Engineering Service Center (NFESC). The three bases were:

- U.S. Army Base Fort Dix, New Jersey,
- Naval Air Station Miramar, San Diego, California, and
- Marine Corps Air Ground Combat Command, Twentynine Palms, California.

Due to a critical time requirement, existing equipment was used to fabricate the two treatment systems. The equipment had been used in previous soil washing demonstrations conducted by the BOM. Hence, it was necessary to determine through treatability studies which soil characteristics most closely matched the capabilities of the completed gravity and leach plants.

Based on the bench-scale treatability studies, the firing range at NAS Miramar was selected as the site for the pilot plant demonstration. Representative soil samples were collected from the small arms firing ranges at the selected bases. Particle size separation, gravity separation using a jig and shaking table, and deagglomeration tests were evaluated by CMRI. Additional vat or percolation and agitation leaching studies were performed on coarse and fine fractions of NAS Miramar and Twentynine Palms soils where the total lead concentrations exceeded 500 mg/kg of soil.

1.3 GENERAL DESCRIPTION OF PILOT PLANT

The treatment system included gravity separation and vat leach pilot plants. Each plant was installed on a flatbed trailer. The equipment within the gravity separation plant performed particle deagglomeration and sizing, gravity separation, and fines and slimes treatment. The leaching plant included vat leaching and lead recovery unit operations. Agitation leaching was not used in the field demonstration at NAS Miramar. Flow diagrams for both plants are shown in Figures 1.1 and 1.2.

The pilot plant was designed to initially deagglomerate firing range soil in the attrition scrubber and trommel. Particle sizing occurred in the attrition scrubber, trommel, spiral classifiers, and hydrocyclone. Gravity separation of the lead particles from soil particles occurred in the attrition scrubber, mineral jig, Reichert spiral concentrators, and Knelson bowl concentrator. The separation equipment components included:

- A soil feed system that consisted of a bottom discharge hopper with belt feeder, a weigh belt feeder, and a bucket elevator. The bucket elevator fed the soil to the attrition scrubber.
- One attrition scrubber with an inclined auger conveyor that was attached to the bottom of the scrubber tank. The auger removed coarse soil and lead particles, +4.76 mm or +4 mesh, that settled to the bottom of the tank.
- One trommel with a 1.19 mm (16 mesh) screen and spray wash system.
- One mineral jig that removed coarse lead particles from trommel oversize material.
- Two spiral classifiers separated slimes particles from the sand particles, −105 μm from +105 μm (−150 mesh from +150 mesh).
- Two Reichert spiral concentrators that removed lead particles from the sand fraction, −1.19 mm by +105 μm (−16 mesh by +150 mesh).
- One hydrocyclone that removed −37 to −44 μm (-325 to −400 mesh) soil particles from the −105 μm (−150 mesh) soil.
- One Knelson bowl concentrator that cleaned the fine lead particles from the hydrocyclone underflow fraction, +37 to +44 μm (+325 to +400 mesh).
- Slimes treatment for all −105 μm (−150 mesh) soil materials in two dewatering roll-off bins.

Supplemental equipment consisted of conveyors, pumps, tanks, and samplers. Water was removed from all non-slimes particles using vibrating Sweco (dewatering) or sieve bend screens.

The vat leach plant treated the dewatered gravel and sand soil fractions leaving the jig and Reichert spiral concentrators. The plant included a vat leach to extract lead from the soil, iron cementation to remove lead from the leach solution, and ion exchange to clean process solutions. The gravel and sand mixture was treated with acetic acid solution. Calcium peroxide was added to oxidize the lead metal. The primary vat leaching components were:

- Two stainless steel vat leach tanks to extract lead from the sized soil samples,
- One Plexiglas™ iron cementation column to remove soluble lead from the leach solution as lead metal particulate,
- One chelating resin column to remove residual lead and other base metals from the leach solution,
- One anion resin column to decrease the acetate, chloride, or nitrate concentration in the recycled spent leach solution,
- Two large stainless steel tanks to hold the pregnant and spent leach solutions, and
- Pumps and mixing drums.

Once on site, equipment installation was completed and the plant readied for operation in four days. On-site assembly requirements are minimal as long as the site is prepared, utilities are in place, and all equipment and supplies are available. Demobilization of the pilot plant, including decontamination of equipment, was accomplished in two days.

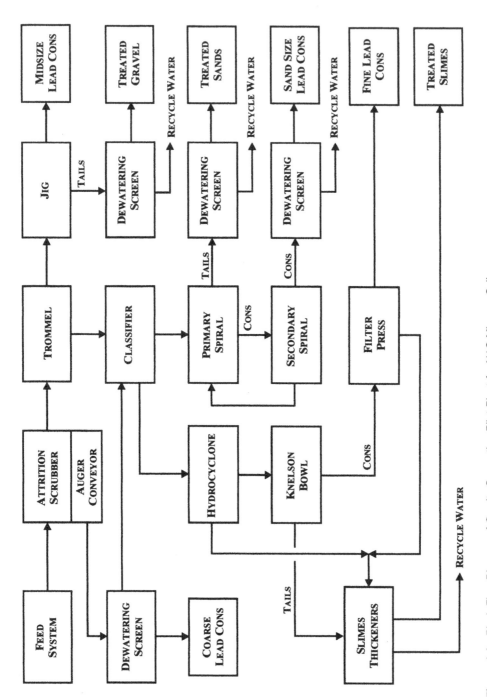

Figure 1.1 Block Flow Diagram of Gravity Separation Pilot Plant for NAS Miramar Soil.

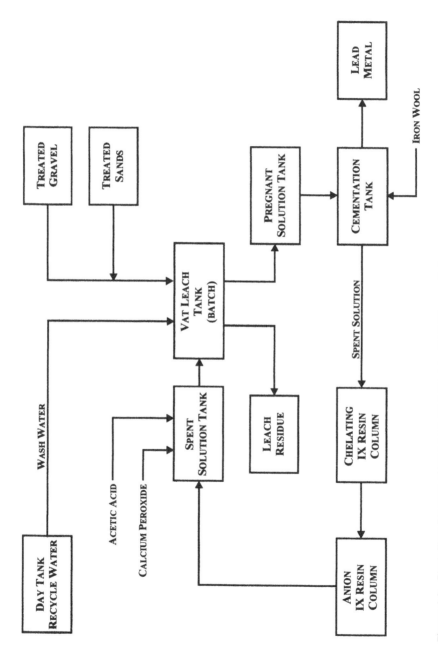

Figure 1.2 Block Flow Diagram of Vat Leach Pilot Plant for NAS Miramar Soil.

1.4 FULL-SCALE PLANT ECONOMICS

A mobile soil washing process design for lead recovery from firing range soils was shown to be feasible in the demonstration at NAS Miramar. The pilot plant and treatability studies provided the basis for the design of a commercial scale mobile plant capable of processing a nominal capacity of 13.60 t/hr (15 tons/hr) of lead contaminated soil from a hypothetical site. The effects of changing operational parameters on the costs were also investigated.

CHAPTER 2

Soil Washing Treatability Studies

In the treatability studies and pilot plant demonstration, the ability of gravity separation and leaching processes in combination to remediate a lead contaminated firing range were evaluated. Lead particles were physically separated from the soil matrix by breaking up the agglomerated material. The deagglomerated contaminated soil was settled or screened to create fractions of particle sizes that were uniform and, therefore, more easily handled and treated. Gravity separation techniques were used to concentrate lead particles and soil particles with associated lead for disposal or further treatment. The coarse and sand soil fractions with lead concentrations greater than 500 mg/kg soil were treated using a vat leaching process. Agitation leaching was also evaluated to treat the contaminated slimes fraction of soil.

Many DOD facilities were screened before three final sites were selected for conducting treatability studies. Treatability studies were conducted on soils from firing ranges on the Fort Dix, New Jersey; Twentynine Palms, California; and NAS Miramar military bases. The firing range to be selected for the pilot plant demonstration had to have sufficient lead contamination of more than 1000 mg/kg of soil. Since the demonstration was designed to test gravity separation and leaching, the lead contamination had to be found as both larger particles and smears distributed throughout the coarse and sand soil fractions, and as oxides in the slimes fractions. The lead particles had to have sufficient size and mass to warrant gravity separation rather than leaching of all of the soil. Contamination of some coarse fractions and the sand and slimes fractions of the soil allowed demonstration of both vat and agitation leaching processes.

The objectives of the treatability studies were to develop the data that could be used to select the demonstration site and to optimize and evaluate pilot plant operations during the field demonstration. Methods of physical and chemical treatment were tested. The physical separation of lead contaminated soils consisted of soil deagglomeration, particle size separation, and gravity (density) separation. Chemical treatment used leaching techniques to determine the solution chemistry for extraction of lead from the soils and to recover lead product from leach solution.

Six samples of firing range soils weighing 20 to 32 kg were furnished to CMRI in Golden, Colorado for treatability tests. Two samples each were furnished from NAS Miramar, Twentynine Palms, and Fort Dix. The bench scale studies were completed during May 1996. The detailed results were included in Appendix A.

The results for the NAS Miramar soil samples were used to prepare a detailed flowsheet and mass balance for the gravity separation and vat leach units of the pilot plant (Appendices B and C). General flowsheets for the gravity separation and vat leaching units are shown in Figures 1.1 and 1.2. The mass balance was used to calculate design and operational capacities (Appendix D) of the major gravity plant equipment. The mass balance and design capacities were also used to determine the effectiveness of the pilot plant demonstration.

2.1 PARTICLE SIZING AND GRAVITY SEPARATION

The objective of particle sizing and gravity separation was to determine which size fractions could be considered "clean," i.e., total lead content less than 500 mg/kg, and to produce a concentrate containing the lead contamination. During the treatability studies, the firing range soil was initially sieved to produce six size fractions. Each fraction was analyzed for lead content and then separated by gravity methods into concentrate, middlings, and tailings fractions. Lead and lead contaminated soil particles were the densest particles and accumulated in the concentrates. In this manner, the soil size fraction(s) with lead contamination was (were) identified and the volume of contaminated soil that required chemical treatment was decreased. Detailed descriptions of the particle sizing and gravity separation tests are included in Appendix A Section A.3. Averaged results for particle size and gravity separation tests on NAS Miramar, Twentynine Palms, and Fort Dix samples are summarized in Tables 2.1, 2.2, and 2.3. The tables also list the gravity methods used to treat each fraction. In these tables, the soil fractions were divided into three size categories:

- Coarse fraction (gravel): +1.19 mm (+16 mesh);
- Sand fraction: –1.19 mm by +105 μm (–16 mesh by +150 mesh); and
- Fine fraction (slimes): –105 μm (–150 mesh).

Figure 2.1 shows the plots of the screen analyses for the three soils. The figure compares graphically the amounts of each particle size in the three firing range soils. The NAS Miramar soil contained some gravel particles with a high percentage of sand and fine material. The Twentynine Palms soil consisted of coarse (gravel) and sand particles with a small amount of fine material. The Fort Dix soil contained sand particles with small amounts of gravel and fine material.

The "soil before treatment" data in the tables showed that particle size separation did not produce clean soil fractions containing less than 500 mg Pb/kg soil. However, gravity separation processes recovered most of the larger metal particles from the coarse and sand fractions so only very fine free lead remained to be leached. Soil particles with heavy lead coatings also behaved like solid metal. Most of these particles were removed from the soil by the gravity circuit before leaching. The remaining particles with or without coatings that required leaching were very fine and had a high surface area to volume ratio. When gravity separation of a fraction was not performed before leaching, the percent lead extracted per leach stage decreased significantly, indicating that most of the free lead particles are removed by the gravity circuit.

2.1.1 NAS Miramar

Results of particle sizing and gravity separation are presented in Table 2.1. The tailings from the gravity separation operations indicate two size fractions required further treatment:

- –4.76 mm by +1.19 mm (–4 mesh by +16 mesh) and
- –105 μm (–150 mesh).

The two tailings fractions had average total lead concentrations of 3045 and 3870 mg/kg, respectively. These fractions represented about 44% of the overall soil by weight in the firing range.

The treatability study evaluated two density separation technologies for treatment of the –105 μm material, a shaking table, and a Knelson bowl concentrator. Both the shaking table and the Knelson bowl concentrator produced a concentrate and tailings. The shaking table also produced a middlings stream with a total lead concentration of 3850 mg/kg. The middlings represented about 8% of the soil by weight. The Knelson concentrator produced no middlings stream. Therefore, for

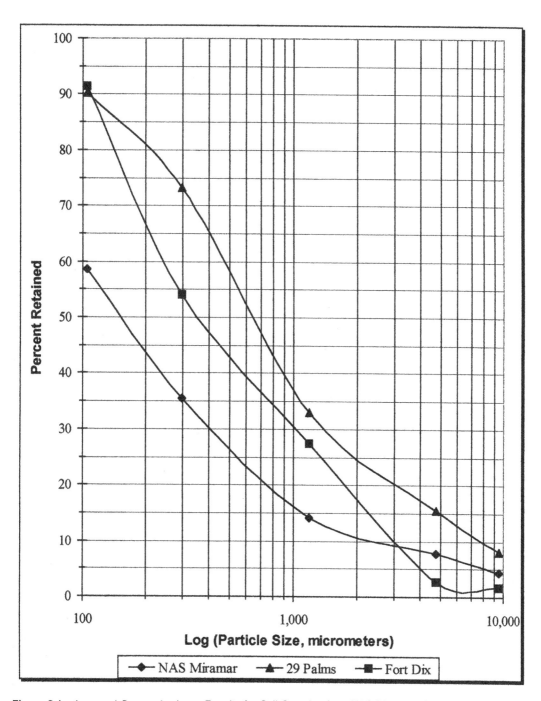

Figure 2.1 Averaged Screen Analyses Results for Soil Samples from NAS Miramar, Twentynine Palms, and Fort Dix.

the pilot plant demonstration, a Knelson bowl concentrator was installed, thus eliminating the middlings stream. The concentrates from the gravity separation operations showed very high concentrations of lead averaging 92,500 mg/kg (9.25% by weight). The concentrate fractions represented about 11% of the soil by weight.

Table 2.1   Average Screen and Lead Analyses For Samples A and B From NAS Miramar

	Soil Before Treatment		Concentrate		Tailings/Middlings	
	Size Fraction Weight %	Total Lead mg/kg	Size Fraction Weight %	Total Lead mg/kg	Size Fraction Weight %	Total Lead mg/kg
Soil Sample Total	100.00	11,806	10.77	92,524	89.23	2068
Coarse Fraction Total	14.20	58,101	4.05	200,173	10.15	1442
+9.50 mm (+3/8 in.)	4.45	74,263	1.00	329,622	3.45	354
+4.76 mm (+4 mesh)	3.25	73,063	0.80	296,228	2.45	195
+1.19 mm (+16 mesh)	6.49	39,543	2.25	108,487	4.25	3045
Sand Fraction Total	44.46	3535	5.25	26,988	39.22	398
+297 μm (+50 mesh)	21.28	4333	2.70	31,603	18.58	374
+105 μm (+150 mesh)	23.18	2802	2.55	22,102	20.63	419
Fine Fraction (Slimes) Total	41.34	4809	1.47	30,209	39.87	3870
−105 μm (−150 mesh)	41.34	4809	1.47	30,209	39.87	3870

The particle sizes use the U.S. Standard Sieve series.

Gravity Treatment Methods

All results for the middlings fractions were added to the results for the tailings fractions.
Humates − skimmed from +3/8-in. tailings
+9.50 mm (+3/8-in.) − Jig tailings and concentrate
−9.50 mm by +4.76 mm (−3/8-in. by +4 mesh) − Jig tailings and concentrate
−4.76 mm by +1.19 mm (−4 mesh by +16 mesh) − Jig tailings and concentrate
−1.19 mm by +297 μm (−16 mesh by +50 mesh) − Shaking table tailings, middlings, and concentrate
−297 μm by +105 μm (-50 mesh by +150 mesh) − Shaking table tailings, middlings, and concentrate
−105 μm (−150 mesh) for Sample A − Shaking table tailings, middlings, and concentrate
−105 μm (−150 mesh) for Sample B − Knelson bowl tailings and concentrate

2.1.2   Twentynine Palms

Results of particle sizing and gravity separation are presented in Table 2.2. The data showed that the coarse tailings fractions were clean with a combined lead content less than 500 mg/kg. These fractions represented 25% of the soil by weight. The combined tailings-middlings from the gravity separation operations indicate that three fractions required further treatment:

- −1.19 mm by +297 μm (−16 mesh by +50 mesh);
- −297 μm by +105 μm (−50 mesh by +150 mesh); and
- −105 μm (−150 mesh).

The three tailings fractions had average total lead concentration of 744, 588, and 19,894 mg/kg, respectively. The two sand fractions in the tailings represented about 51% of the soil by weight. The −105 μm tailings fraction only represented about 9% of the soil by weight. The combined concentrates from the gravity separation operations showed high lead concentrations of about 65,400 mg/kg (6.54%). The concentrates represented about 15% of the soil by weight.

2.1.3   Fort Dix

The treatability results for the Fort Dix soil samples are presented in Table 2.3. The tailings from the gravity separation operations indicated that not all size fractions greater than 105 μm (150 mesh) required further treatment as total lead concentrations were below 500 mg/kg. The −105 μm fraction, representing about 9% of the soil by weight, had a total lead concentration of 2590 mg/kg. The concentrates from the gravity separation operations had an average lead concentration of 33,400 kg/kg, and represented 9% of the soil by weight. The Fort Dix soil samples were eliminated from

Table 2.2 Average Screen and Lead Analyses for Samples LR-23 and LR-30 from Twentynine Palms

	Soil Before Treatment		Concentrate		Tailings/Middlings	
	Size Fraction Weight %	Total Lead mg/kg	Size Fraction Weight %	Total Lead mg/kg	Size Fraction Weight %	Total Lead mg/kg
Soil Sample Total	100.00	12,147	15.30	65,365	84.70	2535
Coarse Fraction Total	33.05	15,329	7.50	66,346	25.55	358
+9.50 mm (+3/8 in.)	8.15	3903	0.40	78,336	7.75	64
+4.76 mm (+4 mesh)	7.35	24,423	2.15	82,760	5.20	313
+1.19 mm (+16 mesh)	17.55	16,827	4.95	58,248	12.60	559
Sand Fraction Total	57.15	8165	6.55	65,821	50.60	701
+297 µm (+50 mesh)	40.15	6469	3.45	67,371	36.70	744
+105 µm (+150 mesh)	17.00	12,169	3.10	64,096	13.90	588
Fine Fraction (Slimes) Total	9.80	24,638	1.25	57,089	8.55	19,894
−105 mm (−150 mesh)	9.80	24,638	1.25	57,089	8.55	19,894

The particle sizes use the U.S. Standard Sieve series.

Gravity Treatment Methods

All results for the middlings fractions were added to the results for the tailings fractions.
Humates − skimmed from +3/8-in. tailings
+9.50 mm (+3/8-in.) − Jig tailings and concentrate
−9.50 mm by +4.76 mm (−3/8-in. by +4 mesh) − Jig tailings and concentrate
−4.76 mm by +1.19 mm (−4 mesh by +16 mesh) − Jig tailings and concentrate
−1.19 mm by +297 µm (−16 mesh by −50 mesh) − Shaking table tailings, middlings, and concentrate
−297 µm by +105 µm (−50 mesh by +150 mesh) − Shaking table tailings, middlings, and concentrate
−105 µm (−150 mesh) for Sample LR-23 − Shaking table tailings, middlings, and concentrate
−105 µm (−150 mesh) for Sample LR-30 − Knelson bowl tailings and concentrate

Table 2.3 Average Screen and Lead Analyses for Samples BK-1 and BK-3 from Fort Dix

	Soil Before Treatment		Concentrate		Tailings/Middlings	
	Size Fraction Weight %	Total Lead mg/kg	Size Fraction Weight %	Total Lead mg/kg	Size Fraction Weight %	Total Lead mg/kg
Soil Sample Total	100.00	3277	8.95	33,449	91.05	313
Coarse Fraction Total	27.39	9773	2.70	97,541	24.69	178
+9.50 mm (+3/8 in.)	1.65	127,802	1.50	140,532	0.15	520
+4.76 mm (+4 mesh)	1.00	31,817	0.15	211,745	0.85	73
+1.19 mm (+16 mesh)	24.74	1013	1.05	19,824	23.69	180
Sand Fraction Total	63.97	589	5.10	6356	58.87	89
+297 µm (+50 mesh)	26.63	1086	1.95	13,546	24.69	103
+105 µm (+150 mesh)	37.33	234	3.15	1906	34.19	80
Fine Fraction (Slimes) Total	8.65	2589	1.15	3113	7.50	2509
−105 µm (−150 mesh)	8.65	2589	1.15	3113	7.50	2509

The particle sizes use the U.S. Standard Sieve series.

Gravity Treatment Methods

All results for the middlings fractions were added to the results for the tailings fractions.
Humates − skimmed from +3/8-in. tailings
+9.50 mm (+3/8-in.) — Jig tailings and concentrate
−9.50 mm by +4.76 mm (−3/8-in. by +4 mesh) — Jig tailings and concentrate
−4.76 mm by +1.19 mm (−4 mesh by +16 mesh) — Jig tailings and concentrate
−1.19 mm by +297 µm (−16 mesh by +50 mesh) — Shaking table tailings, middlings, and concentrate
−297 µm by +105 µm (−50 mesh by +150 mesh) — Shaking table tailings, middlings, and concentrate
−105 µm (−150 mesh) for Sample BK-1 — Shaking table tailings, middlings, and concentrate
−105 µm (−150 mesh) for Sample BK-3 — Knelson bowl tailings and concentrate

the leaching studies since there were no coarse size fractions requiring leaching and the weight percent for the fine fraction was less than 9%.

2.2 ATTRITION SCRUBBING OF MIRAMAR SOIL

The attrition scrubber tests were designed to physically separate agglomerated soil particles and to concentrate lead contaminated soils through gravity separation (Appendix A Section A.4). In the treatability study, the –4.76 mm by +1.19 mm (–4 mesh by +16 mesh) fraction of tailings from two samples of NAS Miramar soil (A and B) were treated separately in an attrition scrubber. The scrubber slurry was then wet screened through a 0.59 mm (30 mesh) opening. Both +0.59 mm fractions were treated in a hand jig. The jig tailings and concentrates were wet screened through 1.19 mm (16 mesh) openings. The products were the -0.59 mm fraction of the scrubbed soil and the +1.19 mm and –1.19 mm fractions of jig tailings and concentrates. The results were analyzed to determine if the lead contamination was associated with one or more particle size(s). The details of the scrubber results are discussed in Appendix A Section A.4.

In Table 2.1, the tailings for the treated fractions of samples A and B averaged 3100 mg/kg lead. From the scrubber tests, the averaged calculated value for the tailings lead content of samples A and B was 770 mg/kg. The lead content of the combined +1.19 mm jig fractions was decreased to about 500 mg/kg. The combined fractions represented 68% of the total weight. The combined –1.19 fractions contained about 1350 mg Pb/kg. The –1.19 mm concentrates and the -0.59 mm screen fractions were quite high in lead content, averaging 2290 and 1950 mg/kg, respectively. Although the treated and calculated values for the tailings lead analysis were quite different, attrition scrubbing did appear to be an effective process for deagglomeration and cleaning the coarse fractions of NAS Miramar soil.

2.3 HYDROCYCLONE TESTS

The hydrocyclone was used to prepare the fine tailings or slimes fraction from the classifier as an acceptable feed for the Knelson bowl concentrator. Proper separation of the lead particles from the soil particles in the concentrator required a feed slurry that contained particle sizes in the range from –105 μm (–150 mesh) to about 30 μm. Accordingly, the –105 μm tailings from the two NAS Miramar samples were evaluated for particle size separation using a Krebs D4B-12 hydrocyclone during the treatability study (Appendix A Section A.5). A 5 to 10% solids slurry of each sample was prepared and run at three or four different pressures. The averaged results for the hydrocyclone tests at an operating gauge pressure of 69 kPa (10 psig) are given in Table 2.4. The hydrocyclone overflow was screened at 37 μm (400 mesh) and the underflow was screened at 53 μm (270 mesh),

Table 2.4   Hydrocyclone Data for NAS Miramar Samples A and B

	Weight grams	Lead mg/kg	% Distribution Weight	% Distribution Lead
Feed	762.5	3870	100.0	100.0
Overflow Total	273.1	4925	35.8	47.6
+37 μm (+400 mesh)	5.4	3890	0.7	0.7
–37 μm (–400 mesh)	267.7	4945	35.1	46.9
Underflow Total	489.3	3020	64.2	52.3
+53 μm (+270 mesh)	189.2	965	24.8	6.5
+44 μm (+325 mesh)	23.8	1450	3.1	1.2
+37 μm (+400 mesh)	70.2	2185	9.2	5.4
–37 μm (–400 mesh)	206.2	5370	27.0	39.2

44 µm (325 mesh), and 37 µm (400 mesh). Table 2.4 lists the weight and lead contents in each fraction and the weight and lead distributions within the hydrocyclone overflow and underflow.

The most significant result was that the smallest size fraction, −37 µm, had the greatest percentage of lead contamination in both the overflow and underflow. The overflow contained mostly −37 µm particles representing about 36% of the mass in the hydrocyclone feed. The overflow also contained about 48% of the total lead mass. The hydrocyclone underflow contained considerable −37 µm material, about 27% of the feed mass with 39.2% of the lead. However, it can be expected that the overflow would contain most of the −10 µm particles in the −105 µm feed. If the Knelson bowl could effectively treat the underflow slurry producing clean tailings, about 64% of the −105 µm fraction would not require chemical treatment.

2.4 SUMMARY OF PHYSICAL TREATMENT STUDIES

The first phase of the treatability studies conducted on the Fort Dix, Twentynine Palms, and Miramar soils included deagglomeration, particle sizing, and gravity separation while determining contaminant distribution. The second phase included vat and agitation leach, attrition scrubber, and hydrocyclone tests on selected samples from the first phase.

The ability to physically separate lead contamination from the bulk soil was determined during the treatability studies. Deagglomeration, particle size separation, and density or gravity separation tests were performed on soil samples from NAS Miramar using screens, a mineral jig, a shaking table, a Knelson bowl concentrator, an attrition scrubber, and a hydrocyclone.

- Particle size separation by screening did not produce sufficient separation of the lead from all of the soil fractions. However, particle sizing and gravity separation removed most of the lead from +105 µm (+150 mesh) particles, resulting in clean products.
- The combined lead concentrates produced during the tests represented about 11% of the soil mass and had an average lead content of 9.2%.
- About 50% of the soil mass, the +105 µm (+150 mesh) tailings fractions, were nearly clean with an average lead content of about 610 mg/kg. However, gravity separation produced one tailing fraction, −4.76 mm by +1.19 mm (−4 mesh by +16 mesh), that had a lead content above 3000 mg/kg. This fraction, representing only 4% of the total soil mass, had to be vat leached.
- About 40% of the soil contained soil and lead particles with sizes less than 105 µm (150 mesh). This fraction had a lead content of 3900 mg/kg and required agitation leaching.
- Gravity concentration of the −105 µm (−150 mesh) soil using the Knelson bowl concentrator was inconclusive. However, it was decided to use the concentrator in the pilot plant after the slimes fraction was treated by a hydrocyclone to remove the −37 µm (−400 mesh) particles.
- Attrition scrubbing tests were performed on the −4.76 mm by +1.19 mm (−4 mesh by +16 mesh) fraction. Deagglomeration of the soil particles was demonstrated by the increase in the amount of the −1.19 mm fraction. Furthermore, gravity concentration of the scrubbed soil produced a clean fraction, representing 68% of the scrubber feed.
- No trend was observed in the hydrocyclone tests. The overflow slurry contained 55% of the −37 µm (−400 mesh) fraction, which was somewhat upgraded in lead content. The −37 µm (−400 mesh) material in the underflow slurry was not decreased to the desired low weight percentage. However, the hydrocyclone probably removed most of the −10 µm soil particles from the underflow, producing a cleaner feed for the Knelson concentrator. The soil in the underflow represented about 25% of the total soil sample. If the Knelson concentrator removed the lead from this fraction, about 15% of the total soil mass required agitation leaching.

2.5 CHEMICAL TREATMENT - LEACHING

Vat leaching of soil usually is carried out in a specially designed roll-off bin or container in which the leach solution percolates down or flows up through the soil. The solution contacts the soil in a near flooded condition. Vat leaching is most effective on coarse soil with particle sizes greater than 1.19 mm (16 mesh). Sandy soil with particle sizes down to +105 μm (+150 mesh) can be mixed with the coarse soil; however, coarse soil fractions should be the predominant material. The box should be designed for ease of filling and removing the soil. The number of roll-off bins required is dependent on the ton per day of contaminated material produced, the cycle time, and the capacity of the roll-off bin.

Agitation leaching of soil usually is carried out in a tank in which the slurry is vigorously mixed with turbine agitator (impeller). The impeller and reactor tank are designed specifically to suspend soil and lead particles as described in Appendix D.2. The design addresses the suspension of the coarsest soil and lead particles in the leach slurry. The suspended particles should easily flow from the reactor into the reactor discharge port. Properly designed leaching systems can be used on slurries containing fine sand or slimes fractions, –1.19 mm (16 mesh) or –105 μm (–150 mesh) soil particles.

2.5.1 Leaching Chemistry

Vat and agitation leach processes use oxidation-reduction and acid-base reactions to dissolve lead soil contamination into aqueous solutions. The major lead contaminants in the soil are lead metal (Pb) and lead oxide (PbO). Acceptable lixiviants that have been used to dissolve the lead compounds are nitric acid (HNO_3), hydrochloric acid (HCl), acetic acid (CH_3COOH), and citric acid [$COOHCH_2COHCOOHCH_2COOH$ or $(CH_2)_2COH(COOH)_3$]. Dilute acidic solutions of nitric acid can also act as an oxidizing agent for lead metal probably forming nitric oxide gas [NO(g)]; however, this oxidizing reaction could be quite slow. Other oxidizing agents that have been used are hydrogen peroxide (H_2O_2) and calcium peroxide (CaO_2). The main chemical reactions and their stoichiometry that dissolve the lead metal and lead oxide are as follows:

Oxidation-Reduction Reactions

$$3Pb^\circ + 8HNO_3 = 3Pb(NO_3)_2 + 4H_2O + 2NO(g) \qquad \text{R1}$$

$$Pb^\circ + H_2O_2 = PbO + H_2O \qquad \text{R2}$$

$$Pb^\circ + CaO_2 = PbO + CaO \qquad \text{R3}$$

Acid-Base Reactions

$$PbO + 2HNO_3 = Pb(NO_3)_2 + H_2O \qquad \text{R4}$$

$$PbO + 2HCl = Pb(Cl)_2 + H_2O \qquad \text{R5}$$

$$PbO + 2CH_3COOH = Pb(CH_3COO)_2 + H_2O \qquad \text{R6}$$

$$3PbO + 2(CH_2)_2COH(COOH)_3 = Pb_3[(CH_2)_2COH(COO)_3]_2 + 3H_2O \qquad \text{R7}$$

$$CaO + 2HNO_3 = Ca(NO_3)_2 + H_2O \qquad \text{R8}$$

$$CaO + 2HCl = Ca(Cl)_2 + H_2O \qquad \text{R9}$$

$$CaO + 2CH_3COOH = Ca(CH_3COO)_2 + H_2O \qquad \text{R10}$$

The minimum concentration of reagent required to react with and solubilize lead can be calculated for reactions R1 through R10 in grams reagent per 1000 mg of lead in the soil. While determining the stoichiometric quantitiy of acid theoretically required to remove the lead contamination, this calculation actually results in an underestimate of the quantity of reagent required in the field.

Oxidation-Reduction Reactions (grams/gram Pb)

$$R1: \frac{HNO_3}{Pb} = 0.811 \qquad R2: \frac{H_2O_2}{Pb} = 0.164 \qquad R3: \frac{CaO_2}{Pb} = 0.348$$

Acid-Base Reactions (grams/gram Pb)

$$R4: \frac{HNO_3}{Pb} = 0.608 \qquad R5: \frac{HCl}{Pb} = 0.352 \qquad R6: \frac{CH_3COOH}{Pb} = 0.580$$

$$R7: \frac{(CH_2)_2COH(COOH)_3}{Pb} = 0.618$$

The use of an oxidizing agent, such as CaO_2, in addition to an acid to dissolve the lead metal and oxide can result in an additional acid requirement. The CaO formed during the oxidation of the lead metal (R3) consumes acid in an acid-base reaction (R8 to R10).

Combined Reactions (grams/gram Pb)

$$R3+R8: \frac{HNO_3}{Pb} = 0.608 \qquad R3+R9: \frac{HCl}{Pb} = 0.352 \qquad R3+R10: \frac{CH_3COOH}{Pb} = 0.580$$

Clays and other silicate minerals in the soil can also consume considerable quantities of the nitric, hydrochloric, acetic, and citric acids. Typically, clay minerals have particle sizes less than 10 μm. These fine particles have a high surface area per unit volume which promotes high reaction rates, thus high acid consumption. Typical clay minerals are kaolinite $[Al_2(Si_2O_5)(OH)_4]$ and montmorillonite $[(Al,Mg)_8(Si_4O_{10})_3(OH)_{10} \times 10H_2O]$. Other silicate minerals in the soil could be from the feldspar groups, for example:

- Orthoclase $[1/2K_2O \times 1/2Al_2O_3 \times 3(SiO_2)]$,
- Albite $[1/2Na_2O \times 1/2Al_2O_3 \times 3(SiO_2)]$, and
- Anorthite $[CaO \times Al_2O_3 \times 2(SiO_2)]$.

The following equations list the typical acid-base reactions between nitric and hydrochloric acids and anorthite and kaolinite minerals:

$$CaO \times Al_2O_3 \times 2(SiO_2) + 8HNO_3 = Ca(NO_3)_2 + 2Al(NO_3)_3 + 2SiO_2 + 4H_2O \qquad \text{R11}$$

$$CaO \times Al_2O_3 \times 2(SiO_2) + 8HCl = Ca(Cl)_2 + 2Al(Cl)_3 + 2SiO_2 + 4H_2O \qquad \text{R12}$$

$$Al_2(Si_2O_5)(OH)_4 + 6HNO_3 = 2Al(NO_3)_3 + 2SiO_2 + 5H_2O \qquad \text{R13}$$

$$Al_2(Si_2O_5)(OH)_4 + 6HCl = 2Al(Cl)_3 + 2SiO_2 + 5H_2O \qquad \text{R14}$$

The following mass relationships can be calculated for reactions R11 through R14 in grams acid reagent per gram of mineral in the soil:

$$R11: \frac{HNO_3}{anorthite} = 1.81 \qquad R12: \frac{HNO_3}{anorthite} = 1.05$$

$$R13: \frac{HNO_3}{kaolinite} = 1.46 \qquad R14: \frac{HCl}{kaolinite} = 0.85$$

2.5.2 Acidic, Basic, and Salt Solutions and pH

(Butler, 1964; Freiser, 1963; Stumm, et al., 1970)

In equilibrium chemistry, the terms strong and weak refer to the degree of dissociation of a compound in solution. Strong acids, such as HNO_3 and HCl, and bases, like NaOH, are almost completely dissociated in aqueous solutions. The salts of strong acids and bases, NaCl for example, are similarly dissociated in aqueous solutions. Sulfuric acid, H_2SO_4, is a combination of a strong and weak acid. The first hydrogen ion dissociates from the HSO_4^{-1} ion while the second is partially dissociated from the sulfate ion, SO_4^{-2}. The dissociation can be described by the following equilibrium relationships:

$$H_2SO_4 = H^{+1} + HSO_4^{-1} \qquad \text{R15}$$

where $[H^{+1}] \approx [HSO_4^{-1}]$

$$HSO_4^{-1} = H^{+1} + SO_4^{-2} \qquad \text{R16}$$

$$K_{a2} = \frac{[H^{+1}][SO_4^{-2}]}{[HSO_4^{-1}]} = 1.00 \times 10^{-2} \qquad (1)$$

K_{a2} is the equilibrium constant for the dissociation of the bisulfate ions, HSO_4^{-1}, and $[H^{+1}]$, $[HSO_4^{-1}]$, $[SO_4^{-1}]$ are the aqueous concentrations of hydrogen, bisulfate, and sulfate ions in mole/L.

Weak acids like acetic and citric acids, and bases such as ammonium hydroxide are only partially dissociated in aqueous solutions. The dissociation of weak acids and bases can be described by the following equilibrium relationships:

For acetic acid

$$CH_3COOH = H^{+1} + CH_3COO^{-1} \qquad \text{R17}$$

For ammonium hydroxide

$$NH_4OH = NH_4^{+1} + OH^{-1} \qquad \text{R18}$$

where

$$K_a = \frac{[H^{+1}][CH_3COO^{-1}]}{[CH_3COOH]} = 1.75 \times 10^{-5} \qquad (2)$$

$$K_b = \frac{[NH_4^{+1}][OH^{-1}]}{[NH_4OH]} = 1.78 \times 10^{-5} \tag{3}$$

K_a and K_b are the equilibrium constant for the dissociation of acetic acid and ammonium hydroxide, NH_4OH or NH_3. $[H^{+1}]$, $[OH^{-1}]$, $[CH_3COO^-]$, and $[NH_4^{+1}]$ are the aqueous concentrations of hydrogen, hydroxyl, acetate, and ammonium ions in mole/L. $[CH_3COOH]$ and $[NH_4OH]$ are the aqueous concentrations of acetic acid and ammonium hydroxide in mole/L. The equilibrium equations for citric acid are more complex since the acid can readily provide three hydrogen ions for acid-base reactions. The ionization constants for the three hydrogen ions have decreasing values as each ion is released from the acid:

$$(CH_2)_2COH(COOH)_3 = H^{+1} + (CH_2)_2COH(COOH)_2(COO)^{-1}$$

$$K_{a1} = \frac{[H^{+1}][(CH_2)_2COH(COOH)_2(COO)^{-1}]}{[(CH_2)_2COH(COOH)_3]} = 1.15 \times 10^{-3} \tag{4}$$

$$(CH_2)_2COH(COOH)_2(COO)^{-1} = H^{+1} + (CH_2)_2COH(COOH)(COO)_2^{-2}$$

$$K_{a2} = \frac{[H^{+1}][(CH_2)_2COH(COOH)_2(COO)_2^{-2}]}{[(CH_2)_2COH(COOH)_2(COO)^{-1}]} = 7.24 \times 10^{-5} \tag{5}$$

$$(CH_2)_2COH(COOH)(COO)_2^{-2} = H^{+1} + (CH_2)_2COH(COO)_3^{-3}$$

$$K_{a3} = \frac{[H^{+1}][(CH_2)_2COH(COO)_3^{-3}]}{[(CH_2)_2COH(COOH)(COO)_2^{-2}]} = 1.51 \times 10^{-6} \tag{6}$$

In leaching with strong acids, pH is a measure of the residual acid concentration. The pH of a solution is indicative of the quantity of acid available for reaction and is often used as an operating parameter in leaching processes. The pH of an aqueous solution can be defined by the equation:

$$pH = -\log[H^{+1}] \tag{7}$$

Another relationship for dilute solutions at 25°C is the water ionization equilibrium for water:

$$K_w = [H^{+1}][OH^{-1}] = 1.00 \times 10^{-14} \tag{8}$$

The value of K_w can change with temperature. Finally, a general equation for pH of a strong acid with an analytical concentration of C, mole/L, can be calculated:

$$C = [H^{+1}] - \frac{K_w}{[H^{+1}]} \tag{9}$$

For pH values less than 5.00, the equation simplifies to $C = [H^{+1}]$, and at infinite dilution of the strong acid the pH approaches 7.00.

Calculation of relationships between pH and the concentration of a weak acid like acetic acid, $[CH_3COOH]$, is more complex. An example provides the best illustration of the relationship. If the acetic acid concentration in water is 43.2 grams/L, 0.720 mole/L, the concentration of all species and the pH can be calculated as shown in Example 1:

Example 1

$$[H^{+1}][CH_3COO^{-1}] = 1.75 \times 10^{-5}[CH_3COOH] \tag{10}$$

$$[H^{+1}][OH^{-1}] = 1.00 \times 10^{-14} \tag{11}$$

The charge balance shows the proton condition:

$$[H^{+1}] = [CH_3COO^{-1}] + [OH^{-1}] \tag{12}$$

Since the concentration of acetic acid is 0.720 mole/L, the mass balance on acetate ions is

$$[CH_3COOH] + [CH_3COO^{-1}] = 0.720 \text{ mole/L} \tag{13}$$

Since the solution is acidic and the acetic acid is slightly dissociated, the hydroxyl and acetate ion concentrations are negligible. With these assumptions, the charge and mass balances can be simplified:

$$[H^{+1}] = [CH_3COO^{-1}] \tag{12a}$$

$$[CH_3COOH] = 0.720 \text{ mole/L} \tag{13a}$$

Substituting (12a) and (13a) into (10) produces:

$$\begin{aligned}
[H^{+1}]^2 &= (1.75 \times 10^{-5})(0.720) = 1.26 \times 10^{-5} \\
[H^{+1}] &= 3.55 \times 10^{-3} \text{ mole/L} \\
pH &= 2.45
\end{aligned} \tag{10a}$$

From (11):

$$[OH^{-1}] = 2.82 \times 10^{-12} \text{ mole/L}$$

From (12)

$$[CH_3COO^{-1}] = 3.55 \times 10^{-3} - 2.82 \times 10^{-12} = 3.55 \times 10^{-3} \text{ mole/L}$$

For a check, the values for hydroxyl and acetate ion concentration are added into (13):

$$[CH_3COOH] + [CH_3COO^{-1}] = 0.720 + 3.55 \times 10^{-3} = 0.7236 \text{ mole/L}$$

This is a less than 1% error.

2.5.3 Activity Coefficients

(Butler, 1964; Freiser, 1963; Stumm, et al., 1970)

The above equilibria equations, (1), (2), (3), (4), (5), and (6), are for very dilute solutions. In the more concentrated solutions that are encountered in leaching processes, the equilibrium relationships must use activity concentrations or activity for all ionic species. For example, the relationship between hydrogen ion activity and hydrogen ion concentration can be defined by:

$$\{H^{+1}\} = [H^{+1}] \times \gamma_{H^{+1}} \tag{14}$$

where $\{H^{+1}\}$ is the hydrogen ion activity, $[H^{+1}]$ is the hydrogen ion concentration, and $\gamma_{H^{+1}}$ is the mean hydrogen ion activity coefficient.

Activity coefficients are dependent upon the ionic strength, I, of the solution, which is defined by:

$$I = \frac{1}{2}\Sigma_i C_i z_i^2 \tag{15}$$

where I is the product of the concentration C_i for each ion (i) and the square of its charge Z_i. The mean activity coefficient $\gamma_{\pm}$ for 1-1 and 1-2 electrolytes can be estimated by using the Davies equation:

$$-\log\gamma_{\pm} = A \times Z_+ Z_- \left(\frac{\sqrt{I}}{1 + \sqrt{I}}\right) - 0.2 \times I \tag{16}$$

For a single ion, the equation becomes:

$$-\log\gamma_{\pm} = A \times Z^2 \left(\frac{\sqrt{I}}{1 + \sqrt{I}}\right) - 0.2 \times I \tag{17}$$

The value of A is defined by:

$$A = 1.825 \times 10^6 (\varepsilon T)^{-3/2} = 0.509 \text{ at } 25°C$$

where εT is the product of the dielectric constant of water and the absolute temperature in K.

In terms of activity coefficients, the equation (2) for acetic acid can be written as follows:

$$K_a = \frac{[H^{+1}]\gamma_{H^{+1}}[CH_3COO^{-1}]\gamma_{Ac^{-1}}}{[CH_3COOH]\gamma_{HAc}} \tag{18}$$

$$K_a = \frac{[H^{+1}][CH_3COO^{-1}]}{[CH_3COOH]} \times \frac{\gamma_{H^{+1}}\gamma_{Ac^{-1}}}{\gamma_{HAc}} = K_a^o \times \frac{\gamma_{H^{+1}}\gamma_{Ac^{-1}}}{\gamma_{HAc}} \tag{18a}$$

K_a^o is the equilibrium constant at infinite dilution, which has the previously defined value of 1.75×10^{-5}; HAc and Ac^{-1} represent acetic acid and acetate ion, respectively.

Example 2 shows the procedures for using the activity to correct equilibrium calculations for the previous acetic acid example.

Example 2

For Example 1 above, the activity coefficients for hydrogen and acetate ions can be estimated assuming that the equilibrium constant is independent of the ionic strength. Hence, the values in Example 1 can be used:

$$[H^{+1}] = 3.55 \times 10^{-3} \text{ mole/L}$$

$$[CH_3COO^{-1}] = 3.55 \times 10^{-3}$$

$$I = \frac{1}{2}\Sigma_i C_i Z_i^2 = \frac{1}{2}[3.55 \times 10^{-3}(+1)^2 + 3.55 \times 10^{-3}(-1)^2] = 3.55 \times 10^{-3}$$

Once the ionic strength is known, the activity coefficients for the hydrogen and acetate ions can be calculated using Equation (17):

$$-\log \gamma_{H^{+1}} = 0.509 \times (+1)^2 \left(\frac{\sqrt{3.55 \times 10^{-3}}}{1 + \sqrt{3.55 \times 10^{-3}}} \right) - 0.2 \times 3.55 \times 10^{-3}$$

$\gamma_{H^{+1}} = 0.937$ and, conversely, $\gamma_{Ac^{-1}} = 0.937$. Hence, K_a is calculated:

$$K_a = \frac{[H^{+1}][CH_3COO^{-1}]}{[CH_3COOH]} \times \frac{\gamma_{H^{+1}}\gamma_{Ac^{-1}}}{\gamma_{HAc}} = 1.75 \times 10^{-5} \times \frac{(0.937)(0.937)}{(1.0)} \qquad (18b)$$

$$K_a = 1.54 \times 10^{-3}$$

The activity of a neutral molecule is assumed to 1.0. Using this corrected value for the equilibrium constant in Equation (10a), the pH is recalculated:

$$[H^{+1}]^2 = (1.54 \times 10^{-5})(0.720) = 1.109 \times 10^{-5}$$

$$[H^{+1}] = 3.33 \times 10^{-3} \text{ mole/L and the pH} = 2.48.$$

In an aqueous solution, nitric acid is completely dissociated. Example 3 shows the effect of the activity coefficient on the solution pH.

Example 3

   For a solution that contains 0.10 mole/L of nitric acid, the pH without the correction for the activity coefficient will be:

$$[H^{+1}] = 1.00 \times 10^{-1} \text{ mole/L}$$

$$pH = -\log[H^{+1}] = -\log[1.00 \times 10^{-1}] = 1.00$$

For the 0.10 mole/L nitric acid solution, the ionic strength for the solution and the activity coefficients for the hydrogen and nitrate ions are:

$$I = \frac{1}{2}\Sigma_i C_i Z_i^2 = \frac{1}{2}[1.0 \times^{-1}(+1)^2 + 1.0 \times 10^{-1}(-1)^2] = 1.0 \times 10^{-1}$$

$$-\log \gamma_{H^{+1}} = -\log \gamma_{NO_3^{-1}} = 0.509 \times (+1)^2 \left(\frac{\sqrt{1.0 \times 10^{-1}}}{1 + \sqrt{1.0 \times 10^{-1}}} \right) - 0.2 \times 1.0 \times 10^{-1}$$

$$\gamma_{H^{+1}} = \gamma_{NO_3^{-1}} = 0.7725$$

$$pH = -\log\{H^{+1}\} = -\log\{[H^{+1}] \times \gamma_{H^{+1}}\} = -\log\{[1.0 \times 10^{-1}] \times 0.7725\} = 0.923$$

2.5.4 Complex Ion Chemistry

(Butler, 1964; Sillen and Martell, 1964; Stumm, et al., 1970)

Table 2.5 gives values for the equilibrium constants of complex ions.

Table 2.5 Equilibrium Constants for Complex Ions and Insoluble Compounds

Ligand	Central Ion	Logarithm of Equilibrium Constant K_{SO}	K_1	K_2	K_3	K_4
Cl^{-1}	H^{+1}	NA	−1.51			
Cl^{-1}	Pb^{+2}	−4.67	1.10	1.16	-0.40	−1.05
CH_3COO^{-1}	H^{+1}	NA	−4.76			
CH_3COO^{-1}	Pb^{+2}	NA	2.43	1.52		
SO_4^{-2}	H^{+1}	NA	1.99			
SO_4^{-2}	Pb^{+2}	−7.78	2.62	0.85		

The ligand is the complexing ion, like chloride (Cl^{-1}) and acetate (CH_3COO^{-1}) ions. Lead ions, Pb^{+2}, can form complex ions with the chloride and acetate ions. Nitrate from nitric acid does not form complex ions with lead ions. In many cases, complex ion formation can increase the solubility of salts that are produced by acid-base reactions.

Lead chloride, $PbCl_2$, dissolves in water by the following reaction:

$$PbCl_{2,\,solid} = Pb^{+2}_{aq} + 2Cl^{-1}_{aq}.$$

As given in the above table, the solubility product, K_{SO}, for lead chloride is 2.14×10^{-5}. The solubility product for this reaction is given by the equation:

$$K_{SO} = [Pb^{+2}][Cl^{-1}]^2 = 2.14 \times 10^{-5}$$

where $[Pb^{+2}]$ and $[Cl^{-1}]$are the aqueous concentrations of lead and chloride ions in mole/L. Without complex ion formation between the lead and chloride ions, the maximum amount of lead chloride dissolved would be as follows:

$$[PbCl_2] = X, \text{ moles/L}; [Pb^{+2}] = X, \text{ moles/L}; \text{ and } [Cl^{-1}] = 2X, \text{ moles/L}.$$

Hence, the solubility product becomes:

$$[X][2X]^2 = 4 \times [X]^3 = 2.14 \times 10^{-5}$$

where $[PbCl_2] = X = 0.0175$ moles/L = 4.86 grams/L and $[Pb^{+2}] = 3.62$ grams/L.

The equilibrium relationships between the soluble lead and chloride ions are as follows:

$$H^{+1} + Cl^{-1} = HCl$$

$$K_a = \frac{[HCl]}{[H^{+1}][Cl^{-1}]} = 3.09 \times 10^{-2} \tag{20}$$

$$Pb^{+2} + Cl^{-1} = PbCl^{+1}$$

$$K_1 = \frac{[PbCl^{+1}]}{[Pb^{+2}][Cl^{-1}]} = 12.59 \tag{21}$$

$$PbCl^{+1} + Cl^{-1} = PbCl_2^{o}$$

$$K_2 = \frac{[PbCl_2^{o}]}{[PbCl^{+1}][Cl^{-1}]} = 14.45 \tag{22}$$

$$PbCl_2^{o} + Cl^{-1} = PbCl_3^{-1}$$

$$K_3 = \frac{[PbCl_3^{-1}]}{[PbCl_2^{o}][Cl^{-1}]} = 3.98 \times 10^{-1} \tag{23}$$

$$PbCl_3^{-1} + Cl^{-1} = PbCl_4^{-2}$$

$$K_4 = \frac{[PbCl_4^{-2}]}{[PbCl_3^{-1}][Cl^{-1}]} = 8.92 \times 10^{-2} \tag{24}$$

In an aqueous solution, hydrochloric acid is almost completely dissociated. Example 4 shows the methods to calculate the concentrations of complex ions.

Example 4

For agitation leach test, LR-23-5, in Table 2.8, the initial lead content of the soil fraction, the solution volume, and the quantity of soil and reagents added during the first stage had the following values:

$$[Pb_{Start}] = 15,100 \text{ mg/kg}$$

$$V = 0.400 \text{ L}$$

$$Dry \text{ Soil} = 0.100 \text{ kg}$$

$$HNO_3 = 56 \text{ kg/1000 kg soil} \equiv 5.60 \text{ grams} = 8.89 \times 10^{-2} \text{ mole}$$

$$[HNO_3] = 14.0 \text{ g/L} = 2.222 \times 10^{-1} \text{ mole/L}$$

$$HCl = 29.6 \text{ kg/1000 kg soil} \equiv 2.96 \text{ grams} = 8.110 \times 10^{-2} \text{ mole}$$

$$[HCl] = 7.40 \text{ g/L} = 2.027 \times 10^{-1} \text{ mole/L}$$

$$H_2O_2 = 6.0 \text{ kg/1000 kg soil} \equiv 0.600 \text{ grams} = 1.765 \times 10^{-2} \text{ mole}$$

$$[H_2O_2] = 1.50 \text{ g/L} = 4.412 \times 10^{-2} \text{ mole/L}$$

At the end of the first stage leach, the final pH and lead concentrations in the soil and solution (Soln) were:

$$[Pb_{End}] = 1117 \text{ mg/kg}$$

$$Pb_{Soln} = \frac{(15,100 \text{ mg/kg} - 1117 \text{ mg/kg})(0.100 \text{kg})}{1000 \text{ mg/g}} = 1.398 \text{ grams} = 6.75 \times 10^{-3} \text{ mole}$$

$$[Pb_{Soln}] = \frac{(1.398 \times 10^{-3} \text{ mole})}{(0.400 \text{ L})} = 1.687 \times 10^{-2} \text{ mole/L or } 3.495 \text{ grams/L}$$

$$pH = 1.33 \text{ or } [H^{+1}] = 4.68 \times 10^{-2} \text{ mole/L}$$

Assuming all the lead is present as metal, Pb^o and major clay mineral is kaolinite, the probable reactions are:

$$Pb^o + H_2O_2 = PbO + H_2O \qquad\qquad R2$$

$$PbO + 2HNO_3 = Pb(NO_3)_2 + H_2O \qquad\qquad R4$$

$$PbO + 2HCl = Pb(Cl)_2 + H_2O \qquad\qquad R5$$

$$Al_2(Si_2O_5)(OH)_4 + 6HNO_3 = 2Al(NO_3)_3 + 2SiO_2 + 5H_2O \qquad\qquad R13$$

$$Al_2(Si_2O_5)(OH)_4 + 8HCl = 2Al(Cl)_3 + 2SiO_2 + 5H_2O \qquad\qquad R14$$

The solution is quite acidic; hence the hydrolysis of the metal ion is quite small and does not complicate the calculations. For the Pb-Cl system, the equilibrium equations are:

$$[HCl] = 3.09 \times 10^{-2} \times [H^{+1}][Cl_{-1}] \qquad\qquad (25)$$

$$[PbCl^{+1}] = 12.59 \times [Pb^{+2}][Cl^{-1}] \qquad\qquad (26)$$

$$[PbCl_2^0] = 14.45 \times [PbCl^{+1}][Cl^{-1}] \qquad\qquad (27)$$

$$[PbCl_3^{-1}] = 3.98 \times 10^{-1} \times [PbCl_2^0][Cl^{-1}] \qquad\qquad (28)$$

$$[PbCl_4^{-2}] = 8.91 \times 10^{-2} \times [PbCl_3^{-1}][Cl^{-1}] \qquad\qquad (29)$$

The mass balances for the lead and chloride ions are:

$$Cl_{Total} = [HCl] + [Cl^{-1}] + [PbCl^{+1}] + 2[PbCl_2^0] + 3[PbCl_3^{-1}] + 4[PbCl_4^{-2}]$$

$$Cl_{Total} = 2.027 \times 10^{-1} \text{ mole/L} \qquad\qquad (30)$$

$$Pb_{Total} = [Pb^{+2}] + [PbCl^{+1}] + [PbCl_2^0] + [PbCl_3^{-1}] + [PbCl_4^{-2}] = 1.68 \times 10^{-2} \text{ mole/L} \quad (31)$$

Equations (25) to (31) are solved by an iterative calculation using Microsoft Excel® by alternately changing $[Cl^{-1}]$ and $[Pb^{+2}]$ until both:

$$Cl_{Total} = Cl_{Calc} = 2.027 \times 10^{-1} \text{ and}$$

$$Pb_{Total} = Pb_{Calc} = 1.680 \times 10^{-2}.$$

The results of the calculation are in the following table:

Lead Complex Ion Concentration, mole/L (Calculated)						
Pb^{+2}	$PbCl^{+1}$	$PbCl_2{}^0$	$PbCl_3{}^{-1}$	$PbCl_4{}^{-2}$	Pb_{Total}	Pb_{Calc}
0.00182	0.00403	0.01023	0.00072	0.00011	0.01680	0.01680

Chloride Concentration in Lead Complex Ions, mole/L (Calculated)						
HCl	Cl^{-1}	$PbCl^{+1}$	$2PbCl_2{}^0$	$3PbCl_3{}^{-1}$	$4PbCl_4{}^{-2}$	Cl_{Total}
0.00025	0.1758	0.00403	0.02046	0.00215	0.00045	0.20270

Using the charge balance, the concentration of aluminum ions that were produced by reactions R13 and R14 can be estimated:

$$3[Al^{+3}] + [H^{+1}] + 2[Pb^{+2}] + [PbCl^{+1}] = [NO_3{}^{-1}] + [Cl^{-1}] + 3[PbCl_3{}^{-1}] + 2[PbCl_4{}^{-2}] + [OH^{-1}]$$

where

$$[NO_3{}^{-1}] = [HNO_3] = 2.222 \times 10^{-1} \text{ mole/L}$$

$$[H^{+1}] = 4.68 \times 10^{-2} \text{ mole/L at pH of 1.33}$$

therefore

$$[Al^{+3}] = 1.148 \times 10^{-1} \text{ mole/L}.$$

The concentration of aluminum ions equates to about 5.9 grams of kaolinite dissolved by the acids.

Without complex ion formation, the solubility of lead chloride would be limited in the presence of the high chloride content, Cl^{-1} at 0.2027 mole/L. For example, the solubility product for a lead content of 0.0168 mole/L would be:

$$K_{SO} = [Pb^{+2}][Cl^{-1}]^2 = (0.0168)(0.2027)^2 = 6.90 \times 10^{-4}$$

This value is about 32 times higher than the solubility product in Table 2.5. Therefore, the lead chloride would be insoluble. With complex ion formation, the solubility product decreases to:

$$K_{SO} = [Pb^{+2}][Cl^{-1}]^2 = (0.00182)(0.1758)^2 = 5.62 \times 10^{-5}$$

This value is about three times higher than the expected table value of 2.14×10^{-5}. The small difference is probably caused by not using activity coefficients for all ionic species and the accuracy of solubility product determination. Furthermore, adding a neutral salt, like sodium chloride, would enhance the solubility of lead chloride.

About 89 mol% of the lead ions in the aqueous solution are combined as lead-chloride complex ions. $PbCl^{-1}$ and $PbCl_2{}^0$ are the predominating species. Only a small amount of undissociated hydrochloric acid remains, about 0.13 mol% of the original acid added. About 40 mol% of the hydrogen peroxide is consumed in oxidizing the lead metal, reaction R2. About 4 mol% of the hydrochloric and nitric acid is consumed in dissolving the lead. Greater than 90 mol% of the two acids is consumed dissolving the clay kaolinite.

During the leaching process, most of the acid applied to soil is consumed in the dissolution of the gangue materials rather than the desired metals. Typically, there is less competition for the acid in coarser materials. Clay particles have a high surface area per unit volume and exhibit high reactivity and acid consumption. Therefore, the removal of lead from clays should not be attempted through leaching, particularly by agitation leaching which increases the contact between solution and soil particles. Vat leaching of coarser materials generally proves more effective.

2.5.5 Lead Recovery from Leach Solution

The vat and agitation leaching in the treatability studies dissolved the lead metal and oxide using nitric and hydrochloric acids and hydrogen peroxide. Acetic acid and calcium peroxide were used in the vat leaching tests during the demonstration. Lead was not recovered from solution in the CMRI treatability study. However, laboratory and demonstration tests by BOR personnel showed that dissolved lead can be removed from the leach solution by iron cementation to form a lead metal precipitate. To provide clean solution for leaching, the ion exchange reactions can remove copper and residual lead ions from the leach solution using a chelating ion exchange resin, and anions like acetate, chloride, and nitrate ions, using anion exchange resins.

Lead Cementation

According to the electrochemical series table (Freiser, 1963; Stumm, 1970), any dissolved metal ions with a higher reduction potential (less negative) will react with a metal that has lower reduction potential (more negative).

Electrochemical Series (As Standard Reduction Potentials)

Reduction Reaction	Potential Volts	Reduction Reaction	Potential Volts
$Zn^{+2} + 2e^- = Zn°$	−0.763	$Pb^{+2} + 2e^- = Pb°$	−0.126
$Fe^{+2} + 2e^- = Fe°$	−0.444	$2H^{+1} + 2e^- = H_2°$	0.000
$Cd^{+2} + 2e^- = Cd°$	−0.402	$Cu^{+2} + 2e^- = Cu°$	+0.337

Hence, copper and lead acetate dissolved in an aqueous solution can react with zinc or iron metal to form a copper and lead metal plating or precipitate while forming zinc and iron acetate in the solution. The overall reactions are:

$$Pb(CH_3COO)_2 + Fe° = Pb° + Fe(CH_3COO)_2 \qquad \text{R19}$$

$$Pb(CH_3COO)_2 + Zn° = Pb° + Zn(CH_3COO)_2 \qquad \text{R20}$$

$$Cu(CH_3COO)_2 + Fe° = Cu° + Fe(CH_3COO)_2 \qquad \text{R21}$$

$$Cu(CH_3COO)_2 + Zn° = Cu° + Zn(CH_3COO)_2 \qquad \text{R22}$$

The above reactions are usually called oxidation-reduction reactions. The reduction reactions are written as in the above table, Oxid + ne⁻ = Red, where 'Oxid' is the oxidized form and 'Red' is the reduced form. The corresponding oxidation reactions are written as follows: Red = Oxid + ne⁻. The resulting oxidation and reduction reaction can be written using the half-cell reactions and reduction potentials, E, from the table above. For example, the half-cell reactions for reaction R19 are as follows:

$Fe° = Fe^{+2} + 2e^-$	$E_{ox} = +0.409$ volts	Oxidation Reaction
$Pb^{+2} + 2e^- = Pb°$	$E_{red} = -0.126$ volts	Reduction Reaction
$Fe° + Pb^{+2} = Pb° + Fe^{+2}$	$E_{net} = +0.283$ volts	Net Reaction

where $E_{ox} = -E_{red}$. Iron metal is the reducing agent and lead ions are the oxidizing agent. A positive net reaction potential, E_{net}, of +0.283 volts means that the reaction will proceed toward completion. Using the same procedure, it can be shown that lead ions will react spontaneously with any precipitated copper by the reaction:

$$Pb(CH_3COO)_2 + Cu^o = Pb^o + Cu(CH_3COO)_2 \qquad \text{R23}$$

Ion Exchange Chemistry

Ion exchange is a process for removing ions from solution and replacing them with different ions producing a solution with a more desirable composition (Kunin, 1982). The process is cyclic. The solution being treated flows through an exchange bed until the resin is nearly loaded with the ion(s) being removed from the solution. For example, the basic reactions for the removal of lead and copper ions from a leach solution using a chelating resin (R_z) are as follows:

$$2R_zH + Pb(CH_3COO)_2 = Pb(R_z)_2 + 2CH_3COOH \qquad \text{R24}$$

$$2R_zH + Cu(CH_3COO)_2 = Cu(R_z)_2 + 2CH_3COOH \qquad \text{R25}$$

Similarly, the basic reactions for the removal of acetate, chloride, or nitrate ions from a leach solution using an anion exchange (R_z) resin are as follows:

$$R_zOH + CH_3COOH = R_zCH_3COO + H_2O \qquad \text{R26a}$$

$$R_zOH + HCl = R_zCl + H_2O \qquad \text{R26b}$$

The next step is the removal of the ions from the resin using a regenerant solution containing a high concentration of an acid (HCl), base (NaOH), or salt (NaCl). The reactions for this step are:

$$Pb(R_z)_2 \text{ (or Cu } (R_z)_2) + HCl = 2R_zH + PbCl_2 \text{ (or CuCl}_2) \qquad \text{R27}$$

$$R_zCH_2COO + NaOH = R_zOH + CH_2COONa \qquad \text{R28a}$$

$$R_zCl + NaOH = R_zOH + NaCl \qquad \text{R28b}$$

Consequently, ion exchange is used to concentrate waste ions into the regenerant. The regeneration of the resin allows for it to be used indefinitely. The ions can be removed from the regenerant by processes such as precipitation and electrowinning or disposed at a waste treatment plant.

The ion exchange resins are usually a spherical, porous organic solids containing fixed cations or anions capable of reversible reactions with ions of the opposite sign in the solution. Some exchange resins have a high affinity or selectivity for certain ions, while others do not. The broad classification of commercial resins is:

- **Strongly Acidic Cation Exchange Resins** — Sodium form for the removal of divalent cations, mainly calcium and magnesium; hydrogen form for the removal of all cations.
- **Strongly Basic Anion Exchange Resins** — Strong hydroxide form is used to remove all anions, like acetate, chloride, and sulfate ions.
- **Weakly Acidic Cation Exchange Resins** — Hydrogen (acid) form for the removal of cations associated with alkaline anions. Water hardness cations associated with bicarbonate alkalinity are removed; whereas cations associated with chloride and sulfate are not removed.
- **Weakly Basic Anion Exchange Resins** — Weak-base form removes entire strong acid ions, including hydrochloric acid or sulfuric acid.

- **Chelating Cation Exchange Resins** — Crosslinked macroreticular resin with a particular affinity for heavy metal cations over the more common alkali or alkaline earth cations such as sodium, potassium, calcium, and magnesium. The resin has a unique chelating functionality that is chemically bonded to the resin. The chelating functionality has several active sites for cations that coordinate with the metal ions.

2.6 LEACHING TREATABILITY STUDIES

Following physical separation and concentration in the treatability studies, the lead contaminated concentrates and slimes were chemically treated in vat and agitation leaching studies (Appendix A Sections A.6 and A.7). The leach tests were conducted to determine whether or not lead removal was possible and were not meant to be an optimization process. The initial leach tests were for a two hour duration. Where more aggressive leaching conditions were required, the leaching time was increased to four hours. It was determined that the final leaching tests in each series approached equilibrium as there was little or no change between the final two pH measurements. Buffering capacity of the soils was not measured directly, although some measure of soil alkalinity was inferred from the leaching tests. However, these were not done on the total soil. Kinetics studies were not performed.

2.6.1 Vat or Percolation Leaching

Tests to evaluate vat leaching were performed on two combined tailings and middlings samples that exceeded the total lead concentration of 500 mg/kg:

- Tailings and middlings for the −1.19 mm by +297 μm (−16 mesh by +50 mesh) fraction of Twentynine Palms LR-23 with a total lead content of 780 mg/kg, and
- Tailings and middlings for the −9.50 mm by 297 μm (−3/8 in. by +50 mesh) fractions of Twentynine Palms LR-30 with a total lead content of 710 mg/kg.

The test results are summarized in Table 2.6.

Three sets of percolation leaching tests were completed. In these tests, enough acid was added to attain a given pH value. The initial concentration of acid (g/L) was calculated from the grams of acid added and the volume of water used in each leach test.

In the first set of tests, LR23-P1 and LR30-P1, a leach solution containing 25 g nitric acid/L was used to maximize the lead extraction in a 24 hour leach. In a 250 mL bottle, 100 g of the combined soil was added to 200 mL of the acid solution. The bottle was placed on a rolling device. With nitric acid additions of 50 kg/t of soil (100 lb/ton) at the beginning of the tests, the final solution pH was maintained at about 0.5. This aggressive leach decreased the total lead content in the soil by at least 96% to a value less than 30 mg/kg.

The second set of leach tests, LR23-P2 and LR30-P2, used substantially less nitric acid, about 15 kg/t (31 lb/ton). The tests were conducted by adding 400 g of the combined soil to a vertical column and percolating the acid solution downward at a flow of 0.5 mL/min. At the beginning of the leach tests, the nitric acid concentration was about 5 g/L. After 72 hours of leaching, the total lead concentration was decreased by about 95% to less than 50 mg/kg in both tests.

In the final set of percolation leach tests, LR23-P3 and LR30-P3, the nitric acid was added incrementally over a 72 hour period. The total acid used was about 9 kg/t (18 lb/ton), resulting in a nitric acid concentration of about 2 g/L in the leach solution. These tests were also conducted in a vertical column where acid solution was percolated downward at a flow of 0.5 mL/min through 400 g of the combined soil. In the first test, the lead concentration decreased by 95% to about 30 mg/kg, while in the second test it decreased by 79% to about 150 mg/kg.

Table 2.6 Vat Leach Results for Twentynine Palms – Coarse, High Lead Fractions

Test Number	Sample: Fraction	Initial Lead Concentration (mg/kg)	Initial HNO₃ Concentration (g/L)	Time (hours)	Final pH	Accumulative HNO₃ Addition (kg/t)	Solubilized Lead	Residue Lead (mg/kg)
LR23P1	LR-23: –1.19 mm by +297 μm	780	25.0	0 24	0.50	49.7 49.7	0% 96%	<30
LR23P2	LR-23: –1.19 mm by +297 μm	780	5.0	0 72		0.0 15.3	0% 93%	46
LR23P3	LR-23: –1.19 mm by +297 μm	780	2.0	0 24 48 72		0.0 2.8 4.7 8.9	0% 46% 76% 95%	30
LR30P1	LR-30: –9.50 mm by +297 μm	710	25.0	0 24	0.46	49.7 49.7	0% >97%	<30
LR30P2	LR-30: –9.50 mm by +297 μm	710	5.0	0 72		0.0 15.3	0% 95%	53
LR30P3	LR-30: –9.50 mm by +297 μm	710	2.0	0 24 48 72		0.0 2.8 4.7 8.9	0% 31% 60% 88%	150

The leach results from the Twentynine Palms soil samples showed that contaminated fractions coarser than 105 μm (150 mesh) were amenable to percolation leaching with solutions containing about 2 g/L of nitric acid. The residues after leaching contained total lead concentrations ranging from 30 to 150 mg/kg, which were considered clean or remediated.

2.6.2 Agitation Leaching

Five sets of agitation leach tests were performed on the –105 μm (–150 mesh) fractions from the NAS Miramar (Miramar) and Twentynine Palms (TNP) samples. The total lead concentrations of the four samples are listed below:

Sample	Lead, mg/kg	Sample	Lead, mg/kg
Miramar A	5570	Miramar B	2370
TNP LR23	15,100	TNP LR30	24,300

The first set of tests used a single stage leach of Miramar and Twentynine Palms samples in which only nitric acid was added (Table 2.7). In the tests, 100 g of the –105 μm soil samples were added to 200 mL of leach solution in 400 mL beaker, which formed a 33% solid slurry. Before addition of the soil, the solution pH was adjusted to 2.0, 1.5, and 1.0 with nitric acid. The slurry was agitated with an impeller stir bar. During the 240 minute test, acid was added periodically to maintain the pH near the starting value. The results were:

- Miramar A samples seemed to be more refractory to leaching than the Miramar B sample. However, similar acid additions produced similar lead extractions: tests A-1 and A-2 versus B-1.
- The maximum acid additions were 35 kg/t for Miramar samples and 67 kg/t for TNP samples. These acid quantities decreased the total lead content to a minimum of 970 mg/kg and 2090 mg/kg, respectively. The target lead content was 500 mg/kg.
- Most of the acid was probably consumed in dissolving clay minerals as discussed in Example 4 in Section 2.5.4. Since the particle size of clay minerals is usually less than 10 μm, they will be quite reactive in strong acid solutions.

The second set of agitation leach tests examined the effects of increasing hydrochloric acid additions and adding hydrogen peroxide as a lead metal oxidant (Table 2.7). The tests also were a single-stage leach of the Miramar B sample using a 16.5% solid slurry. The test results showed the following effects:

- The lead content of the Miramar B sample was decreased to about 400 mg/kg by adding 50 kg nitric acid/t, 30 kg hydrochloric acid/t, and 6 kg hydrogen peroxide/t.
- As expected, the presence of chloride ions increased the solubility of the lead: tests B-3 versus B-14, and B-5 versus B-11.
- Hydrogen peroxide was an effective replacement oxidant for nitric acid: tests B-13 versus B-14.
- A combination of nitric acid and hydrochloric acid increased the extraction of lead: tests B-5, B-6, and B-11. However, eliminating nitric acid and increasing the hydrochloric acid addition could have produced similar results.

The third set of agitation leach tests explored the use of citric and acetic acids as replacements for nitric and hydrochloric acids and sodium chlorate as an oxidant (Table 2.7). The tests again were a single-stage leach of the Miramar B sample using a 16.5% solid slurry. The test results showed several effects:

- The nitric and hydrochloric acids were much better lixiviants than high additions of acetic or citric acids: tests B-7, B-8, and B-16 versus B-10 and B-12.
- Sodium chlorate could be an effective replacement for the hydrogen peroxide oxidant.

Table 2.7   Single-stage Agitation Leach Results for NAS Miramar and Twentynine Palms – Minus 150 Mesh, Middlings and Tailings Fractions

Test Number	Sample	Feed Lead (mg/kg)	Dry Weight Sample (grams)		Lixiviant Components	Lixiviant Added (kg/t)	Final pH	Solubilized Lead	Lead Residue (mg/kg)
			Middlings	Tailings					
A-1	Miramar A[1]	5570	79	21	HNO_3 / H_2O_2 / HCl	22 / 0 / 0	2.04	32%	3850
A-2	Miramar A[1]	5570	79	21	HNO_3 / H_2O_2 / HCl	23 / 0 / 0	1.56	42%	3230
A-3	Miramar A[1]	5570	79	21	HNO_3 / H_2O_2 / HCl	21 / 0 / 0	1.26	29%	3940
B-1	Miramar B[1]	2370	0	100	HNO_3 / H_2O_2 / HCl	25 / 0 / 0	2.15	38%	1510
B-2	Miramar B[1]	2370	0	100	HNO_3 / H_2O_2 / HCl	30 / 0 / 0	1.53	52%	1150
B-3	Miramar B[1]	2370	0	100	HNO_3 / H_2O_2 / HCl	35 / 0 / 0	1.24	59%	970
LR23-1	TNP LR-23[1]	15,100	45	55	HNO_3 / H_2O_2 / HCl	60 / 0 / 0	2.47	76%	3840
LR23-2	TNP LR-23[1]	15,100	45	55	HNO_3 / H_2O_2 / HCl	65 / 0 / 0	1.57	84%	2650
LR23-3	TNP LR-23[1]	15,100	45	55	HNO_3 / H_2O_2 / HCl	70 / 0 / 0	1.25	88%	2090
LR30-1	TNP LR-30[1]	24,300	0	100	HNO_3 / H_2O_2 / HCl	60 / 0 / 0	2.05	87%	3310
LR30-2	TNP LR-30[1]	24,300	0	100	HNO_3 / H_2O_2 / HCl	64 / 0 / 0	1.51	90%	2600
LR30-3	TNP LR-30[1]	24,300	0	100	HNO_3 / H_2O_2 / HCl	67 / 0 / 0	1.12	92%	2200
B-9	Miramar B[2]	2370	0	100	HNO_3 / H_2O_2 / HCl	14 / 6 / 7	2.10	45%	1270
B-12	Miramar B[2]	2370	0	100	HNO_3 / H_2O_2 / HCl	14 / 6 / 15	1.44	68%	770
B-10	Miramar B[2]	2370	0	100	HNO_3 / H_2O_2 / HCl	28 / 6 / 15	1.18	75%	599
B-13	Miramar B[2]	2370	0	100	HNO_3 / H_2O_2 / HCl	14 / 6 / 30	1.19	75%	596
B-14	Miramar B[2]	2370	0	100	HNO_3 / H_2O_2 / HCl	0 / 6 / 30	1.28	75%	608
B-5	Miramar B[2]	2370	0	100	HNO_3 / H_2O_2 / HCl	50 / 7 / 0	1.19	70%	704
B-6	Miramar B[2]	2370	0	100	HNO_3 / H_2O_2 / HCl	50 / 7 / 31	0.78	84%	389
B-11	Miramar B[2]	2370	0	100	HNO_3 / H_2O_2 / HCl	56 / 7 / 31	0.95	83%	434
B-7	Miramar B[2]	2370	0	100	acetic acid	182	3.83	50%	1160
B-8	Miramar B[2]	2370	0	100	citric acid	154	2.67	56%	1030
B-16	Miramar B[2]	2370	0	100	citric acid/H_2O_2/HCl	40 / 12 / 15	1.79	55%	1060
B-12	Miramar B[2]	2370	0	100	HNO_3 / H_2O_2 / HCl	14 / 6 / 15	1.44	65%	770
B-10	Miramar B[2]	2370	0	100	HNO_3 / H_2O_2 / HCl	28 / 6 / 15	1.18	75%	599
B-15	Miramar B[2]	2370	0	100	HNO_3 / $NaClO_3$ / HCl	28 / 20 / 15	1.21	73%	664

[1] 100 grams of middlings and tailings added to 200 mL of water and leached for 240 min.
[2] 100 grams of tailings added to 400 mL of water and leached for 240 min.

The fourth set of tests used two-stage leaching of Miramar and Twentynine Palms samples, in which varying amounts of nitric and hydrochloric acids and hydrogen peroxide were added (Table 2.8). In the first stage leach, 100 g of the −105 μm soil samples were mixed with 400 mL or 600 mL of solution, forming 23% or 15% solid slurries, respectively. In the second stage leach, about one-half the first stage residue, or about 50 g, was mixed with 200 mL or 300 mL of solution to create 20% or 15% solid slurries, respectively. Before adding the soil, an initial amount of nitric acid and all of the hydrochloric acid and hydrogen peroxide were added to water. The slurry was agitated with an impeller stir bar. During the 120 or 240 minute test, nitric acid was added periodically to maintain the pH near the starting value. The test results showed the following effects:

- For the Miramar A, adding 85 kg/t nitric acid, 45 kg/t hydrochloric acid, and 8 kg/t hydrogen peroxide to each stage decreased the lead content to 600 mg/kg: tests A-7 and A-8.
- Aggressive leaching of Miramar A samples decreased the lead content to less than 500 mg/kg. In these tests, the stage additions of nitric acid, hydrochloric acid, and hydrogen peroxide were 105 kg/t, 55 kg/t, and 9 kg/t respectively. The leach temperature was maintained at 40 or 60°C: tests A-11 and A-12.
- For Miramar B, adding 55 kg/t nitric acid, 30 kg/t hydrochloric acid, and 7 kg/t hydrogen peroxide to each stage decreased the lead content to 300 mg/kg: test B-17.
- For both Twentynine Palms samples, adding to each stage 60 kg/t nitric acid, 30 kg/t hydrochloric acid, and 7 kg/t hydrogen peroxide decreased the lead content to less than 500 mg/kg: tests LR23-6 and LR30-8.

2.6.3 Leaching Conclusions

Vat (Percolation) Leaching

- The contaminated coarse fraction from NAS Miramar that represented about 5% of the firing range soil had a total lead content of 3000 mg/kg.
- The contaminated sand fractions from Twentynine Palms that represented 50% of the firing range soil had a total lead content of 700 mg/kg.
- The sand fractions were easily leached, achieving lead concentrations that were less than the target of 500 mg/kg.
- Nitric acid additions for vat leaching of coarse and sand fractions would be about 10 kg/t.
- Based on the total soil mass, the nitric acid additions for the NAS Miramar fraction would be less than 3 kg/t.
- For the coarse and sand fractions of Twentynine Palms, the nitric acid additions based on the total soil mass would be less than 5 kg/t.

Soil fractions coarser than 105 μm (150 mesh) with a total lead content greater than 500 mg/kg showed good amenability to vat or percolation leaching with a 2 g/L nitric acid solution. During these tests, the leach solution pH was the targeted operating parameter. The residues after leaching contained total lead concentrations that ranged from 30 to 150 mg/kg. The nitric acid requirements for the leach averaged about 3 kg/t of firing range soil. When gravity separation was not performed before leaching, the percent lead extracted from the coarse fractions decreased significantly, indicating that most of the free lead particles were removed by gravity treatment.

Agitation Leaching

- Since chloride ions form lead-chloride complex ions, hydrochloric acid has the greatest effect on the extraction of lead from contaminated soil.
- The contaminated −105 μm (−150 mesh) fraction represented about 40% and 9% of the firing range soil from NAS Miramar and Twentynine Palms, respectively.

Table 2.8 Two-stage Agitation Leach Results for NAS Miramar and Twentynine Palms – Minus 150 Mesh, Middlings and Tailings Fractions

Test Number	Sample	Feed Lead (mg/kg)	Dry Weight Sample (grams)		Lixiviant Components	Lixiviant Added (kg/t)	Final pH	Solubilized Lead	Lead Residue (mg/kg)
			Middlings	Tailings					
A-5	Miramar A[1]	5570	21	79	HNO₃ / H₂O₂ / HCl	56 / 6 / 30	1.22	74%	1570
	2nd Stage [1]	1570	1/2 Residue	37	HNO₃ / H₂O₂ / HCl	153 / 16 / 80	1.24	89%	590
A-6	Miramar A[1]	5570	21	79	HNO₃ / H₂O₂ / HCl	56 / 6 / 30	1.14	67%	1710
	2nd Stage[3]	1710	1/2 Residue	47	HNO₃ / H₂O₂ / HCl	60 / 6 / 32	1.04	84%	875
A-7	Miramar A[2]	5570	21	79	HNO₃ / H₂O₂ / HCl	84 / 9 / 44	1.50	79%	1210
	2nd Stage[4]	1210	1/2 Residue	47	HNO₃ / H₂O₂ / HCl	90 / 7 / 47	1.00	91%	565
A-8	Miramar A[2]	5570	21	79	HNO₃ / H₂O₂ / HCl	84 / 9 / 45	1.65	77%	1290
	2nd Stage[4]	1290	1/2 Residue	50	HNO₃ / H₂O₂ / HCl	84 / 6 / 45	1.32	94%	622
A-9	Miramar A[1]	5570	21	79	HNO₃ / H₂O₂ / HCl	105 / 9 / 56	0.70	84%	995
	2nd Stage[3]	995	1/2 Residue	44	HNO₃ / H₂O₂ / HCl	119 / 10 / 63	0.80	91%	580
A-10	Miramar A[1]	5570	21	79	HNO₃ / H₂O₂ / HCl	126 / 12 / 67	0.60	85%	935
	2nd Stage[3]	935	1/2 Residue	44	HNO₃ / H₂O₂ / HCl	144 / 14 / 76	0.70	91%	565
A-11	Miramar A[1]	5570	21	79	HNO₃ / H₂O₂ / HCl	105 / 9 / 56	1.10 @ 40°C	85%	885
	2nd Stage[3]	885	1/2 Residue	48	HNO₃ / H₂O₂ / HCl	109 / 9 / 58	0.85 @ 40°C	92%	473
A-12	Miramar A[1]	5570	21	79	HNO₃ / H₂O₂ / HCl	105 / 9 / 56	1.00 @ 60°C	88%	691
	2nd Stage[3]	690	1/2 Residue	50	HNO₃ / H₂O₂ / HCl	105 / 9 / 56	0.90 @ 60°C	94%	327
B-17	Miramar B[1]	2370	0	100	HNO₃ / H₂O₂ / HCl	56 / 6 /30	1.10	71%	682
	2nd Stage[3]	682	1/2 Residue	48	HNO₃ / H₂O₂ / HCl	59 / 7 / 31	0.99	87%	313
LR23-5	TNP LR-23[1]	15,100	45	55	HNO₃ / H₂O₂ / HCl	56 / 6 / 30	1.33	93%	1120
	2nd stage[3]	1120	1/2 Residue	51	HNO₃ / H₂O₂ / HCl	111 / 12 / 59	1.25	99%	170
LR23-6	TNP LR-23[1]	15,100	45	55	HNO₃ / H₂O₂ / HCl	56 / 6 / 30	1.17	74%	3200
	2nd stage[3]	3200	1/2 Residue	47	HNO₃ / H₂O₂ / HCl	60 / 7 / 32	0.98	98%	360
LR30-5	TNP LR-30[1]	24,300	0	100	HNO₃ / H₂O₂ / HCl	56 / 6 / 30	1.33	88%	3090
	2nd stage[3]	3090	1/2 Residue	44	HNO₃ / H₂O₂ / HCl	56 / 6 / 30	1.30	99%	250
LR30-6	TNP LR-30[1]	24,300	0	100	HNO₃ / H₂O₂ / HCl	127 / 14 / 67	1.13	75%	6043
	2nd stage[3]	6043	1/2 Residue	46	HNO₃ / H₂O₂ / HCl	56 / 6 / 30	0.96	97%	640
LR30-7	NP LR-30[2]	24,300	0	100	HNO₃ / H₂O₂ / HCl	84 / 9 / 45	1.69	98%	1250
	2nd stage[4]	1250	1/2 Residue	42	HNO₃ / H₂O₂ / HCl	100 / 7 / 53	1.30	+99%	176
LR30-8	TNP LR-30[2]	24,300	0	100	HNO₃ / H₂O₂ / HCl	56 / 6 / 30	1.73	92%	2010
	2nd stage[4]	2010	1/2 Residue	46	HNO₃ / H₂O₂ / HCl	61 / 7 / 32	1.27	98%	242

[1] 100 grams of middlings and tailings added to 400 mL of water and leached for 120 min; A-5, LR23-5 and LR30-5 for 240 min.
[2] 100 grams of middlings and tailings added to 600 mL of water and leached for 120 min; A-7 for 240 min.
[3] One-half first stage residue, ~50 grams, added to 200 mL of water and leached for 120 min; LR23-5 and LR30-5 for 240 min.
[4] One-half first stage residue, ~50 grams, added to 300 mL of water and leached for 120 min; A-7 for 240 min.

- Treating the −105 μm fraction for NAS Miramar with a hydrocyclone and Knelson bowl concentrator could have decreased the amount that required chemical treatment to about 24% of the total soil mass.
- Two-stage agitation leaching of the average −105 μm fraction for NAS Miramar required an average of 70 kg/t, 40 kg/t, and 7 kg/t per stage for the nitric acid, hydrochloric acid, and hydrogen peroxide additions, respectively.
- Based on the total soil mass, the total nitric acid, hydrochloric acid, and hydrogen peroxide additions for NAS Miramar were 34 kg/t, 19 kg/t, and 4 kg/t, respectively.
- Heating the slurry containing the −105 μm fraction for NAS Miramar might be required to achieve the target lead concentration.
- Two-stage agitation leaching of the average −105 μm fraction for Twentynine Palms required nitric acid, hydrochloric acid, and hydrogen peroxide additions of 60 kg/t, 30 kg/t, and 7 kg/t per stage, respectively.
- Based on the total soil mass, the total nitric acid, hydrochloric acid, and hydrogen peroxide additions for Twentynine Palms would be 11 kg/t, 6 kg/t, and 1 kg/t, respectively.

Agitation leach tests were performed on NAS Miramar samples from the −105 μm (−150 mesh) fraction, which had total lead concentrations up to 5600 mg/kg. The high surface area to volume ratio of the particles enabled this fraction to be easily leached. The lixiviant which yielded the greatest extraction of lead was a combination of nitric acid (34 kg/t), hydrochloric acid (19 kg/t), and hydrogen peroxide (4 kg/t). The initial series of tests used a single-stage leach, which was unsuccessful in decreasing the total lead concentration to less than 500 mg/kg. The second series of agitation leach tests used two stages, which decreased the total lead concentration to less than 400 mg/kg. Increasing the leach temperature improved lead extraction.

2.7 FLOWSHEET DESIGN

For each of the three soils tested in the treatability studies, removal of the large gravel-sized particles should precede introduction of the soil into the treatment plant. Pretreatment of the soil should involve use of a double-deck vibrating screen to remove the +50.8 mm (+2 in.) and −50.8 mm by +19.0 mm (−2 in. by +3/4 in.) soil fractions. Both fractions were clean, with lead content less than 500 mg/kg. These materials should be placed in soil stockpiles, mixed with the clean treated soil, and returned to the original excavation area. In most applications, the treatment plant should require a feed system consisting of a storage hopper and feed conveyor, a weigh-belt conveyor, and a bucket elevator or conveyor to feed the first unit operation.

The particle size and gravity separation treatability studies demonstrated that specific unit operations were needed to process each soil in a treatment plant. For the NAS Miramar soils, nine major pieces of equipment were needed for the treatment plant: single-deck wash screen, primary jig, attrition scrubber, trommel wash screen, secondary jig, spiral classifier, Reichert spiral concentrator, hydrocyclone, and Knelson bowl concentrator. For the Twentynine Palms soil, five pieces of equipment were needed: double-deck wash screen, primary jig, secondary jig, spiral classifier, and Reichert spiral concentrator. The Fort Dix soil needed only four major pieces of equipment: single-deck wash screen, primary jig, spiral classifier, and Reichert spiral concentrator. Ancillary equipment for all treatment plants included a clarifier, centrifuge, and plate and frame filter. Several soil fractions from NAS Miramar and Twentynine Palms required vat and agitation leaching. Leach processes were not required for any fractions from the Fort Dix soil.

NAS Miramar
Based on the treatability study results, the recommended gravity plant flowsheet for the treatment of NAS Miramar firing range soil is shown in Figure 2.2. The major unit operations for the treatment process are as follows:

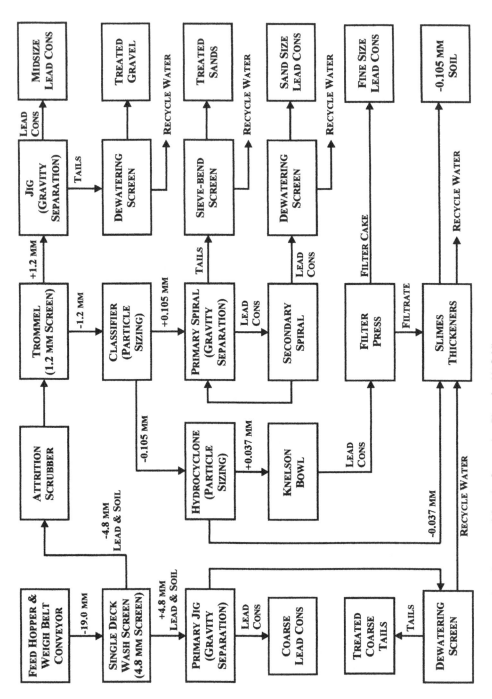

Figure 2.2   Block Flow Diagram of Gravity Separation Plant for NAS Miramar.

- Single-deck vibrating screen with high-pressure water spray nozzles for initial soil deagglomeration and particle sizing.
  - Screen opening at 4.76 mm (4 mesh).
  - Screen products:
    - Oversize: −19.0 mm by +4.76 mm fraction, 7.7% of total soil by weight containing 73,800 mg Pb/kg.
    - Undersize: −4.76 mm fraction, 92.3% of total soil by weight containing 6650 mg Pb/kg.
- Primary jig for gravity separation of the −19.0 mm by +4.76 mm fraction.
  - Jig concentrate: 1.8% of total soil by weight containing 314,800 mg Pb/kg
  - Jig tailings: 5.9% of total soil by weight containing 290 mg Pb/kg
- Attrition scrubber for deagglomeration of −4.76 mm fraction.
- Single-deck vibrating screen or trommel screen with low-pressure water spray nozzles for particle sizing of attrition scrubber overflow.
  - Screen opening at 1.19 mm (16 mesh).
  - Screen products:
    - Oversize: −4.76 mm by +1.19 mm fraction, 6.5% of total soil by weight containing 39,500 mg Pb/kg.
    - Undersize: −1.19 mm fraction, 85.8% of total soil by weight containing 4150 mg Pb/kg.
- Secondary jig for gravity separation of the −4.76 mm by +1.19 mm fraction.
  - Jig concentrate: 2.25% of total soil by weight containing 108,500 mg Pb/kg.
  - Jig tailings: 4.25% of total soil by weight containing 3045 mg Pb/kg.
  - Jig tailings require vat leaching to decrease lead content to less than 500 mg/kg.
- Spiral classifier for particle sizing of 1.19 mm (16 mesh) screen undersize.
  - Maximum particle size of overflow: 105 μm (150 mesh).
  - Classifier products:
    - Underflow (Sand): 44.5% of total soil by weight containing 3500 mg Pb/kg.
    - Overflow (Slimes): 41.3% of total soil by weight containing 4800 mg Pb/kg.
- Primary and secondary Reichert spiral concentrators for gravity separation of the −1.19 mm by +105 μm fraction.
  - Spiral concentrate: 5.3% of total soil by weight containing 27,000 mg Pb/kg.
  - Spiral tailings: 39.2% of total soil by weight containing 400 mg Pb/kg.
- Hydrocyclone for particle sizing of the −105 μm slimes.
  - Particle size and lead splits are based on the hydrocyclone treatability tests.
  - Removal of 57% of the −37 μm and +95% of the −10 μm particles into the overflow.
  - Overflow, −37 μm particles: 14.9% of total soil by weight containing 6700 mg Pb/kg.
  - Underflow, fine sand particles: 26.5% of total soil by weight containing 3750 mg Pb/kg.
  - The overflow is fed into the fines (slimes) settling basin.
  - Before discharge into settling basin, the overflow slurry requires agitation leaching to decrease the lead content below 500 mg/kg.
- Knelson bowl concentrator for gravity separation of hydrocyclone underflow slurry.
  - Knelson bowl concentrate: 1.5% of total soil by weight containing 59,000 mg Pb/kg.
  - Knelson bowl tailings: 25.0% of total soil by weight containing <500 mg Pb/kg.
- Settling basins (roll-off bins or lamella clarifiers for fines, or slimes products).
  - Hydrocyclone overflow: 14.9% of total soil by weight containing 6700 mg Pb/kg.
  - Knelson bowl tailings: 25.0% of total soil by weight containing <500 mg Pb/kg.
- Final products:
  - Concentrates (primary jig, secondary jig, Reichert spiral, and Knelson bowl):
    - 10.8% of total soil by weight containing 96,500 mg Pb/kg.
  - Clean soil (primary jig, Reichert spiral, and Knelson bowl tailings):
    - 70.1% of total soil by weight containing about 400 mg Pb/kg.
- Leaching:
  - Vat leach feed (secondary jig tailings):
    - 4.2% of total soil by weight containing 3045 mg Pb/kg.
  - Agitation leach feed (hydrocyclone overflow):
    - 14.9% of total soil by weight containing 6700 mg Pb/kg.

- Ancillary equipment consists of:
  - High and low pressure water pumps, slurry pumps,
  - Dewatering screens for coarse and sand products,
  - Filter press for Knelson bowl concentrate,
  - Centrifuge for settling basin underflow slurry, and
  - Conveyors for dewatered coarse, sand, and slimes products.

An attrition scrubber was used in the demonstration pilot plant (Figure 1.1) in place of the wash screen, mineral jig, and commercial scrubber due to availability.

Twentynine Palms

Based on the treatability study results, the recommended gravity plant flowsheet for the treatment of Twentynine Palms firing range soil is shown in Figure 2.3. The major unit operations for the treatment process are as follows:

- Double-deck vibrating screen with high-pressure water spray nozzles for initial soil deagglomeration and particle sizing.
  - Screen openings at 4.76 mm (4 mesh) and 1.19 mm (16 mesh)
  - 4.76 mm screen products:
    - Oversize: –19.0 mm by +4.76 mm fraction, 15.5% of total soil by weight containing 13,600 mg Pb/kg.
    - Undersize: –4.76 mm fraction, 84.5% of total soil by weight containing 11,900 mg Pb/kg.
  - 1.19 mm screen products:
    - Oversize: –4.76 mm by +1.19 mm fraction, 17.6% of total soil by weight containing 16,800 mg Pb/kg.
    - Undersize: –1.19 mm fraction, 66.9% of total soil by weight containing 10,600 mg Pb/kg.
- Primary jig for gravity separation of the –19.0 mm by +4.76 mm fraction.
  - Jig concentrate: 2.5% of total soil by weight containing 82,100 mg Pb/kg
  - Jig tailings: 13.0% of total soil by weight containing 165 mg Pb/kg
- Secondary jig for gravity separation of the –4.76 mm by +1.19 mm fraction.
  - Jig concentrate: 5.0% of total soil by weight containing 58,300 mg Pb/kg.
  - Jig tailings: 12.6% of total soil by weight containing 550 mg Pb/kg.
  - Jig tailings might require vat leaching to decrease lead content to less than 500 mg/kg.
- Spiral classifier for particle sizing of 1.19 mm (16 mesh) screen undersize.
  - Maximum particle size of overflow: 105 μm (150 mesh).
  - Classifier products:
    - Underflow (sand): 57.1% of total soil by weight containing 8200 mg Pb/kg.
    - Overflow (slimes): 9.8% of total soil by weight containing 24,600 mg Pb/kg.
- Primary and secondary Reichert spiral concentrators for gravity separation of the –1.19 mm by +105 μm fraction.
  - Spiral concentrate: 6.5% of total soil by weight containing 65,800 mg Pb/kg.
  - Spiral tailings: 50.6% of total soil by weight containing 701 mg Pb/kg.
  - Spiral tailings might require vat leaching to decrease lead content to less than 500 mg/kg.
- Final products:
  - Concentrates (primary jig, secondary jig, and Reichert spiral):
    - 14.0% of total soil by weight containing 66,100 mg Pb/kg.
  - Clean soil (primary jig tailings):
    - 13.0% of total soil by weight containing about 165 mg Pb/kg.
- Leaching:
  - Vat leach feed (secondary jig and Reichert spiral tailings):
    - 63.2% of total soil by weight containing 670 mg Pb/kg.
  - Agitation leach soil (spiral classifier overflow):
    - 9.8% of total soil by weight containing 19,900 mg Pb/kg.

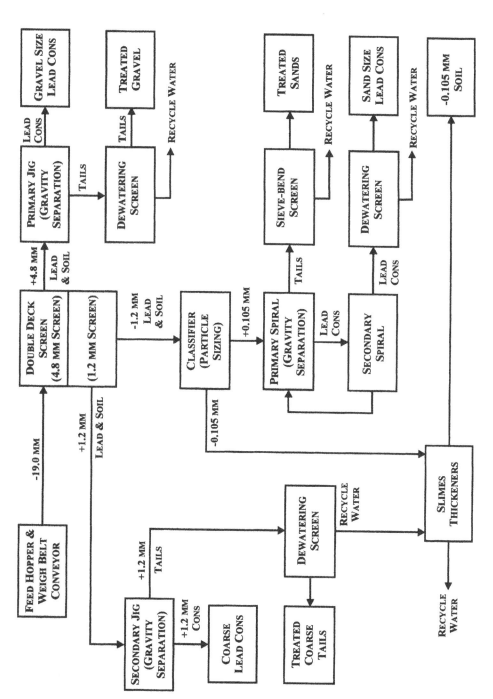

Figure 2.3   Block Flow Diagram of Gravity Separation Plant for Twentynine Palms Soil.

- Ancillary equipment consists of:
  - High and low pressure water pumps, slurry pumps,
  - Dewatering screens for coarse and sand products,
  - Centrifuge for settling basin underflow slurry, and
  - Conveyors for dewatered coarse, sand, and slimes products.

Fort Dix

Based on the treatability study results, the recommended gravity plant flowsheet for the treatment of Fort Dix firing range soil is shown in Figure 2.4. The major unit operations for the treatment process are as follows:

- Single-deck vibrating screen with high-pressure water spray nozzles for initial soil deagglomeration and particle sizing.
  - Screen openings at 1.19 mm (16 mesh)
  - 1.19 mm screen products:
    - Oversize: +1.19 mm fraction, 27.4% of total soil by weight containing 9800 mg Pb/kg.
    - Undersize: −1.19 mm fraction, 72.6% of total soil by weight containing 825 mg Pb/kg.
- Primary Jig for gravity separation of the −19.0 mm by +4.76 mm fraction.
  - Jig concentrate: 2.7% of total soil by weight containing 97,500 mg Pb/kg
  - Jig tailings: 24.7% of total soil by weight containing 180 mg Pb/kg
- Spiral classifier for particle sizing of 1.19 mm (16 mesh) screen undersize.
  - Maximum particle size of overflow: 105 μm (150 mesh).
  - Classifier products:
    - Underflow (sand): 64.0% of total soil by weight containing 589 mg Pb/kg.
    - Overflow (slimes): 8.6% of total soil by weight containing 2600 mg Pb/kg.
- Primary and secondary Reichert spiral concentrators for gravity separation of the −1.19 mm by +105 μm fraction.
  - Spiral concentrate: 5.1% of total soil by weight containing 6350 mg Pb/kg.
  - Spiral tailings: 58.9% of total soil by weight containing 89 mg Pb/kg.
- Final Products:
  - Concentrates (primary jig and Reichert spiral):
    - 7.8% of total soil by weight containing 37,900 mg Pb/kg.
  - Clean soil (primary jig and Reichert spiral tailings and spiral classifier overflow):
    - 92.2% of total soil by weight containing about 315 mg Pb/kg.
- Ancillary equipment consists of:
  - High and low pressure water pumps, slurry pumps,
  - Dewatering screens for coarse and sand products,
  - Centrifuge for settling basin underflow slurry, and
  - Conveyors for dewatered coarse, sand, and slimes products.

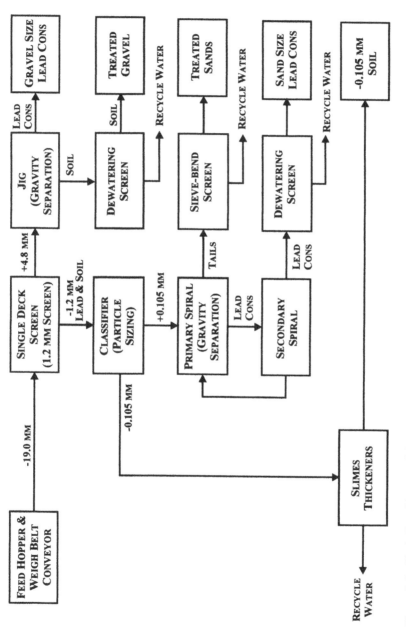

Figure 2.4 Block Flow Diagram of Gravity Separation Plant for Fort Dix Soil.

CHAPTER 3

Technology Demonstration

3.1 TEST SITE DESCRIPTION

3.1.1 Site Selection

Soil samples from the small arms ranges of three DOD military bases — Fort Dix, New Jersey; Twentynine Palms, California; and NAS Miramar, California — were provided for treatability studies. The treatability results were used to select the site for the field demonstration. The selection was based on the criteria that most closely met the capability of the pilot plant equipment and included:

- About 25% of the soil as coarse fractions, +1.19 mm (+16 mesh), and at least 25% as sand fractions, −1.19 mm by +105 µm (+150 mesh);
- At least 50% of the coarse and sand fractions requiring vat leaching;
- At least 25% as slimes fraction, −105 µm, for testing the Knelson bowl concentrator; and
- At least 25% of the total soil mass as −105 µm soil requiring agitation leaching.

Other factors used to aid in the selection of the site were:

- U.S. Navy support for the demonstration,
- Climatic conditions,
- Proximity to material and equipment vendors,
- Large flat isolated working area to set up pilot plant, and
- Minimized travel for operating crew from the BOR in Salt Lake City.

Based on these factors, NAS Miramar was selected as the demonstration site.

3.1.2 Site Description

The field demonstration for the lead remediation project was conducted at the former Camp Elliott small arms range located on NAS Miramar, San Diego, California. The range was about 50 stations across with shooting locations at 92 m (100 yd), 183 m (200 yd), and 457 m (500 yd) from the berm. The total berm length was approximately 152 m (500 ft) with a height of 9.1 m (30 ft).

The soil excavation area for the pilot demonstration was directly behind stations 27 through 48. The soil excavated was taken from the face along the toe of the berm, to a height of about 3.65 m (12 ft), a depth of about 0.45 m (1.5 ft), and a distance of about 64 m (210 ft). The excavation provided about 105 m^3 (140 yd^3) of soil for the demonstration.

3.1.3 Site Location

NAS Miramar is located in the City of San Diego, California, approximately 20.9 km (13 miles) north of downtown San Diego. The air station contains over 9300 hectares (23,000 acres) and is one of the largest and most active military airfields in the country.

Interstate 15 (I-15) bisects NAS Miramar into western and eastern parcels. The 3065 hectares (7573 acres) of West Miramar, also known as the Main Station, is the site of major operations. The 6204 hectares (15,330 acres) in East Miramar are mostly undeveloped. The boundaries of NAS Miramar may be generally described as follows. The western boundary is formed by a portion of Interstate Highway 805 (I-805), approximately 9.6 km (6 miles) east of La Jolla. About 19 km (12 miles) to the northeast, the eastern boundary is formed by a portion of Sycamore Canyon. Miramar and Pomerado Roads roughly define the northern boundary, and State Highway 52 approximates the southern boundary of West Miramar. No significant natural or manmade features clearly demarcate the southern boundary of East Miramar (NEESA, 1984).

3.1.4 Site History

The present site of the Main Station has been utilized by the U.S. government since May 23, 1917. In 1917, about 5300 hectares (13,000 acres) were leased for the establishment of an Army Infantry Training Center known as Camp Kearny for practice and drill maneuvers. Shortly after World War I, Camp Kearny was closed and most of the buildings were demolished.

During the time between World Wars I and II, the site was used for a variety of military functions, including a target bombing range. A portion of Camp Kearny was developed by the U.S. Navy as a lighter-than-air aircraft base in 1929, with plans to make it the major Pacific coast dirigible base. In May 1932, the first steerable lighter-than-air aircraft, *U.S.S. Akron*, arrived at Camp Kearny.

Approximately 170 hectares (420 acres) of land were procured in 1939 for the establishment of Kearny Mesa Naval Air Station. The station was again expanded in 1942 through the acquisition of an additional 401 hectares (991 acres) of land. In 1943, a total of 446 hectares (1101 acres) of land was obtained, and the Station was subsequently divided into southern and northern halves. The southern half of the Station was commissioned as an auxiliary air station of Naval Air Station, North Island, with a mission to administer operations and maintain facilities, and to provide services in support of fleet units. The northern half of the station was designated as Marine Corps Air Depot, Miramar, and was used as a staging area for pilots and crewmen. However, after V-J Day the Marine Corps Air Depot was used as a separation center where 25,000 men were processed for separation from the Marine Corps. On May 1, 1946, the northern and southern halves were combined and designated as Marine Corps Air Station Miramar.

Following the transfer of the Marines to El Toro on June 30, 1947, the Air Station was ordered into a reduced operational status by the Chief of Naval Operations (CNO) and was unofficially designated NAS Miramar. In July 1949, funds were appropriated to develop NAS Miramar as a Master Jet Air Station, in response to the upsurge in the utilization of jets in war. During the next few years, a major construction and rehabilitation program was undertaken on the Air Station. On April 1, 1952, Miramar was officially commissioned as a Naval Air Station (NEESA, 1984).

East Miramar has been part of the Navy's land holdings since 1941, with the acquisition of the 7867 hectare (19,438 acre) Marine Corps training base known as Camp Elliott. At this time, the base functioned as the Pacific Coast Marine Training Base. After World War II, the majority of Camp Elliott was declared excess to military needs and sold. However, the Marine Corps retained a part that included Sycamore Canyon.

In 1955, the Atlas Missile Facility was established in the Sycamore Canyon property of Camp Elliott. This secure facility was operated by General Dynamics under the auspices of the National Aeronautics and Space Agency (NASA) and the U.S. Air Force.

Following the disestablishment of Camp Elliott in 1960, a total of 3120 hectares (7709 acres) of land, including 175 buildings and structures, was transferred to the custody of NAS Miramar. In the same year, the 2994 hectares (7399 acres) of Sycamore Canyon property was transferred to the U.S. Air Force and subsequently to NASA for further testing of missiles. This property was declared excess to NASA's needs in 1970, and was again transferred to the Department of the Navy in 1971. NAS Miramar assumed responsibility for the property beginning in 1972. Portions of this area are currently used for training and ordnance storage.

3.2 PILOT PLANT DESCRIPTION AND DESIGN

3.2.1 Process Flow Design

The treatment systems used to process firing range soil at NAS Miramar included the gravity separation and vat leach pilot plants. The gravity separation plant contained unit operations that:

- Prescreened excavated soil to produce −19.0 mm (−3/4 in.) feed for the gravity plant,
- Deagglomerated and scrubbed soil particles,
- Sized soil and lead particles for gravity separation,
- Separated the lead from the soil by density, and
- Separated slimes particles from the water by thickening.

The vat leach plant included unit operations that dissolved or leached the lead from coarse and sand-sized soil fractions and recovered lead from the leach solution.

Using the treatability results for NAS Miramar soil in Chapter 2, simplified flow diagrams of the two treatment plants were prepared to show all unit operations and the major flows (Figures 3.1 and 3.2). An abbreviated mass balance for an operating rate at 907 kg/hr (2000 lb/hr) of −19.0 mm (−3/4 in.) soil was also prepared (Table 3.1). Detailed flow diagrams for the gravity plant and detailed mass balances for both the gravity and leach plants are given in Appendices B and C, respectively.

The major unit operations of the gravity separation pilot plant in Figure 3.1 included:

- Double-deck screen for prescreening soil (not shown),
- Soil feed system: feed hopper and conveyor belt, weigh-belt conveyor, and bucket elevator,
- Attrition scrubber,
- Trommel,
- Two spiral classifiers,
- Mineral jig,
- Two Reichert spiral concentrators,
- Hydrocyclone,
- Knelson bowl concentrator, and
- Slimes thickening system.

Deagglomeration and scrubbing of the firing range soil was accomplished in the attrition scrubber and trommel. Particle sizing occurred in the double-deck screen, attrition scrubber, trommel, spiral classifiers, and hydrocyclone. Gravity separation occurred in the attrition scrubber, mineral jig, Reichert spiral concentrators, and Knelson bowl concentrator.

The major unit operations of the leach pilot plant in Figure 3.2 included:

- Batch vat leach tank,
- Lead cementation column,
- Chelating ion exchange (IX) column, and
- Anion ion exchange column.

Table 3.1 Mass Balance for the Design of the NAS Miramar Demonstration Plant

Flow	Description	Mass Flows		Slurry Solids wt %	Solids Lead mg/kg	Comments
		Solid kg/hour	Liquid kg/hour			
Gravity Treatment Plant						
1	Feed: Soil from Firing Range Berm	1209.58	134.40	90.00	8979	Mass Split, +/−19.0 mm: 25% & 75%
2	+19.0 mm Soil	302.39	33.60	90.00	500	**To Clean Soil Stockpile**
3	−19.0 mm Soil	907.18	100.80	90.00	11,805	Feed to Attrition Scrubber
4	Dilution Water from Day Tank	0.00	474.89	0.00	0	Feed to Attrition Scrubber
5	Overflow: −4.76 mm (−4 mesh) Soil	837.35	558.23	60.00	6638	Feed to Trommel Screen
6	Auger Product: Bullets/Stones	69.84	17.46	80.00	73,757	Feed to Vibrating Screen #1
7	Oversize: +1.19mm (+16 mesh) Product	69.84	3.68	95.00	73,757	**Lead Product (to storage barrel)**
8	Undersize: −1.19 mm (−16 mesh) Slurry	0.00	13.78	0.00	0	Feed to Spiral Classifier
9	Wash Water from Day Tank	0.00	235.67	0.00	0	Feed to Trommel Screen (S/L=4/1)
10	Oversize: −4.76 mm, +1.19 mm (−4 mesh, +16 mesh) Soil	58.92	14.73	80.00	39,543	Feed to Jig: +1.19 mm (+16 mesh)
11	Undersize: −1.19 mm (−16 mesh) Soil	758.97	759.69	49.98	4147	Feed to Spiral Classifier
12	Dilution Water from Day Tank	0.00	95.55	0.00	0	Feed to Mineral Jig
13	Jig Concentrate	20.39	20.39	50.00	108,487	**Lead Product (to storage barrel)**
14	Trommel Overflow −1.19 mm (−16 mesh)	19.46	19.48	49.98	4147	Feed to Spiral Classifier
15	Jig Tailings	38.52	89.89	30.00	3045	Feed to Vibrating Screen #2
16	Oversize: Jig Tailings	38.52	16.51	70.00	3045	**To Vat Leach Feed Stockpile**
17	Undersize: Water	0.00	73.38	0.00	0	Feed to Classifier Recycle Sump
18	Sands: −1.19 mm, +105 µ (−16 mesh, +150 mesh)	413.72	177.31	70.00	3535	Feed to Gravity Separator: +105 µ (+150 mesh)
19	Fines: −105 µ (−150 mesh Soil)	384.67	792.20	32.69	4821	Feed to Hydrocyclone −105 µ (−150 mesh)
20	Gravity Separator Dilution Water	0.00	3157.42	0.00	0	Feed Water to Sands Sump at +52 lpm
21	Sump Overflow	10.34	83.37	11.04	3535	Feed to Recycle Sump at 2.50% Total Flow
22	Sands Sump Slurry	403.37	3251.36	11.04	3535	Feed to Reichert Spiral at ~55 lpm
23	Gravity Separator Concentrate	47.59	269.66	15.00	26,988	Feed to Screen #3
24	Separator Tailings	355.79	2981.70	10.66	398	Feed to Sieve Bend Screen
25	Undersize: Water	0.00	2918.91	0.00	0	Feed to Bowl Pump
26	Oversize: +105 µ (+150 mesh) Soil	47.59	11.90	80.00	26,988	**Lead Product (to storage barrel)**
27	Undersize: Water	0.00	257.76	0.00	0	Feed to Dewatering Tank
28	Fines Sump Overflow	9.62	19.80	32.69	5392	Feed to Recycle Sump at 2.50% of Total Flow

No.	Stream					Destination
29	Recycle to Spiral Classifier	19.96	176.55	10.16	4430	Recycle to Spiral Classifier
30	Classifier Slimes: –105 µ (–150 mesh)	375.05	772.39	32.69	5392	Feed to Hydrocyclone
31	Overflow: –37 µ (–400) mesh Soil	134.33	575.44	18.93	6395	Feed to Dewatering Tank
32	Underflow: +37 µ (+400) mesh Soil	240.72	196.95	55.00	3920	Feed to Knelson Bowl
33	Knelson Bowl Fluidizing Water from Day Tank	0.00	6813.00	0.00	0	Feed to Knelson Bowl at ~115 lpm
34	Knelson Bowl Concentrate	8.58	3.68	70.00	30,209	Feed to Filter Press
35	Knelson Bowl Tailings	232.14	7006.28	3.21	2948	Feed to Dewatering Tank
36	Filter Cake	8.58	0.95	90.00	30,209	**Lead Product (to storage barrel)**
37	Filtrate Water	0.00	2.72	0.00	0	Feed to Dewatering Tank
38	Sieve-Bend Underflow Water from Bowl Pump	0.00	3176.68	0.00	0	Feed to Dewatering Tank
39	Dewatering Tank Overflow (<1 micron)	0.00	10,516.80	0.00	0	Feed to Day Tank
40	Dewatering Tank Underflow	366.47	244.32	60.00	4211	**Feed to Agitation Leach**
50	System Make-up Water	0.00	259.73	0.00	0	Feed to Gravity System

Vat Leach Plant

No.	Stream					Destination
16	Jig Tailings	38.52	16.51	70.00	3045	Feed to Vat Leach
41	Reichert Separator Tailings	355.79	62.79	85.00	398	Feed to Vat Leach
44	Leach Solution	0.00	157.29	0.00	0	Feed to Vat Leach; Pb Content = 0.00 g/L
45	Wash Water from Day Tank	0.00	236.59	0.00	0	
46	30% Acetic Acid (Solution)	0.00	2.39	0.00	0	Acetic Acid = 0.593 kg/1000 kg Feed (Flow #1)
47	30% Calcium Peroxide (Solution)	0.00	0.36	0.00	0	Ca Peroxide = 0.089 kg/1000 kg
48	Vat Leach Solution	0.00	377.34	0.00	164	Leach Solution Pb Content = 0.251 g/L
49	Vat Leach Tailings	394.31	98.58	80.00	500	Pb in Solid (Wet Tailing) = 500 mg/kg

Lead Precipitation

No.	Stream					Destination
48	Feed: Vat Leach Solution	0.0000	377.34	0.00	164	Leach Solution Pb Content = 0.281 g/L
Fe	Feed: Scrap Iron or Iron Wool	0.0185	0.00	100.00	0	Iron (Fe) Utilization = 90%
51	Precipitate: Lead Metal	0.0636	0.00	100.00	970,924	**To Lead Product Storage Barrels**
52	Treated Solution	0.0000	377.29	0.00	0	Recycled to Vat Leach
53	Bleed Solution	0.0000	220.00	0.00	0	Leach Bleed Solution

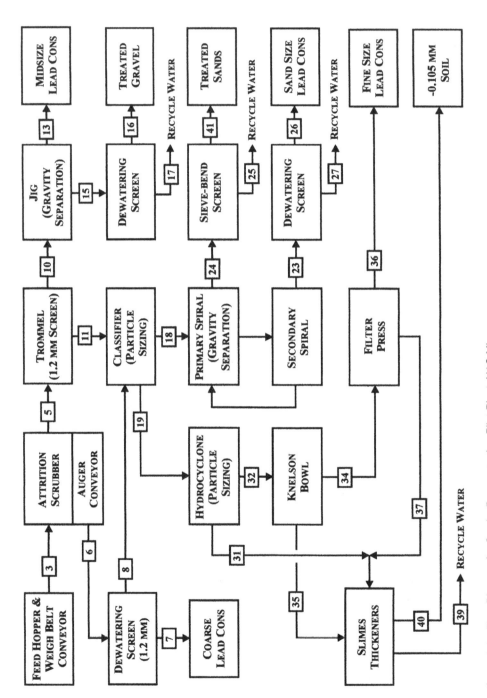

Figure 3.1 Flow Diagram for Gravity Demonstration Pilot Plant: NAS Miramar.

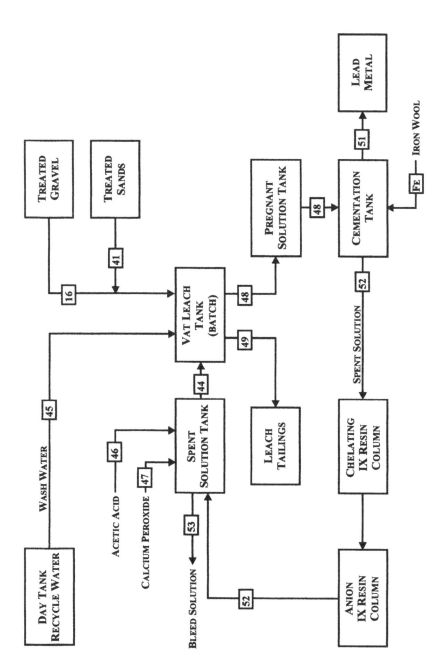

Figure 3.2 Flow Diagram for Vat Leach Demonstration Pilot Plant: NAS Miramar.

In the vat leach, the lead in the soil was dissolved using acetic acid and calcium peroxide. The lead was removed from the leach solution by cementation with iron wool, forming a lead metal precipitate. To provide clean solution for leaching, ion exchange columns removed copper and residual iron ions from the leach solution using a chelating ion exchange resin. An anion ion exchange column can be used to remove acetate, chloride, and nitrate ions. The treated solution was recycled to the leach.

3.2.1.1 *Soil Prescreen and Feed Systems*

Screening is a mechanical process that separates particles based on size. Particles are presented to apertures in a screen surface and rejected if larger than the opening or accepted and passed through if smaller.

The NAS Miramar firing range soil was prescreened before feeding the firing range soil into the pilot plant. A feed hopper, a double-deck vibrating screen, and a belt conveyor comprised the prescreening system. The screen removed oversize stone and debris on the upper screen and +19.0 mm (+3/4 in.) material on the second screen. These fractions were clean and did not require additional processing or treatment. The −19.0 mm material was stockpiled as feed to the gravity treatment plant.

A mobile feed unit was used to add the soil into the pilot plant feed hopper. The feed unit consisted of a hopper and an inclined belt conveyor. The −19.0 mm soil was added to the hopper using a front-end loader. The conveyor carried the soil to the pilot plant feed hopper. A belt conveyor under this hopper fed the weigh-belt conveyor. The weigh-belt conveyor transported the soil to a bucket elevator, which lifted the soil into the feed chute on the attrition scrubber.

3.2.1.2 *Deagglomeration*

Deagglomeration is the process of separating particles from each other by mechanical means, which allows for accurate particle sizing. The attrition scrubber and the trommel were used in this process to separate the lead particles from the soil matrix in order to recover the lead particles.

Attrition Scrubber

The attrition scrubber consisted of two cells. Each scrubber cell used an agitator with a double set of opposing turbine blades to mix the soil-water slurry. The effective diameter of the cells, the slurry height, and the agitator blade diameter, spacing, and rpm were selected to suspend −4.76 mm (−4 mesh) soil particles (Appendix D.1). The +4.76 mm soil and heavy lead particles settled to the bottom of the scrubber. The scrubber's mixing parameters were also selected to aid the breakup of agglomerated soil particles and the removal of lead smears on soil particles. Review of attrition-scrubber literature showed that optimum deagglomeration occurs with a 60 to 70 % solids; therefore, sufficient water would have to be added to the scrubber to produce a 60 to 70% solids slurry. The attrition scrubber can also grind soil particles into finer particles when operated with high solids slurries.

The scrubber overflow slurry containing the −4.76 mm soil flowed through a chute into the trommel. An auger conveyor was attached to the bottom of the scrubber to remove the coarse soil and lead particles. The spiral auger conveyed these particles to a double-deck Sweco screen for dewatering. The +4.76 mm (+4 mesh) and +1.19 mm (+16 mesh) fractions from this screen were stored as coarse lead concentrate. The screen slurry containing the −1.19 mm soil and lead particles discharged into the spiral classifiers.

Trommel

The trommel was a horizontal rotating drum consisting of two sections. A blind section was at the front end of the drum, and a screening section was at the rear end. Additional deagglomeration of the soil particles in the attrition scrubber overflow probably occurred in the blind section. The soil was sized by the 1.19 mm (16 mesh) wire screen in the rear section. Two water spray bars were placed in the trommel to enhance the deagglomeration of particles and screening of material. The trommel's wet oversized material (+1.19 mm) discharged into the mineral jig and the slurry containing the undersized soil particles (−1.19 mm) flowed through a chute into the spiral classifiers.

3.2.1.3 Particle Sizing

Particle sizing occurred in the attrition scrubber, trommel, spiral classifiers, and hydrocyclone. Initially, the +4.76 mm (+4 mesh) soil and lead particles settled to the bottom of the attrition scrubber. The spiral auger transported these coarse particles to a Sweco screen for dewatering. The slurry containing the −4.76 mm particles flowed into the trommel. The trommel performed the second particle-sizing step at 1.19 mm (16 mesh). The +1.19 mm soil and lead discharged into the mineral jig and the slurry containing the −1.19 mm material flowed through a chute into the spiral classifiers.

Spiral Classifiers

Mechanical classification in a spiral classifier is a two-step separation process. Initially, the feed slurry was added to the overflow pool area or horizontal overflow zone (Appendix D.4). The slimes particles were separated from the coarse or sand particles by a cross flow of water. The water carried the slimes out of the overflow pool area toward the overflow weir. Secondly, the sand particles settled out in the compression pool zone. A spiral or auger conveyor removed and drained the settled sand particles. Sand removal by the auger was a dewatering process; therefore, the majority of the water added to the classifiers left with the slimes through the overflow weir.

In the pilot plant, the two spiral classifiers were operated in parallel to achieve particle sizing. The spiral classifiers produced a −1.19 mm by +105 μm (−16 mesh by +150 mesh) sand fraction and a −105 μm (−150 mesh) slimes fraction. The dewatered sand material discharged from auger conveyors into the feed sump for the Reichert spiral concentrators. The overflow slurry containing the slimes fraction flowed from the overflow weirs into a feed sump where a pump fed the slurry to the hydrocyclone.

Hydrocyclone

The hydrocyclone had a cylindrical feed chamber attached to a conical section (Weiss, 1985), as shown in Appendix D.5. The feed slurry entered tangentially into the cylindrical chamber. A vortex finder in the center of the chamber removed the fine and light particles into the overflow. The coarse and heavy particles were removed as a slurry, which flowed out of the adjustable apex valve at the bottom of the cone section. For proper operation, the hydrocyclone required a centrifugal pump to feed the slurry at constant pressure and flow rate.

In the hydrocyclone, flow velocity and pressure were changed to angular and linear accelerations, which produced a cyclone effect. The angular acceleration was increased as the feed fluid moved from the outside wall toward the axis of rotation. As the angular acceleration increased, the centrifugal forces increased, which caused the separation of particles by size and/or gravity. Water and fine particles moved inward and upward as a vortex toward the overflow, while the coarse particles were driven toward the wall and downward in an accelerating spiral toward the apex valve.

The final particle-sizing step in the pilot plant occurred in the hydrocyclone, which prepared the feed for the Knelson bowl concentrator. For proper operation, the bowl concentrator needed a feed slurry that minimized −37 μm (−400 mesh) particle concentration. Accordingly, the hydrocyclone vortex-finder diameter and apex opening were selected to make a cut at approximately 37 μm. The hydrocyclone essentially removed the clay fractions from the soil. The −37 μm overflow slurry flowed to the slimes thickeners. The +37 μm underflow slurry was processed through the Knelson bowl concentrator.

3.2.1.4 Gravity Separation

Gravity separation occurred in the attrition scrubber, mineral jig, Reichert spiral concentrators, and Knelson bowl concentrator. The unit operations selected for gravity separation of the bullets and the coarse lead particles were the attrition scrubber and mineral jig. Deagglomeration, particle sizing, and gravity separation occurred simultaneously in the attrition scrubber. The mineral jig treated the +1.19 mm (+16 mesh) particles from the trommel. The Reichert spiral concentrators were used to treat the −1.19 mm by +105 μm (−16 mesh by +150 mesh) sand fraction from the spiral classifier. The Knelson bowl concentrator treated the −105 μm by +37 μm (−150 mesh by +400 mesh) slimes fraction from the hydrocyclone.

Mineral Jig Concentrator

The mineral jig produced an alternating rising and falling (pulsating) flow of water in which all particles with a high specific gravity, like lead, settled through the jig bed (rag) and collected in the bottom container or hutch (Appendix D.3). The lead product was periodically removed from the hutch. The particles of lower specific gravity collected and overflowed at the top of the jig bed.

The mineral jig treated the +1.19 mm (+16 mesh) soil and lead particles from the trommel. To remove excess water, the jig tailings overflowed onto a Sweco vibrating screen with 1.19 mm (16 mesh) openings. The screen oversize flowed onto a weigh-belt conveyor that monitored the wet weight of the product. The conveyor discharge was stockpiled as a clean product. The jig's lead concentrate was periodically removed from the hutch and placed into a storage drum.

Reichert Spiral Concentrators

The Reichert spiral concentrators applied gravity and centrifugal forces to separate particles based on differences in specific gravity. The concentrators consisted of vertical curved channels arranged in the form of a spiral having a diameter of 61 cm (24 in.) and a pitch of 34 cm (13.5 in.). The primary spiral had 6 turns, and the secondary spiral had 6.5 turns. The feed slurry and wash water were added at the top of the channel. As the combined water and slurry flow down the spiral channel, centrifugal forces made the low specific gravity particles or tailings travel to the outside of the flow. The tailings discharged at the bottom of the concentrator. The high specific gravity particles or lead concentrate were held to the inside of the flow by gravity. A mixed flow of light and heavy particles or middlings usually occurred next to the concentrate flow. The concentrate and middlings flows were withdrawn through ports located on the inside near the bottom of the channel.

The Reichert spiral concentrators treated the sand flow from the spiral classifiers. Tailings from the rougher or first concentrator flowed into a sieve bend screen for dewatering. The dewatered sand was placed into vat leach storage. The rougher middlings were recirculated to the rougher spiral for reprocessing. The concentrate was pumped to the cleaner spiral. Tailings from the cleaner concentrator were recirculated to the rougher spiral. The cleaner middlings were recirculated through the cleaner spiral. The cleaner concentrate, or lead product, was fed onto a Sweco screen for dewatering. After dewatering, the lead product was placed into storage drums.

Knelson Bowl Concentrator

A Knelson bowl concentrator can treat a slurry containing heavy particles with sizes from about 4.76 mm (4 mesh) down to about 1 μm. The optimum lower limit is about 30 μm. The Knelson bowl separated lead particles from soil particles using strong water flow and gravitational enhancement created by centrifugal forces (Appendix D.8). When operating, the slurry was fed near the bottom of the concentrator's inner cone, which was rotated at a high speed to produce about 60 g's of gravitational force. The rotating cone had ribs on the inside that formed a concentrating bed for the heavy lead particles. Injecting water through small holes in the cone walls prevented the bed from compacting. The bed was kept fluidized, enabling lead particles as small as 1 μm to enter the bed. The bed continuously exchanged light particles for heavy particles to produce an enriched concentrate. The combined wash water and tailings slurry flowed over the top of the inner cone and into the space between the cone and the concentrator's outer cylindrical shell. The slurry flowed out through a pipe in the outer shell.

The underflow from the hydrocyclone contained -105 μm by $+37$ μm (-150 mesh by $+400$ mesh) soil and lead particles, which were processed through the Knelson bowl concentrator. The concentrate collected on the Knelson bowl's rotating cone. At the end of each collection cycle, the accumulated concentrate was washed from the cone into a concentrate drum. The wash water and tailings slurry flowed out of the concentrator and into the slimes thickeners.

3.2.1.5 Leaching Systems

Vat Leaching

The treatability studies were sufficient to develop an efficient vat leach flowsheet for the demonstration. The batch vat leach, using a 2.5% acetic acid leach solution and calcium peroxide, was designed to extract lead from soil fractions larger than 105 μm ($+150$ mesh) by the reactions:

$$Pb^{\circ} + CaO_2 = PbO + CaO \qquad\qquad R3$$

$$PbO + 2CH_3COOH = Pb(CH_3COO)_2 + H_2O \qquad\qquad R6$$

The acetic acid leach solution was mixed in a solution storage tank and was distributed over the coarse and sand fractions in the leach tank. The leach solution percolated downward through the soil. If plugging or channeling became a problem, up-flow leaching would have been used. The pregnant leach solution was collected in another storage tank for recycle to the vat leach column or for treatment to remove the dissolved lead and other metals. Wash water was used to clean the leached solid, and it was collected in a second storage tank. The leach cycle time and the number of batch stages were determined by the quantity of soil, the lead content, and the clean-up standards.

Lead Recovery

The dissolved lead was precipitated from the leach solution as powdered metal in a cementation process by pumping the solution through a column containing iron wool. After the lead was precipitated, the solution was treated with a cation selective resin (chelating resin) that removed both residual lead and copper ions. If acetic acid or nitric and hydrochloric acids are used as leaching reagents, the acetate ions or nitrate and chloride ions can be removed or the concentration minimized using an anion exchange resin. The spent solution from the treatment system flowed into a holding tank where fresh acid was added as required. The ion exchange resins were regenerated with strong hydrochloric acid or sodium hydroxide solutions. The reactions for each step are as follows:

Lead Cementation

$$Pb(CH_3COO)_2 + Fe^{\circ} = Pb^{\circ} + Fe(CH_3COO)_2 \qquad\qquad R19$$

$$Cu(CH_3COO)_2 + Fe^{\circ} = Cu^{\circ} + Fe(CH_3COO)_2 \qquad\qquad R21$$

Leach Solution Treatment Using a Chelating Resin

$$2R_zH + Pb(CH_3COO)_2 = Pb(R_z)_2 + 2CH_3COOH \qquad\qquad R24$$

$$2R_zH + Cu(CH_3COO)_2 = Cu(R_z)_2 + 2CH_3COOH \qquad\qquad R25$$

Acetate Ion Removal Using an Anion Exchange Resin

$$R_zOH + CH_3COOH = R_zCH_3COO + H_2O \qquad\qquad R26a$$

Regeneration of Ion Exchange Resins

$$Pb(R_z)_2 \text{ (or } Cu\ (R_z)_2) + HCl = 2R_zH + PbCl_2 \text{ (or } CuCl_2) \qquad\qquad R27$$

$$R_zCH_3COO + NaOH = R_zOH + CH_3COONa \qquad\qquad R28a$$

After leaching was completed, the hydrochloric acid and sodium hydroxide regenerant solutions had to be treated to remove lead and copper. In the pilot plant demonstration, only one or two batches of material were leached so solution treatment by ion exchange resin was not required.

Agitation Leaching

For slimes soil fractions with particle sizes of −105 μm (−150 mesh) or −37 μm (−400 mesh), an agitation leach in batch stirred reactors would be used to dissolve the lead contamination. The tailings from the leach would be dewatered and washed using centrifuges and/or filters. The leach and wash solutions would be handled in the same manner as those in the vat leach system.

Resin-In-Pulp Leaching

If the slimes, −37 μm fraction, required leaching, then a resin-in-pulp (RIP) leach could also be utilized. The dewatered slimes or clay product from one of the gravity plant thickeners would be mixed with acetic acid and an oxidant. The slurry would be pumped through a series of columns containing an ion exchange resin. The slurry would flow counter-current to the resin in the columns. A screen would be used to keep the resin in a column, while allowing the slurry to pass onto the next column. Once clean, the solids would be settled and mixed with other solids from the site. Natural bacterial action would be used to decompose any residual acetic acid in the soil, and all resins would be treated off-site to recover the metal values.

3.2.2 Equipment Design Specifications and Capacities

Table 3.2 lists the major pieces of equipment that were selected for the demonstration pilot plant treating NAS Miramar firing range soil. Using the mass balance data developed in Section 3.2.1, the design specifications and capacities for the attrition scrubber, trommel, mineral jig, spiral classifiers, hydrocyclone, Reichert spiral and Knelson bowl concentrators, and sieve bend screen were calculated (Appendix D). These design and operating requirements were compared in Section 3.4 to the calculated results for the equipment that was installed in the demonstration gravity plant.

The feed belt, weigh belt, and bucket elevator were selected to handle up to 7260 kg/hr of soil. The three vibrating Sweco screens were used to remove water from the auger product, mineral jig concentrate, and Reichert spiral concentrate. The screen underflow water was less than 260 kg/hr (1.15 gpm). Review of the screen sizing methods in Appendix J.1 showed that the 46 cm (18 in.) diameter Sweco screens were adequate to remove the water from the coarse and sand products. The five vertical bowl (sand) pumps manufactured by Svedala Pump & Process had a rated slurry capacity of about 6.0 m³/hr (26 gpm) at gauge pressures up to 195 kPa (28 psig or 58 ft of water).

Table 3.2 Equipment Design Capacities for Pilot Scale Treatment Plant

Equipment Description	Number of Units	Equipment Specifications	Design Capacity kg/hr (tons/hr)
Hopper & Feed Belt	1	V_{HOPPER} = 7.65 m³ (10 yd³), 46 cm (18 in.) Feed Belt	7260 (8.00)
Weigh Belt	1	46 cm (18 in.) Flat Belt	3630 (4.00)
Bucket Elevator	1	No Size Data Available	4540 (5.00)
Attrition Scrubber	1	V_T = 570 L, RT = 15 min, Solids = 60%, rpm = 495, kW = 15.1	2150 (2.40)
Trommel Scrubber (−1.19 mm Soil)	1	L_T = 2.53 m; Screen: ID = 1.22 m, L_S = 0.56 m; A_{eff} = 0.227 m²	855 (0.94)
Mineral Jig Concentrator	1	Simplex 4 × 6, Bed = 0.016 m², Media Flow = 0.20 m³/hr	120 (0.13)
Spiral Classifiers	2	300 mm (12 in.) Diameter, S/Simplex/150, A_{pool} = 0.64 m²	1275 (1.41)
Hydrocyclone (High Capacity)	1	50 mm (2 in.) Diameter, 138 kPa, Flow = 1.35 m³/hr	550 (0.61)
Reichert Spiral Concentrators	2	Std. Design; Units in Series; Solids = 25%; Wash = 0.68 m³/hr	1295 (1.42)
Sieve Bend Screen	1	Rapped Screen; R=1625 mm, L=2300 mm; W=610 mm; Gravity Feed, Fixed Surface; Bar Spacing = 100 μm	45 (195)[1]
Knelson Bowl Concentrator	1	KD12 Concentrator, 30.4 cm D, Fluidizing Water = 8.16 m³/hr	1800 (1.98)
Sweco Screens	3	Vibrating Screen, 45.7 cm (18 in.) Diameter, Maximum Capacity	405 (0.45)
Vertical Bowl (Sand) Pumps	4	Sala SPV 181-40, Flow rate at operating gauge pressure of 195 kPa (28 psig)	
• Reichert Spiral Feed (10% Solids)			6.0 (26)[2]
• Sieve bend Feed (11% Solids)			6.0 (26)[2]
• Hydrocyclone Feed (33% Solids)			6.0 (26)[2]
• Hydrocyclone Underflow (55% Solids)			6.0 (26)[2]

1. Underflow capacity, m³/hr (gpm).
2. Pump capacity, m³/hr (gpm).

3.2.2.1 Attrition Scrubber

The –19.0 mm (–3/4 in.) stockpiled soil from the prescreening process was fed to the attrition scrubber (Figure 3.3). The scrubber consisted of two tanks combined into one unit, with a total volume of 570 L (20 ft³). A 12.7 cm (5 in.) diameter auger conveyor was attached to the bottom of the scrubber to remove the coarse soil and lead particles that settled to the bottom of the scrubber. Each tank measured 0.61 m (2.0 ft) square with a slurry depth of 0.762 m (2.50 ft). The soil and water was fed into the top of the first tank. The mixed slurry flowed into the second tank through a bottom opening in the common tank wall. The slurry overflowed the scrubber unit through a weir at the top of the second tank. The design retention time for the two tanks was 15 minutes.

Each tank had an agitator with two opposed pitched-blade turbines. Each turbine had four blades at a 45° pitch. The agitators were belt driven by one 3-phase motor. The following impeller diameter and height, spacing, shaft speed, and power requirements were selected to suspend –4.76 mm (–4 mesh) soil particles in a 60% solids slurry (Appendix D.1):

- Impeller Diameter = 30.5 cm (1.0 ft)
- Impeller Spacing = 31.7 cm (1.04 ft)
- Impeller Height = 10.1 cm (0.33 ft)
- Shaft Speed = 495 rpm
- Power Input to Mix Slurry = 15.1 kW (20.2 Hp)

Figure 3.3 Bucket Elevator and Attrition Scrubber.

Figure 3.4 Trommel.

3.2.2.2 *Trommel*

The trommel scrubber (Figure 3.4) was a horizontal-rotating drum with a blind section at the front end of the drum and a screening section at the rear end. The rear screening section used a 1.19 mm (16 mesh) wire screen with the following dimensions (Appendix D.2):

- Screen Inside Diameter (ID) = 1.219 m (4.00 ft or 48 in.)
- Screen Length (L_S) = 0.559 m (1.83 ft or 22 in.)

The overall length (L_T) of the trommel scrubber was about 2.53 m (~8.30 ft or ~100 in.). The length (L_B) of the blind section was 1.52 m (5.00 ft). The effective area of the trommel screen was defined in Appendix D.2 as the product of the "effective screen width" times the "screen length." The effective screen width was equal to one-third the inside diameter of the trommel screen:

$$\text{Effective Screen Width} = \frac{\text{Screen ID, m}}{3} = \frac{1.219 \text{ m}}{3} = 0.407 \text{ m (1.33 ft or 16 in.)}$$

$$\text{Effective Screen Area} = (0.407 \text{ m})(0.559 \text{ m}) = 0.227 \text{ m}^2 \text{ (2.43 ft}^2\text{)}$$

A trommel with this screen area had a capacity for –1.19 mm soil of about 855 kg/hr (0.94 t/hr). This screen throughput was greater than the mass balance requirement of 760 kg/hr (0.84 t/hr).

3.2.2.3 *Mineral Jig*

The mineral jig (Figure 3.5) treated the coarse soil fractions and lead particles from the trommel. A Simplex 4×6 (10.2 cm $\times$ 15.2 cm) mineral jig was needed to handle the low mass flow of solid (Appendix D.3). The design specifications for the mineral jig were:

Figure 3.5   Mineral Jig.

- Mineral Jig Bed Area = 0.016 m² (0.17 ft²)
- Mass Balance Media Flow = 0.1 to 0.3 m³/hr (0.4 to 1.3 gpm)

3.2.2.4 Spiral Classifiers

Two 300 mm (12 in.) spiral classifiers (Figure 3.6), operated in parallel were required for processing the slurry from the trommel (Appendix D.4). The overflow pool area and the compression zone pool area for the classifiers were based on the settling rate of the 105 µm (150 mesh) particles. The area requirements were:

- Overflow Pool Area = 0.35 m² (3.7 ft²)
- Compression Zone Pool Area = 0.40 m² (4.3 ft²)

Figure 3.6   Spiral Classifiers.

The total pool area of 0.640 m² (6.89 ft²) for both classifiers exceeded these pool area requirements.

The feed to the spiral classifier contained about 45% solids. Because of the sand removal, the percent solids in the classifier overflow, –105 µm (–150 mesh) soil, decreased to about 30% solids. The minimum amount of sand that could be transported by the spiral auger in the pilot plant, the raking capacity, was about five times the mass balance requirement

3.2.2.5 *Hydrocyclone*

The –105 µm fraction from the spiral classifier was pumped through a 50 mm (2 in.) hydrocyclone (Figure 3.7) to decrease the amount of –37 µm (–400 mesh) material in the feed to the Knelson bowl concentrator. The Knelson bowl concentrator required a feed with a minimum amount of clay-sized, less then 10 µm, particles. The following vortex-finder diameter, spigot opening, and operating pressure were calculated for the hydrocyclone making a cut at about 37 µm particles (Appendix D.5):

- Vortex-finder Diameter = 8 mm (0.31 in.),
- Spigot Opening = 3.2 mm to 4.5 mm (0.13 in. to 0.18 in.),
- Operating Gauge Pressure = 69 kPa to 207 kPa (10 psig to 30 psig), and
- Flow = 1.0 to 1.7 m³/hr (4.4 to 7.6 gpm).

By adjusting the spigot opening, the hydrocyclone operating parameters can produce the 30% and 70% by volume split in the underflow and overflow, respectively, required by the mass balance.

3.2.2.6 *Reichert Spiral Concentrators*

The Reichert spiral gravity system consisted of a primary and secondary concentrator (Figure 3.8). The feed particle size, –1.19 mm by +105 µm, was within the range of optimum size requirements. Feed and wash water flows to each concentrator had to be balanced to prevent sandbar formation at low flows. At high flows, the heavy minerals are swept wide of the inside exit ports. For optimum operation, the following operating conditions were suggested (Appendix D.6):

Figure 3.7   Hydrocyclone.

Figure 3.8   Reichert Spiral Concentrators.

- Slurry % Solids: 20% and 30%,
- Solid Feed Rate per Unit: 450 kg/hr (fine material) to 2250 kg/hr (coarse material),
- Feed Slurry: 3.52 m³/hr to 5.64 m³/hr, and
- Wash Water = 0.68 m³/hr to 3.40 m³/hr.

3.2.2.7 Sieve Bend Screen

The sieve bend screen (Figure 3.9) was used to remove water from the Reichert concentrator tailings with a particle size range of −1.19 mm to +105 µm (−16 mesh to +150 mesh). The standard sieve bend screens were gravity fed, which limited the bar spacing of the screen to about 350 µm (~45 mesh). Using a bar spacing of about 100 µm, the design process selected a rapped sieve bend screen, which had the following operating parameters (Appendix D.7):

- Optimum Sieve Bend Screen: Gravity Feed, Fixed Surface plus Rapped, 45°,
- Screen Surface Radius = 1625 mm,
- Screen Surface Length = 2390 mm,
- Width of Screen = 610 mm,
- Feed Falling Height of 102 cm (40 in.),
- Water Passing through Screen (Undersize) = 2.92 m³/hr (12.9 gpm), and
- Maximum Screen Capacity = 45 m³/hr (195 gpm).

Figure 3.9 Sieve Bend Screen.

3.2.2.8 Knelson Bowl Concentrator

The operating specifications for the KD12 Knelson bowl concentrator (Figure 3.10) were discussed in Appendix D.8. The Knelson bowl can operate under the following wide range of conditions:

- Diameter Bowl = 30 cm (12 in.),
- Maximum-Minimum Feed Size: 2.0 mm (10 mesh) to ~37 μm (~400 mesh),
- Feed Capacity: 0.0 kg/hr to 3600 kg/hr (0.0 lb/hr to 8000 lb/hr),
- Feed Pulp Density: 0 to 75% (dwb),
- Feed Slurry Capacity: 0 m^3/hr to 11.4 m^3/hr (0 gpm to 50 gpm), and
- Fluidization Water: 6.78 m^3/hr to 9.54 m^3/hr (30 gpm to 42 gpm).

Using the mass balance results, the cycle time for the collection of the concentrate containing 3% lead was quite low, about 25 minutes. Higher lead contents were needed to increase the concentrate cycle time.

Figure 3.10 Knelson Bowl Concentrator.

3.2.3   Summary of Equipment Design and Operating Capacities

The gravity separation plant used for lead recovery in the NAS Miramar demonstration was designed and the major equipment selected following the treatability study. The mass balance (Appendix C) for the plant was calculated based on the mass split, screen fractions, and gravity treatment values obtained from the treatability study of two soil samples. From the mass balance, the operating capacities for the attrition scrubber, trommel, mineral jig, spiral classifiers, hydrocyclone, Reichert spiral and Knelson bowl concentrators, and sieve bend screen that were used in the demonstration were calculated in Appendix D. These capacities were based on treating 907 kg/hr (1.0 ton/hr) of –19.0 mm (–3/4 in.) as received soil. The following table compares the design and operating capacities.

Equipment Item	Number of Units	Design Capacity kg/hr	Operating Capacity kg/hr
Hopper and Feed Belt Conveyor	1	7260	7260
Weight Belt Conveyor	1	3630	3630
Bucket Elevator	1	4540	4540
Attrition Scrubber	1	2150	907
Trommel Scrubber (−1.19 mm Soil)	1	855	435
Mineral Jig	1	120	485
Spiral Classifiers	2	1275	520
Hydrocyclone (High Capacity)	1	550	1680
Reichert Spiral Concentrators	2	1295	404
Sieve Bend Screen [With Bar Spacing: ~100 μm]	1	45[1]	<7[1]
Knelson Bowl Concentrator	1	1800	240

1. Underflow capacity, m^3/hr.

The capacities of the trommel scrubber and spiral classifiers limited the capacity of the plant to less than 907 kg/hr, as shown in this table. The diameter of the demonstration hydrocyclone was too large and required a high volumetric feed rate. This produced an ineffective separation of the −37 μm fraction from the +37 μm fraction.

The attrition scrubber capacity was limited by the power of the agitator drive motor to suspend −4.76 mm (−4 mesh) soil particles. The slurry percent solids had to be decreased to less than the desired 60%. Hence, the field demonstration capacity at 28% solids was acceptable.

The demonstration capacities for the hopper and belt conveyors, bucket elevator, mineral jig, Reichert spiral concentrator, and Knelson bowl concentrator equaled or exceeded the design capacities. A rapped sieve bend screen with bar spacing of 100 μm had sufficient capacity to remove water from the Reichert spiral tailings. A non-rapped unit was limited to bar spacing greater than 350 μm. If the spacing had been 100 μm, the capacity would have been markedly decreased.

3.3 PILOT SYSTEM FABRICATION

The pilot plant for the lead recovery demonstration at NAS Miramar was composed of a gravity separation and a vat leaching process. The equipment for the gravity separation circuit and the leaching circuit was placed separately onto two trailers. Appendix B shows the detailed flowsheet for the gravity separation system. Appendix E contains photographs of the equipment installed at the demonstration site.

3.3.1 Gravity Separation System

The gravity separation system was designed to maintain, as much as possible, an omnidirectional movement of all of the material through the equipment on the trailer. The maximum vertical height allowed was 2.3 m (7.5 ft) for over-the-road capability. The gravity separation plant consisted of the following systems:

- Soil feed
- Coarse soil treatment
- Sand treatment
- Slimes treatment

Due to the condition of available equipment for this demonstration, modifications to the equipment were required for mounting on the trailer, direction of operation, and discharge location. The custom designed attrition scrubber required the majority of modifications, including:

- Adding structural members to reinforce the motor mounting,
- Replacing the bearings supporting the impellers with much larger bearings and tripling the number of bearings,
- Adding a double set of opposing impellers to the agitator shaft in each cell as the original agitators only had a single impeller,
- Fabricating larger diameter impeller shafts, as the original impeller shafts were undersized,
- Enlarging the clean-out ports,
- Replacing the original rubber covers with metal plates, and
- Adding an electronic soft start to the attrition scrubber's electrical circuits to the drive motors.

3.3.1.1 Soil Feed System

Equipment, installed on the gravity separation trailer, to feed the soil to the attrition scrubber (Figure 3.11) included:

- Soil feed hopper, 7.65 m^3 (10 yd^3), with a feed conveyor belt,
- Weigh belt conveyor #1 to continuously record the mass of feed soil, and
- Bucket elevator to feed the soil to the attrition scrubber.

3.3.1.2 Coarse Soil Treatment

Equipment installed on the gravity separation trailer (Figure 3.12) for treatment of the coarse soil particles in the feed, consisted of:

- Attrition scrubber with auger conveyor,
- Double-deck Sweco screen #1 with 4.76 mm (4 mesh) and 1.19 mm (16 mesh) screens to remove water from the auger product (coarse lead concentrate),
- Trommel scrubber with 1.19 mm screen openings,
- Mineral jig to treat the +1.19 mm soil particles.
- Sweco screen #2 with 1.19 mm openings to remove water from the mineral jig tailings (treated gravel),
- Weigh belt conveyor #2 to continuously record the mass of the mineral jig tailings, and
- Two conveyor belts to transport the coarse lead concentrate and treated gravel into collection wheelbarrows.

3.3.1.3 Sand Treatment

Equipment installed on the gravity separation trailer (Figure 3.13) for the treatment of the sand fractions consisted of:

- Two spiral classifiers with 15.2 cm (6 in.) diameter spiral conveyors,
- An agitated sump for the slurry containing the sand fractions from the spiral classifiers with a centrifugal pump to feed the slurry to the rougher Reichert concentrator,
- Two Reichert spiral concentrators with recirculation sumps and pumps,
- Sump for spiral concentrator tailings with SandPIPER double diaphragm pumps to feed the sieve bend screen,
- Sieve bend screen with ~0.100 mm bar spacing to remove the water from the spiral concentrator tailings, and
- Sweco screen #3 with 0.149 mm (100 mesh) openings to remove the water from the spiral concentrator concentrate.

During the demonstration, the centrifugal pump and SandPIPER pumps that fed the hydrocyclone and sieve bend screen were replaced with vertical bowl (sand) pumps that provided consistent discharge rates.

Figure 3.11 Soil Feed System.

Figure 3.12 Coarse Soil Treatment.

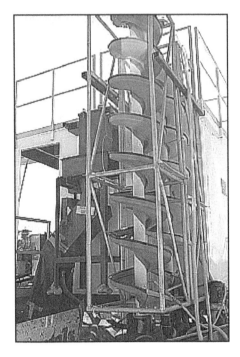

Figure 3.13 Sand Treatment System.

3.3.1.4 Slimes Treatment

The final circuit constructed on the gravity separation trailer was for treatment of the slimes overflow from the spiral classifiers, the −105 μm (−150 mesh) soil fraction. The slimes circuit (Figure 3.14) consisted of:

- Cone bottom sump for the slurry containing the −105 μm (−150 mesh) soil with a centrifugal pump to feed the slurry to the hydrocyclone,
- Hydrocyclone and underflow sump with centrifugal pump to feed the Knelson bowl,
- Knelson bowl concentrator,
- Thickeners or dewatering tanks (Figure 3.15) for collecting and settling the slimes fractions from the hydrocyclone and Knelson bowl concentrator and the water flows from the sieve bend screen and pressure filters, and
- Peristaltic pumps to transfer the thickener overflow to the day tank.

During the demonstration, the centrifugal pumps that fed the hydrocyclone and Knelson bowl concentrator were replaced with a vertical bowl (sand) pump providing consistent discharge rates.

3.3.2 Vat Leaching System

The vat leach equipment was installed on a separate trailer (Figure 3.16) from the gravity separation equipment. The system was designed to leach the −1.19 mm, +105 μm (−16 mesh, +150 mesh) tailings products from the Reichert spiral concentrators and the −4.76 mm, +1.19 mm (−4 mesh, +16 mesh) tailings from the mineral jig. The vat leach plant consisted of:

- Two stainless steel vat leach tanks,
- Plexiglas column containing steel wool for cementation of lead from the vat leach solution,

Figure 3.14 Slimes Treatment.

Figure 3.15 Thickeners (Roll-off Boxes).

Figure 3.16 Vat Leach Plant.

- Plexiglas column containing the chelating ion exchange resin which removed the residual lead and the copper, zinc, and iron ions from the solution leaving the cementation column,
- Two plastic tanks to hold the solutions leaving the cementation and ion exchange columns, and
- Several peristaltic pumps to transfer solutions from between the several unit operations.

The inside diameter of the vat leach tanks was about 61 cm (~2 ft). The soil depth in the vat was about 61 cm (~2 ft). Hence, each leach column held about 180 L (6.3 ft^3) of solids. A perforated plate was placed over the opening in the bottom of the tank and covered with a fabric filter cloth. This kept the solids in the tank while allowing liquid to percolate though the solids and to drain from the tank. Two stainless steel storage tanks were placed on a plastic liner alongside the trailer. One tank held the fresh leach solution and the other held the pregnant leach solution. Since a limited number of batches of the coarse soil fractions were to be leached during the demonstration, the anion exchange resin columns were not included on the trailer.

3.3.3 Water System

The water system was built in two parts. Water was first removed from the day tank, passed through filters, and stored in a clean water tank. Two pumps, each with its own filter to allow back-flushing of one at any time, were used. Then the water from the clean water tank was pressurized for use in the pilot plant. Two pumps were available to draw water from the clean water tank and supply it to a pressurized holding tank connected to the main water lines that fed the plant. The doubling of pumps also provided a backup in the event either of the pumps failed and provided either independent or joint operation. When operated jointly, the dual system provided a surge capacity if added volume was needed. All feed lines in the water systems were 5.08 cm (2 in.) diameter galvanized steel pipe.

3.3.4 Compressed Air System

Air was required to operate the flow control valves and the SandPIPER double diaphragm pumps used to add water and to transport slurries between unit operations. Each of the six air pressure compressors supplied about 0.34 m^3/min (12 ft^3/min). Calculations indicated that only five compressors were needed. The primary air lines consisted of 5.08 cm (2 in.) galvanized steel pipe.

3.3.5 Electrical System

Each piece of equipment had at least two electrical connections: one for primary electrical power, and another for computer control. Several pieces of equipment also had additional electrical connections for feedback circuits to provide data such as feed rate, flow rate, or mass flow. Wherever possible, the conduits for these circuits were placed under the center of the trailer, between the primary support beams to protect from physical damage. All relays for three-phase motors were mounted in a panel in the front of the trailer to provide easy access and for moisture protection. There were four panel or circuit boxes located in an enclosure at the back-end of the trailer, as shown in Figures 3.13 and 3.14. The first circuit box contained the primary fuses and cutoff switch. The second box contained the individual circuit breakers for each of the discrete equipment circuits. The third contained all the automated controls and relays, and the fourth contained relays for the equipment and the three-phase motor.

3.3.6 Process Control System

The process control system was divided into the following three modules:

- Process control platform module,
- Input/output (IO) communications link module, and
- Sensor and controller module.

The process control platform (PCP) was a computer with a 90 MHz Pentium processor, a 256 kb L2 cache, a 1.6 gigabyte hard drive, a 4X CD ROM, a tape backup, a computer case ventilation fan, and a 230 watt power supply.

The process control system included 22 OPTOMUX intelligent digital and analog input/output (I/O) controllers that operated as slave devices to a distributive process control computer called the DCS. Data acquisition and process control were provided by the computer running the industrial automation software, Intellution FIX 5.1. This software provided the supervisory control and data acquisition functions for the process control system. The software, residing in the DCS, required no proprietary hardware to interface with the 22 OPTOMUX I/O hardware and communicated directly with the DCS computer via I/O driver software.

The treatment system was equipped with sensors to acquire process information such as mass, mass flow rate, slurry density, and sump slurry levels. The majority of the process equipment had dedicated mounted controllers that provided the operator with the option of operating the device in either automatic (computer controlled) or manual control modes.

3.3.7 Pilot System Shakedown

The initial shakedown test of the pilot plant was conducted before the pilot plants were transported to the firing range site at NAS Miramar. The shakedown involved four series of tests: initial rotational tests, tests using water, tests using washed sand, and tests using a high clay soil.

The initial rotational tests were conducted upon installation of each piece of equipment to confirm that the equipment was wired correctly and the rotation was in the direction specified on the equipment. If the rotational test indicated the equipment was properly installed and rotating correctly, the piece was further tested by allowing it to run for a short period of time. Any problems encountered were immediately corrected.

The second series of tests was conducted running various parts of the plant, using only water. These tests usually involved a single piece or small grouping of equipment, such as the attrition scrubber, trommel, spiral classifiers, and mineral jig. Any problems encountered were immediately corrected.

The third series of tests was conducted using washed sand from a local sand and gravel company. These tests were designed to analyze the gravity separation process for particles larger than 105 μm (150 mesh). The feed hopper, weigh belt #1, bucket elevator, attrition scrubber, trommel, mineral jig, Sweco screen #1, weigh belt #2, spiral auger on the attrition scrubber, Sweco screen #2, spiral classifiers, centrifugal pump, Reichert spiral concentrators, spiral recirculation sumps and pumps, spiral concentrate dewatering Sweco screen #3, sieve bend screen, sieve bend feed pump, and all associated samplers were tested. This test was considered successful after the plant operated continuously for 4 hours.

The fourth series of tests used a soil collected from a berm near the fabrication facility. This soil was sized by screening to produce about 5% +1.19 mm (+16 mesh), 20 to 25% −1.19 mm to 105 μm (−16 to +150 mesh), and 70 to 75% −105 μm (−150 mesh) soil fractions. This test was considered successful after the plant operated continuously for 4 hours.

3.4 PROJECT EXECUTION

After successfully completing the shakedown tests, the gravity separation and vat leach pilot plant trailers were mobilized. The pilot plant required about 4 days to set up and shake down at the NAS Miramar site. Detailed daily logs are included in Appendix F.

The plant was designed to process 907 kg/hr (2000 lb/hr or 1.0 ton/hr) of firing range soil. During the demonstration, the gravity separation plant processed about 50,000 kg (55 ton) of firing range soil in seven processing days. To test the attrition scrubber and the capacities of the major equipment, the plant was fed with rates as high as 2270 kg/hr (5000 lb/hr). The results of the feed rate tests are also included in Appendix F.

3.4.1   Field Implementation

Before the pilot plant was transported to the site, a lined stockpile cell was constructed and 180 t (200 ton) of the firing range soil was excavated and prescreened. A double-deck vibrating screen with 5.1 cm (2 in.) and 1.9 cm (3/4 in.) openings was used to prescreen the soil. The −1.9 cm (−3/4 in.) fraction was stockpiled as feed for the pilot plant. The other fractions were stockpiled as clean material for use as backfill. A containment cell for the pilot plant was also constructed using a 30 mil HDPE liner. The liner was placed over sandbags stacked around the perimeter of the cell. The cell acted as a catch basin for process water and spilled process soils.

After the pilot plant was brought to the demonstration site, the setup activities included:

- Placing sumps and pumps,
- Connecting electrical wiring,
- Connecting slurry lines,
- Installing a process water storage system,
- Installing the power generator and main electrical feed cables, and
- Testing the plant electrical equipment and the rotation of all motors.

All setup activities were completed in 2 days.

Water was then added to the treatment units, and the pilot plant was operated for 3 hours before water flow equilibrium was achieved. After achieving water flow equilibrium, small amounts of range soil were added to the plant to test each piece of equipment and the slurry flows. During this shakedown period, numerous slurry pumps were replaced because of inadequate performance or failure. Several key peristaltic and SandPIPER pumps were finally replaced with vertical bowl (sand) pumps.

After the shakedown period was completed, the pilot plant was operated for a total of 10 days. For the first 3 days, operating procedures were established that included startup time to achieve water balance and shutdown time to completely flush the system of all soil. Following these disciplined procedures minimized clogging of slurry lines and increased actual operating time. After establishing the operating procedures, the plant was operated for 7 consecutive days, processing about 50 t (55 ton) of soil.

Samples were taken upstream and downstream of each unit during plant operation. These samples were sent to the laboratory for analysis daily. Analytical results were returned within 48 hours to insure a timely measure of effectiveness for each unit operation.

Since the pilot plant was a closed loop system, water treatment was critical. The initial thickeners and pumps were not properly sized to provide sufficient residence time and flow control to adequately handle the flow rate of the pilot plant. Hence, two 15.3 m³ (20 yd³) watertight roll-off bins were installed in series to replace the thickeners. Flocculent was added to the flow stream prior to entry into the roll-off bins to facilitate the settling of the fine soil particles. The water from the sieve bend screen, Knelson bowl tailings slurry, and hydrocyclone overflow were fed into the first roll-off bin to settle out the -105-μm (-150 mesh) soil particles. The second bin increased the retention time of the overflow to allow settling of submicron particles that escaped the first settling bin. The overflow water from the second roll-off bin was fed into the day tank and used as process water for the pilot plant. Since this treatment process consumed water, it was necessary to add water to the day tank throughout the demonstration.

At the completion of the field demonstration, the pilot plant was drained of all water, decontaminated, and transported back to Salt Lake City. The stockpiled firing range soil and treated soils were then placed back on the berm, and the job site was cleaned.

3.4.2 Operating Capacities

Utilizing results from the treatability study, the mass balance in Appendix C was developed based on the process flowsheet in Appendix B. The mass balance information along with detailed equipment specifications is included in Appendix D. As discussed in Section 3.2, the design capacities for the equipment listed in the mass balance were calculated. The specifications and the operating (or demonstration) capacities of primary pieces of equipment installed in the gravity separation plant with supporting calculations were also included. Table 3.3 compares the design and operating capacities for the major equipment. However, during the field demonstration, flow rates at each unit operation were not measured. Therefore, no comparison can be made between the design, expected, and measured capacities.

The comparison of the capacities in Table 3.3 indicates that the 907 kg/hr capacity of the pilot plant was limited by two major unit operations:

- Trommel: The installed trommel had a capability of screening a maximum of 435 kg/hr (0.96 ton/hr), which was 325 kg/hr less than the mass balance requirement of 760 kg/hr (0.84 t/hr).
- Spiral classifiers: The installed classifiers had a capability of separating sand from slimes at a maximum rate of 520 kg/hr (0.57 ton/hr) which was 755 kg/hr less than the required 1275 kg/hr.

The equipment that equaled or exceeded their design capacity were:

- Mineral jig,
- Hydrocyclone,
- Reichert spiral and Knelson bowl concentrators,

Table 3.3 Comparison of Design and Demonstration Equipment Sizes and Capacities

Equipment Description	Capacity Method	Number of Units	Equipment Size	Capacity kg/hr (tons/hr)
Hopper & Feed Belt	Design	1	V_{HOPPER} = 7.65 m³ (10 yd³), 46 cm (18 in.) Feed Belt	**7260 (8.00)**
	Operating	1	Same Size Hopper	7260 (8.00)
Bucket Elevator	Design	1	No Size Data Available	**4540 (5.00)**
	Operating	1	No Size Data Available	4540 (5.00)
Attrition Scrubber	Design	1	V_T = 570 L, RT = 15 min, Solids = 60%, rpm = 495, kW = 15.1	**2150 (2.40)**
	Operating	1	V_T = 570 L, RT = 13 min, Solids = 28%, rpm = 425, kW = 7.2	907 (1.00)
Trommel Scrubber	Design	1	L_T = 2.53 m; Screen: ID = 1.22 m, L_S = 0.56 m; A_{eff} = 0.227 m²	**855 (0.94)**
	Operating	1	L_T = 1.67 m; Screen: ID = 0.47 m, L_S = 0.76 m; A_{eff} = 0.116 m²	435 (0.96)
Mineral Jig	Design	1	Simplex 4 × 6, Bed = 0.016 m², Media Flow = 0.20 m³/hr	**120 (0.13)**
	Operating	1	Simplex 10 ×10, Bed = 0.065 m², Media Flow = 0.10 m³/hr	485 (0.54)
Spiral Classifiers	Design	2	300 mm (12 in.) Diameter, S/Simplex/150, Max A_{pool} = 0.64 m²	**1275 (1.41)**
	Operating	2	150 mm (6 in.) Diameter, S/Simplex/150, Max A_{pool} = 0.22 m²	520 (0.57)
Hydrocyclone (High Capacity)	Design	1	50 mm (2 in.) Diameter, 138 kPa, Flow = 1.35 m³/hr	**550 (0.61)**
	Operating	1	80 mm (3 in.) Diameter, 238 kPa, Flow = 6.00 m³/hr	2400 (2.65)
Reichert Spiral Concentrators	Design	2	Std. Design; Units in Series; Solids = 25%; Wash = 0.68 m³/hr	**1295 (1.42)**
	Operating	2	Std. Design; Units in Series; Solids = 11%; Wash = 0.00 m³/hr	404 (0.45)
Sieve Bend Screen	Design	1	Rapped Screen; R = 1625 mm, L = 2300 mm; W = 610 mm; Gravity Feed, Fixed Surface; Bar Spacing = 100 μm	**45 (195)**[1]
	Operating	1	Non-rapped Screen; R = 915 mm, L = 760 mm; W = 610 mm; Gravity Feed, Fixed Surface; Bar Spacing = 105 μm	<7 (30)[1]
Knelson Bowl Concentrator	Design	1	KD12 Concentrator, 30.4 cm D, Fluidizing Water = 6.78 m³/hr	**1800 (1.98)**
	Operating	1	KD12 Concentrator; 30.4 cm D, Fluidizing Water = ND m³/hr	240 (0.26)
Sweco Screens	Design	3	Vibrating Screen, 45.7 cm (18 in.) Diameter, Maximum Capacity	**405 (0.45)**
	Operating	3	Vibrating Screen, 45.7 cm (18 in.) Diameter, Maximum Capacity	405 (0.45)
Vertical Bowl (Sand) Pumps	Design	4	Sala SPV 181-40, flow rate at operating gauge pressure of 195 kPa (28 psig)	
• Reichert Spiral Feed				**6.0 (26)**[2]
• Sieve bend Feed				**6.0 (26)**[2]
• Hydrocyclone Feed				**6.0 (26)**[2]
• Hydrocyclone Underflow				**6.0 (26)**[2]
Vertical Bowl (Sand) Pumps	Operating	4	Sala SPV 181-40, mass balance flow rates	
• Reichert Spiral Feed				3.4 (15)[3]
• Sieve bend Feed				3.1 (14)[3]
• Hydrocyclone Feed				0.92 (4.0)[3]
• Hydrocyclone Underflow				0.29 (1.4)[3]

1. Underflow capacity, m³/hr (gpm).
2. Capacity, m³/hr (gpm).
3. Flow rate, m³/hr (gpm).

- Sweco vibrating screens,
- Vertical bowl (sand) pumps, and
- Sieve bend screen.

Because of the power limitations of the drive motor, the capacity of the attrition scrubber and the size of soil particles suspended could be affected by the percent solids of the slurry.

3.4.3 Process Design, Operations, and Monitoring

Size and gravity separation was accomplished through a series of treatment processes utilizing the equipment as previously described. Besides calculating the design and operating equipment sizes and capacities in Table 3.3, each piece of equipment was evaluated by plant operating personnel during the demonstration. Analytical data for daily samples from major slurry flows were also measured during field operations. The averaged analytical values for these samples are presented in Table 3.4. In this table, the averaged values were compared to the mass balance values listed in Table 3.1, which were derived from the treatability values for NAS Miramar. The equipment capacities, operational evaluations, and analytical results are discussed in the following sections.

During the project, input and output samples were collected and analyzed to monitor all major unit operations. The total lead analytical procedure for soils was:

- Dry and pulverize each sample to about 149 µm (100 mesh),
- Fuse 0.2 grams of the nominal 149 µm (100 mesh) sample with sodium peroxide,
- Digest the peroxide fusion in nitric acid, and
- Read the concentration by atomic absorption using a diluted nitric acid solution.

About every tenth soil analysis was duplicated. The software for the atomic absorption instrument was set for level "C" that required duplicates to fall within plus or minus 5% of the original value. Spikes were not used with the fusion techniques. All lead concentrations were reported as milligrams per kilogram of dry soil, mg/kg.

The results of the chemical analyses of samples are summarized in Table 3.5. The samples were daily composites, except for samples collected the last two days. The samples taken the last two days were related to feed rates to the plant and were reported as such. Discussions on sample points are included in Appendix G. Additional process monitoring discussion is included in Appendix H.

3.4.3.1 Feed Hopper, Weigh Belts, and Bucket Elevator

Design and Operation — The feed hopper and conveyor belt operated as specified. With the opening set at 5.08 cm (2 in.), the hopper was capable of a feed rate of 3600 kg/hr (8000 lb/hr), significantly higher than required for the pilot plant. Once calibrated, the weigh belts operated as specified. The capacity of the bucket elevator was 4540 kg/hr (10,000 lb/hr). It was run at a rate of 3600 kg/hr for two short periods and worked well.

Feed — The feed is the raw soil transported from the feed hopper, to the bucket elevator. The average lead content of the treatability feed samples was 11,800 mg/kg while the samples from the demonstration averaged 8115 mg/kg. However, for the demonstration samples, the whole bullets were removed before the samples were analyzed. Hence, the demonstration samples were biased toward low lead contents.

3.4.3.2 Attrition scrubber

Design and Operation — The attrition scrubber for this pilot plant was a custom design that provided both particle deagglomeration and particle sizing for a coarse feed containing –19.1 mm

Table 3.4 Comparison of Treatability and Average Demonstration Lead Analyses

Sample	Particle Size	Lead Analyses		Applicable Comments
		Treatability mg/kg	Demonstration mg/kg	
Feed Soil	−19.0 mm (−3/4-in.)	11,810	8115	Two NAS Miramar samples from treatability studies: A = 17,600 mg Pb/kg; B = 5970 mg Pb/kg
Screen #1 (1st Oversize)	+4.76 mm (+4 mesh)	130	75	The bullets were hand-picked out of the field samples. In the treatability studies, the bullets were removed by a jig.
Screen #1 (2nd Oversize)	−4.76 mm, 1.19 mm (−4 mesh, +16 mesh)	36,400	24,955	The particles were combined with 1.19 mm (+16-mesh) trommel fraction and fed to the jig.
Mineral Jig Concentrate	+1.19 mm (+16 mesh)	99,300	255,545	Jig tailings were clean.
Mineral Jig Tailings	+1.19 mm (+16 mesh)	3170	795	
Reichert Spiral Concentrator Feed	−1.19 mm, +105 µm (−16 mesh, +150 mesh)	4220	5900	The −1.19 mm, +105 µm (−16 mesh, +150 mesh) sand fraction from classifiers.
Reichert Spiral Concentrator Concentrate	−1.19 mm, +105 µm (−16 mesh, +150 mesh)	32,700	47,045	
Reichert Spiral Concentrator Tailings	−1.19 mm, +105 µm (−16 mesh, +150 mesh)	410	755	
Hydrocyclone Feed (Spiral Classifier Overflow)	−105 µm (−150 mesh)	5390	5600	The −105 µm (−150 mesh) slime fraction from classifiers.
Hydrocyclone Overflow	−44 µm (−325 mesh)	6010	5380	
Hydrocyclone Underflow (Knelson Bowl Feed)	−105 µm, +44 µm (−150 mesh, +325 mesh)	4930	5160	
Knelson Bowl Concentrator Concentrate	−105 µm, +44 µm (−150 mesh, +325 mesh)	30,200	42,695	Good gravity concentration. Usually, the Knelson bowl does not work on material finer than 37 µm (400 mesh).
Knelson Bowl Concentrator Tailings	−105 µm, +44 µm (−150 mesh, +325 mesh)	4000	1740	Good removal of lead from Knelson bowl feed.
Mixed Slimes	−105 µm (−150 mesh)	4870	4525	Settled slurry in roll offs (thickeners)

Table 3.5 Analytical Results of Samples Collected During Pilot Plant Demonstration

Sample	Daily Composite Samples mg/kg, Lead							Run Date: 19 Sept 1996 mg/kg, Lead			Run Date: 20 Sept 1996 mg/kg, Lead		
	30 Aug	10 Sept	12 Sept	13 Sept	16 Sept	17 Sept	18 Sept	Feed 454 kg/hr	Feed 680 kg/hr	Feed 1361 kg/hr	Feed 907 kg/hr	Feed 2041 kg/hr	Feed 2722 kg/hr
Feed Soil	13,600	11,200	13,400	11,950	3100	3140	3270	11,250	11,300	7050	16,050	NS	NS
Screen #1 (1st Oversize)	311	45	15	41	54	40	NS	NS	38	NS	59	NS	NS
Screen #1 (2nd Oversize)	54,500	56,200	56,100	NS	10,750	59,050	8550	3100	563	437	318	NS	NS
Mineral Jig Concentrate	1843	49,000	28,500	54,300	227,800	415,000	363,500	437,000	575,000	403,500	NS	NS	NS
Mineral Jig Tailings	1543	480	1091	835	520	970	901	868	194	991	990	697	472
Reichert Spiral Feed	NS	1500	8460	1470	3250	7050	7100	6200	6050	8250	9650	NS	NS
Reichert Spiral Concentrate	31,000	13,500	92,900	56,550	36,250	38,150	27,500	51,650	35,600	65,100	69,300	NS	NS
Reichert Spiral Tailings	910	806	715	511	892	496	950	890	451	NA	881	660	923
Hydrocyclone Feed	NS	5600	5700	2900	4500	4700	9050	9000	4250	6550	3750	NS	NS
Hydrocyclone Overflow	NS	4050	5800	7500	5650	5400	6250	6000	3550	6100	3500	NS	NS
Hydrocyclone Underflow	NS	3550	8300	4000	5650	4800	5600	6350	3400	6600	3350	NS	NS
Knelson Bowl Concentrate	36,200	6400	73,000	12,950	11,900	56,470	28,500	55	119,200	75,300	49,650	NS	NS
Knelson Bowl Tailings	NS	NS	NS	1850	344	960	1220	1175	1476	1443	1142	3493	4406
Mixed Slimes	3700	4000	6500	5500	7800	4850	3	2150	4100	6000	5150	NS	NS

NS = No Sample
NA = No Analyses

(–3/4 in.) soil and lead particles. Any agglomerated soil was broken apart, and the particles were scrubbed by the intense agitation within the tank. It was estimated that all of the +4.76 mm (+4 mesh) and some of the +1.19 mm (+16 mesh) soil settled to the bottom of the scrubber tanks. An auger conveyor attached to the bottom of the scrubber removed the settled particles. The slurry containing the –4.76 mm particles flowed through a weir at the top of the tank into the trommel feed chute. A Sweco vibrating screen with 4.76 mm (4 mesh) and 1.19 mm (16 mesh) screens removed the water from the coarse soil product producing the initial lead concentrate. The auger shredded the lead bullets, contaminating the finer soil fractions. Accordingly, any –1.19 mm particles in the auger product flowed into the spiral classifier feed trough and markedly increased the lead content in the feed to the Reichert spiral concentrators.

At a pulp density of 60% solids, the attrition scrubber could treat up to 2150 kg/hr of soil at an acceptable retention time of 15 minutes. Operating the agitators at 495 rpm with a power consumption of 15.1 kW (20.2 Hp) should suspend –4.76 mm (–4 mesh) soil particles. For the demonstration, the scrubber treated 907 kg/hr and suspended –4.76 mm soil particles while operating at 425 rpm and a pulp density of 28% solids. The power consumed was about 7.2 kW (9.7 Hp). The maximum power output of the motor was 14.9 kW (20.0 Hp), which was adequate for the demonstration conditions but not for the design conditions. Hence, with a fixed agitator speed of 425 and a maximum power output of 14.9 kW, the maximum treatment capacity was calculated to be less than 2150 kg/hr.

Because of the problems with the auger grinding the lead bullets and the suspension of –4.76 mm (–4 mesh) solids, the attrition scrubber should have been replaced with a commercial unit that would be easier to operate and maintain. Furthermore, the feed to an attrition scrubber would require prescreening to produce less than 4.76 mm (4 mesh) particles for proper particle suspension in the scrubber. In all applications, the maximum particle size that can be suspended in an attrition scrubber must be estimated for a given set of operating conditions, as shown in Appendix D.1.

Auger Product (+4.76 mm) — The +4.76 mm (+4 mesh) Sweco screen fraction was composed of the bullets, large rocks, and some clods. When this fraction was analyzed, the bullets were removed to determine both the amount of lead smeared on the rocks and if the rocks could be considered clean. The lead content of the bullet-free material was less than 100 mg/kg (Table 3.4, Screen #1, 1st Oversize). After treatment in the mineral jig, the treatability or expected mass balance value was 130 mg/kg.

Auger Product (–4.76 mm, +1.19 mm) — The second split produced by the Sweco screen was the –4.76 mm to +1.19 mm (–4 mesh to +16 mesh) fraction. This fraction contained small gravel, rock chips, and lead fragments. It was also noted that the auger crushed and rolled the lead fragments contained in this fraction into small cigar-shaped cylinders about 0.5 cm (0.25 in.) long. These lead cylinders initially were about 1.2 mm in diameter. However, by the end of the test the diameter expanded to about 1.7 mm due to wear on the steel auger conveyor. During the pilot plant operation, this material had to be hand fed into the trommel feed chute. The material was subsequently treated in the mineral jig where the lead was removed. The lead content of this auger product averaged 24,955 mg/kg (Screen #1, 2nd Oversize). The treatability or mass balance value was 36,000 mg/kg.

Auger Product (–1.19 mm) — The small particles from the attrition scrubber auger were composed primarily of sand and some clay mixed with water. This fraction was combined with the trommel underflow that fed into the spiral classifiers. The single sample of the attrition scrubber fraction that was analyzed indicated that it contained about twice the lead as in the trommel underflow. An accumulation of the small lead particles was also noted in the classifier sand product that flowed into the feed sump for the rougher Reichert spiral concentrator.

Attrition Scrubber Overflow — The scrubber overflow fraction was composed of small gravel, rock chips, small pieces of lead, sand, silt, clay, and some fine organic materials. Screening tests of this material indicated that the largest particles leaving the attrition scrubber were about 5.66 mm to 4.76 mm (3 mesh to 4 mesh) for soil and about 1.69 mm (10 mesh) for the lead. Based on several observations of the small gravel and rock chip particles using a 15X hand lens, it was determined that these particles were well washed. The attrition scrubber overflow was separated into two size fractions by the trommel, −1.19 mm (−16 mesh) and −4.76 mm to +1.19 mm (−4 mesh to +16 mesh) particles.

3.4.3.3 Trommel Scrubber

Design and Operation — The trommel scrubber was used primarily as a rotating screen that made a particle separation at 1.19 mm (16 mesh). The trommel design calculations based on the pilot plant flowsheet and mass balance indicated that the capacity of the pilot plant was limited by the trommel capacity. If the effective design area (A_{eff}) was 0.227 m², the trommel scrubber easily treated up to 855 kg/hr of −1.19 mm soil. For the trommel used in the demonstration, the effective area was only 0.133 m², which had a calculated capacity of 435 kg/hr of −1.19 mm soil. The demonstration trommel scrubber was undersized.

Trommel overflow — This fraction was composed of all the +1.19 mm (+16 mesh) fractions that came from the attrition scrubber overflow and auger products. The overflow was composed of small gravel, rock chips, large sand, and lead particles.

Trommel underflow — The underflow was composed of the −1.19 mm (−16 mesh) soil fractions. Constituents of this fraction were sand, sand-sized lead particles, silt, silt-sized lead particles, clay, clay-sized lead particles, and some fine organic material. Trommel underflow was fed to the spiral classifiers, which made a particle size separation at 105 μm (150 mesh).

3.4.3.4 Mineral Jig

Design and Operation — The mineral jig operated as specified with excess available capacity. The design calculations specified a Simplex 4 × 6 mineral jig that had a bed area of 155 cm² (0.167 ft²) and capacity of 120 kg/hr (0.13 ton/hr). The demonstration mineral jig had a bed area of about 645 cm² (0.694 ft²) with a maximum capacity of 485 kg/hr (0.54 ton/hr). The media (wash water) flow of 0.11 m³/hr was similar to the design requirement. The hutch containing the lead concentrate was monitored and cleaned by opening a valve when it reached capacity. Almost all of the lead that reached the jig was removed by the jig. The average lead content of the feed samples to the mineral jig was an estimated 18,000 mg Pb/kg. The lead concentration of the concentrate and tailings averaged 255,500 mg/kg and 795 mg/kg, respectively. The mineral jig effectively removed about 98% of the coarse lead.

Jig Concentrate — The jig concentrate was a mixture of lead, copper, and heavy minerals. The size range was from 1.19 mm (16 mesh) up to about 1.69 mm (10 mesh). Lead particles larger than 1.69 mm (10 mesh) did not leave the attrition scrubber in the overflow. The lead content of the mineral jig concentrate averaged 255,545 mg/kg, which was much higher than the treatability value of 99,300 mg/kg.

Jig Tailings — The mineral jig tailings are also described as weigh belt #2 product. This product was composed of small rocks, rock chips, gravel, large sand, and some organic material with essentially all of the lead removed. The lead content of the jig tailings averaged 795 mg/kg, which was considerably better than the value of 3170 mg/kg obtained during the treatability test.

3.4.3.5 Spiral classifiers

Design and Operation — Spiral classifiers are a simple and reliable method for separating fine soil particles from coarse soil particles. The expected size fractions for the pilot demonstration were the sand fractions, −1.19 mm to +105 μm (−16 mesh to +150 mesh), and the overflow or slimes fraction, −105 μm (−150 mesh). The sand product was pumped to the Reichert spiral for lead removal. The overflow product was fed into the hydrocyclone for another size fraction split.

The design calculations based on the mass balance specified two spiral classifiers, model S/Simplex/150, with a 300 mm (12 in.) diameter spiral. The total feed and sand-raking capacities were 1275 kg/hr (1.41 ton/hr) and 2270 kg/hr (2.50 ton/hr), respectively. However, the two demonstration classifiers with 150 mm (6 in.) diameter spirals had an estimated total feed capacity of only 520 kg/hr (0.57 ton/hr). Higher capacities could have been achieved if the overflow or slimes particle size cutoff was increased above 105 μm. These classifiers also had a total sand raking capacity of about 1740 kg/hr (1.92 ton/hr), which was adequate for the design mass flow rate. Nevertheless, the amount of the sand fraction would occasionally overload and stall one of the sandscrew motors. Then the classifier sandscrew had to be rotated by hand until the motor resumed operation.

Spiral Classifier Overflow (Hydrocyclone Feed) — The overflow from the spiral classifiers contained the –105 μm (–150 mesh) particles and consisted of silt, silt-sized and clay-sized lead particles, and some fine organic materials. The solid material in the hydrocyclone feed had a lead content of about 5600 mg/kg. This value was close to the value of 5400 mg/kg from the treatability tests.

3.4.3.6 Hydrocyclone

Design and Operation — The hydrocyclone made a particle size split at about 37 μm (400 mesh), separating the silt from the clay particles. The underflow-overflow volumetric split was estimated at 30/70% V. The design calculations specified one 50 mm (2 in.) hydrocyclone. The calculated feed capacity containing 33% solids was about 550 kg/hr (0.61 ton/hr) when operating at a gauge pressure of 138 kPa (20 psig). The 80 mm (3 in.) hydrocyclone used in the demonstration had a maximum feed capacity of about 2400 kg/hr (2.65 ton/hr) while operating with the same percent solids and pressure. The volumetric flow for this unit was about 6.0 m³/hr (26 gpm). For a similar volumetric split to the 50 mm hydrocyclone while operating at the same pressure and solid-mass flow, the feed to the 80 mm unit had to be diluted with recycled water to achieve the correct volumetric flow.

Moreover, the feed pump for the demonstration plant was not properly sized and was unable to provide constant flow at operating pressure. During the first series of test runs, the hydrocyclone pump was a 3.8 cm (1.5 in.) Galigher horizontal centrifugal pump. The Galigher pump was oversized, constantly draining the feed sump, which resulted in surges in the feed to the hydrocyclone. During the last week of the demonstration, the Galigher pump was replaced with a Sala 3.8 cm (1.5 in.) vertical bowl (sand) pump. The Sala vertical bowl pump can produce a feed pressure of at least 195 kPa (28 psig) while maintaining the flow at 6.0 m³/hr (26 gpm).

The concept of using a hydrocyclone to make a size separation at 37 μm (400 mesh) was sound, provided the hydrocyclone was properly sized and the feed was constant with a minimum gauge pressure of 69 kPa (10 psig).

Hydrocyclone Underflow — The hydrocyclone underflow stream consisted of silt-sized particles of soil and lead ranging in size from –105 μm to +37 μm (–150 mesh to +400 mesh). This stream was sent to the Knelson bowl to remove the fine lead particles. The soil in the hydrocyclone underflow had a lead content of about 5160 mg/kg. This value was similar to the 4930 mg/kg value from the treatability test.

Hydrocyclone Overflow — This stream consisted of –37 μm (–400) mesh or clay size particles, which were referred to as slimes. This stream did not undergo any further processing in the gravity plant. A flocculent was added to the slimes and they were discharged into the roll-off boxes for settling. The clean water was returned to the plant. The soil in the hydrocyclone overflow had a lead content of about 5380 mg/kg. This value was again close to the 6010 mg/kg value from the treatability tests. Comparison of the demonstration analyses for feed, underflow, and overflow showed no concentration of the lead in the underflow, which was similar to the treatability results.

3.4.3.7 Reichert Spiral Concentrators

Design and Operation — Each Reichert spiral concentrator had a design capacity of 1295 kg/hr (1.42 ton/hr) at a pulp density of 25% solids. The design volumetric feed rate was 4.38 m³/hr (19.3 gpm) with a minimum wash water rate of 0.68 m³/hr (3 gpm). The minimum design capacity for fine material was 450 kg/hr (0.50 ton/hr) with a volumetric flow of 1.52 m³/hr (6.7 gpm). The operating capacity based on the mass balance was 404 kg/hr with a volumetric rate of 3.40 m³/hr (15.0 gpm) while operating at 11% solids. Wash water was not added to the concentrators. During the demonstration, the rougher concentrator operated close to the minimum design requirements. However, the cleaner concentrator operated below the design requirements. Insufficient dilution water was added to the rougher concentrate that was fed to the cleaner concentrator.

Each spiral made three gravity splits: tailings, middlings, and concentrate. The primary (rougher) spiral produced the clean tailings (Reichert Spiral Concentrator Tailings). The secondary (cleaner) spiral produced the concentrate (Reichert Spiral Concentrator Concentrate).

Reichert Spiral Concentrator Feed — The feed to the Reichert spiral concentrator contained primarily sand and lead particles ranging in size from 1.19 mm to 105 μm (16 mesh to 150 mesh). The grain size was distinct and visible lead was easily observable. The total lead concentration from the demonstration averaged 5900 mg/kg, while that for the treatability samples averaged 4220 mg/kg.

Reichert Spiral Concentrator Tailings — The Reichert concentrator tailings were sand particles with sizes ranging from −1.19 mm to +105 μm. Under a 15X hand lens, this product appeared to be clean sand. The lead content for the tailings from the primary spiral averaged 755 mg/kg, an effective extraction of 87% of the lead in the feed stream. The lead content for the treatability samples averaged 410 mg/kg having an extraction of 90%.

Reichert Spiral Concentrator Concentrate — The Reichert spiral concentrate was a heavy dark sand that was a mixture of fine lead and sand and was amenable to further upgrading. The total lead concentration for the demonstration samples averaged 47,045 mg/kg while that for the treatability samples averaged 32,700 mg/kg.

3.4.3.8 Sieve Bend Dewatering Screen

Design and Operation — The sieve bend screen removed the water from the rougher spiral-concentrator tailings. The mass balance data indicated that the undersize water flow was at least 2.9 m³/hr (13 gpm). If the screen's bar spacing was 100 μm (~150 mesh), a rapped screen easily handled a water flow up to 45 m³/hr (195 gpm). A non-rapped screen with the same bar spacing could have a maximum water flow up to 7 m³/hr (30 gpm). However, the non-rapped screens are usually limited to a bar spacing greater than 350 μm (45 mesh). With a bar spacing of 100 μm, the expected flow would be much less than 7 m³/hr. During the demonstration, the screen provided incomplete dewatering, and large quantities of free water collected in the sand collection container. A pump was installed in the container to remove excess water.

3.4.3.9 Knelson Bowl Concentrator

Design and Operation — This MD12 model Knelson bowl concentrator had a design capacity of 1800 kg/hr (1.98 ton/hr). Although acceptable, the operating feed rated based on the mass balance was considerably lower at 240 kg/hr (0.26 ton/hr). The minimum fluidizing water flow through the concentrator bowl was 6.78 m³/hr (30 gpm). Usually, the Knelson bowl operated as specified. However, when the hydrocyclone was not operating properly, the slimes entering the Knelson bowl concentrator reduced its effectiveness. The hydrocyclone must be designed and operated properly or the Knelson bowl's sensitivity may outweigh its benefit. Constant feed was also necessary to operate the Knelson bowl.

Knelson Bowl Concentrator Feed — During the demonstration, the hydrocyclone underflow fed the Knelson bowl, and had an average total-lead concentration of 5160 mg/kg. In comparison, the feed analysis from the treatability study averaged 4930 mg/kg.

Knelson Bowl Concentrator Concentrates — The concentrates consisted of lead and silt with a particulate size range of −105 μm to + 37 μm (−150 mesh to +400 mesh). The Knelson bowl never reached capacity. The largest lead load in the bowl during this test was approximately 10% or less. While the Knelson bowl removed free lead particles well, the concentrate grade was decreased by dilution with silt. The total lead content for the concentrate samples from demonstration averaged 42,695 mg/kg. The treatability concentrates, which were produced by treating the −105 μm (−150 mesh) fraction, averaged 30,200 mg/kg.

Knelson bowl tailings — In the demonstration, the −105 μm to +37 μm tailings were recombined in the roll-off boxes with the −37 μm slimes from the hydrocyclone. The total lead concentration for the demonstration tailings samples averaged 1740 mg/kg, while those for the treatability samples averaged 4000 mg/kg. Removing the slimes fraction in the hydrocyclone seemed to decrease the lead content of the tailings. However, the results also indicated that the tailings might need further treatment by leaching to remove the attached lead.

3.4.3.10 Vibrating Sweco Screens

Design and Operation — All Sweco vibrating screens had a 45.7 cm (18 in.) diameter. Screen #1 was a double-deck unit, which separated the product discharging from the attrition scrubber auger. This unit used 4.76 mm (4 mesh) and 1.19 mm (16 mesh) screens. The trommel feeding the jig had a 1.19 mm (16 mesh) screen. Thus, a 0.84 mm (20 mesh) screen was installed in Screen #2, which removed water from the jig tailings. Screen #3 for the Reichert spiral concentrate had 149 μm (100 mesh) openings. The following table lists mass balance data from Table 3.1 and the estimated capacities using the calculation methods described in Appendix Figure J.1:

Screen Number	Screen Opening mm (mesh)	Mass Balance Data kg/hr (lb/hr)			Estimated Soil Capacity kg/hr (lb/hr)
		Solid	Water	Undersize	
1	1.19 (16)	70 (154)	18 (40)	14	405 (893)
2	0.84 (20)	38 (84)	90 (198)	73	320 (705)
3	0.149 (100)	48 (106)	270 (595)	258	225 (496)

Overall, the Sweco screens performed within the limits of the estimated capacities. For Screen #1, overloads were not experienced, even at a soil feed rate of 3200 kg/hr (3.5 ton/hr). Screen #2 experienced product overloads when humate material accumulated on the screen. Screen #3 successfully removed the water from the Reichert cleaner concentrate but operated near the upper limit of its capacity.

3.4.3.11 Vertical Bowl (Sand) Pumps

Design and Operation — Vertical bowl (sand) pumps are normally the best pumps for moving slurries. These pumps typically do not plug and can run either dry or under varying slurry loads. However, these pumps develop a head pressure at the discharge of about 195 kPa (28 psig). At these discharge pressures, the bowl pump cannot deliver water to heights above 17.7 m (58 ft). The density of slurries can decrease the maximum head. Consequently, the bowl pump had trouble lifting the 10–20% solids slurry from the spiral classifier's sand sump to the top of the rougher spiral concentrator.

3.4.3.12 Thickeners

Design and Operation — The original thickeners used on this plant were too small and were replaced with two dewatering roll-off boxes connected in series to settle out the slimes particles.

Mixed Slimes — The Knelson tailings and hydrocyclone overflow slurry were mixed with a flocculent and fed into the first of two roll-off boxes. The overflow slurry from the first box was pumped into the second box. Recycled water was pumped from the second box into the day tank. The combined slimes from the hydrocyclone overflow and the Knelson bowl tailings settled out to form the mixed slimes fraction. During the demonstration, the mixed slimes had an average total lead concentration of 4525 mg/kg. In comparison, the slimes from the treatability study averaged 4870 mg/kg.

3.4.4   Ancillary Equipment Operation

3.4.4.1   Conveyor Belts

The conveyor belts used to transport the coarse products to collection containers were flat and smooth. Water and particles leaked beneath the belts. These belts should have been replaced with curved, ribbed belts so that the water and particles would be retained on the belts.

3.4.4.2   Centrifugal pumps

Generally, centrifugal pumps perform well for delivering slurry at high pressures. However, the flow capacities of the pumps were too high for the unit operations in the pilot plant. The pumps would also air lock if allowed to run dry. To avoid operational problems, each centrifugal pump must be sized correctly for its particular application in a soil washing plant.

Two centrifugal pumps supplied water to the pilot plant through 5.08 cm (2 in.) pipes that had delivery trees about every 1.5 m (5.0 ft). The designated delivery capacity with one pump was 189 L/min (50 gpm) and 289 L/min (75 gpm) with both pumps operational. These capacities provided an adequate water supply to the pilot plant.

3.4.4.3   SandPIPER Diaphragm Pumps

The diaphragm pumps tended to plug easily, were unreliable, needed constant attention, required a considerable amount of air, and were not very efficient. Therefore, these pumps had to be replaced in the pilot plant and are not recommended for use in other soil washing plants.

3.4.4.4   Peristaltic Pumps

These peristaltic pumps were adequate for moving liquid or slimes, but the tubing ruptured frequently when used to pump sand. Particles, small rocks, and bullets also prematurely wore the tubing. However, with selected use, as for pumping flocculent, they worked well.

3.4.4.5   Filters

Cloth filters were used to produce clean recirculated water for the plant. However, the original thickeners were too small and allowed fines into the overflow. The filters plugged within a few hours of operation. The fines made it impossible to clean or clear the filter by backflushing. The filters were removed from the treatment plant.

3.4.4.6   "Y" Samplers

The sampler valves were made of plastic and were fragile. All the plastic fittings were found to be cracked or leaking during shakedown testing of the plant. Therefore, all fittings except the "Y" samplers were replaced. During the demonstration, all but one "Y" sampler plugged. The only circuit that did not fail was the slimes circuit. Proper design of the piping network to eliminate sharp bends was necessary. Control of the slurry pulp density had to be maintained, as well as careful monitoring of the percentage of lead particles being fed to the spiral classifier from the attrition scrubber auger.

3.4.4.7 Coriolis Mass Flowmeters

These flowmeters stopped functioning in all the streams that contained large particles. The only Coriolis mass flowmeter that did not completely plug was the one in the slimes circuit. The inside diameter of the flowmeter may have been smaller than the particle diameter, the piping diameter may have been too large, or the pipe angle may have been too sharp.

3.4.4.8 Flowmeters

Flowmeters were used to monitor feed water to various pieces of equipment. They operated adequately throughout the test.

3.4.4.9 Flow Control Valves

Green valves (Badger Company) used a needle constriction to control flow and plugged with fine soil particles. As fine particles are expected to flow through any similar plant, this type of valve is not recommended.

Red valves (Red Valve Company) were used in the feedback circuits to control the levels of the sumps. This type of valve constricted the flow, using a rubber sleeve. When working with slurries, this type of automated valve is recommended.

3.4.4.10 Computer Control

Computer-controlled data gathering was helpful for determining how the plant functioned at various levels and under varying conditions. However, because of interference with the stream flows throughout the operation of the system, a definitive evaluation of effectiveness was not possible. The gravity separation plant functioned as specified when the plant was operated at a constant 907 kg/hr (2000 lb/hr).

3.4.5 Health and Safety

Safety Procedures

The following safety procedures were initiated during the 2 days prior to equipment setup:

- All field personnel for the project underwent an 8-hour refresher course in hazardous waste site worker training presented by Mr. Bill Collier, a Certified Industrial Hygienist (CIH) and health and safety manager for Metcalf & Eddy.
- The site was monitored on four occasions for ambient background levels of lead. The results for all monitoring days were below regulatory minimum standards.
- Site personnel wore lead monitors on two occasions to test for work area levels of lead. Work area levels for lead were found to be below regulatory minimum standards.
- Both an exclusion zone and a decontamination zone were constructed. The exclusion zone barrier was roped with caution flags and warning signs.
- First aid stations were installed in the office trailer and on the gravity separation trailer.
- Eye wash stations were installed on the gravity separation trailer and on the vat leach trailer.
- The vat leach trailer was fitted with railings and barriers to prevent accidental falls.

- Personnel protective equipment (PPE), such as ear plugs, safety glasses, protective glasses, Tyvek suits, respiratory protection, and hard hats, were provided for site personnel and for visitors in the office trailer.
- All electrical outlets on both trailers were equipped with ground fault interrupters (GFI).
- An Occupational Safety and Health Administration (OSHA) poster was positioned visibly in the office trailer.
- The emergency route to the local hospital was verified and posted.
- A drinking water source was established.

Safety Meetings

Formal safety meetings were held as follows:

- August 22, 1996. Participants reviewed site hazards such as rules for heavy lifting, hot weather conditions, prevention of heat stroke, insect and rattlesnake hazards, noise hazards, locations of first-aid kits and eye wash stations, and lock-out/tag-out procedures.
- August 29, 1996. Participants reviewed work zone rules, such as no smoking or eating in the exclusion zone, decontamination procedures, and heat and noise hazards.

Informal safety meetings were held intermittently and as needed.

3.4.6   Pilot Plant Demonstration Results Summary

Initially, the operational data were compared to mass balance data that were calculated using the NAS Miramar treatability results. Each piece of equipment used in the pilot plant was also evaluated during the demonstration runs. A summary of the primary operational characteristics and analytical results for the major treatment units follows.

Attrition Scrubber — The attrition scrubber used in the pilot plant was designed to provide both particle deagglomeration and sizing of a coarse feed using an auger conveyor in conjunction with the scrubber. A slurry containing the sand and fine particles flowed through a weir at the top of the scrubber tank into the trommel feed chute, while the coarse soil and lead particles settled to the bottom of the scrubber. An auger conveyor removed the settled coarse particles. As expected, the coarse product had a lead content of about 25,000 mg/kg. However, visual inspection of this product showed that the auger shredded lead bullets, which markedly increased the lead content in the feed to the Reichert spiral concentrators. Consequently, the current attrition scrubber with auger should be replaced with a vibrating wet screen to remove the coarse particles and a commercial attrition scrubber that would be easier to operate and maintain.

Trommel Scrubber — The trommel scrubber was used as a rotating screen and made a particle separation at 1.19 mm (16 mesh). The feed to the trommel from the attrition scrubber contained –4.76 mm (–4 mesh) soil primarily. The capacity of the gravity circuit was limited by the trommel capacity, as predicted by the particle size distribution from the treatability study and the pilot plant design calculations.

Mineral Jig — The oversize fraction from the trommel was fed to the jig. All of the lead in the soil that was fed to the jig was removed in the sink product. Therefore, the lead concentration in the tailings was low, 800 mg/kg, as compared to the mass balance value, 3200 mg/kg. The jig removed about 98% of the lead from the coarse soil. One of the difficulties encountered with the use of the pilot-scale mineral jig was that the jig hutch had to be cleaned by hand when it reached capacity.

Spiral Classifiers — The undersize fraction from the trommel flowed into two spiral classifiers for separating sand soil particles from slimes soil particles. The expected size fractions were the sand fraction, –1.19 mm by +105 µm (–16 mesh by +150 mesh), and the slimes fraction, –105 µm (–150 mesh). The sand fraction was fed to the Reichert spiral concentrators, while the slimes went to the hydrocyclone. The classifiers had a combined capacity of about 520 kg/hr (0.57 ton/hr) based on the required sand-compression pool area. As expected, the sand fraction occasionally overloaded and stalled the sand-screw motors. For the pilot plant to operate at 1275 kg/hr (1.41 ton/hr), larger classifiers should have been installed.

Reichert Spiral Concentrators — Two Reichert spiral concentrators were used in series to produce a lead concentrate and clean tailings. Each spiral concentrator had a rated solids capacity of 450 to 2250 kg/hr (0.50 to 2.5 ton/hr) with a pulp density between 20 to 30% solids. The spiral concentrators had excess capacity and operated satisfactorily. The total lead concentration for the spiral feed averaged 5900 mg/kg. The lead content for the tailings from the primary spiral averaged 750 mg/kg, an effective decrease of about 87% of the lead in the feed stream. The tailings lead content was higher than the mass balance result of 410 mg/kg. However, the mass balance content for the feed was lower at 4200 mg/kg. The tailings and the lead concentrate were dewatered separately by using screens.

Hydrocyclone — The hydrocyclone was designed to make a particle size separation at 37 μm. The overflow from the hydrocyclone containing −37 μm particles went to a slimes thickener (roll-off bins), while the larger +37 μm particles went to the Knelson bowl concentrator. However, the hydrocyclone used in the pilot plant was oversized to treat the spiral classifier overflow feed. The required slurry capacity had to be between 4000 and 6000 L/hr at 30% solids, but the feed pump was unable to provide constant flow at a minimum operational gauge pressure of 69 kPa (10 psig). The hydrocyclone used in the pilot plant allowed −37 μm particles to pass into the feed to the Knelson bowl concentrator. The concept of using a hydrocyclone to make a size separation at 37 μm (400 mesh) was sound. However, the hydrocyclone had to be sized properly and the slurry feed kept constant with an optimal gauge pressure of 103 kPa (15 psig).

Knelson Bowl — During the demonstration, the solid feed to the Knelson bowl concentrator from the hydrocyclone had an average lead concentration of 5200 mg/kg. The concentrator produced tailings with a lead concentration of about 1700 mg/kg, a reduction in excess of 65%. This reduction was better than the expected mass balance value of 4000 mg/kg for a feed containing about 5000 mg/kg, or about a 20% reduction. When the Knelson bowl concentrator was operated as specified, it had a capacity of about 1800 kg/hr (1.98 ton/hr). However, a constant feed is necessary for proper operation of the Knelson concentrator, just as for a hydrocyclone. Since the hydrocyclone was not designed and operated properly, the increase in the −37 μm (400 mesh) fraction in the +37 μm feed decreased the effectiveness of the concentrator.

CHAPTER **4**

Design and Operation of Hypothetical System

4.1 REMEDIAL OBJECTIVES

A conceptual design of components that might be required for full-scale application of mining and beneficiation techniques to remediate typical small arms firing ranges was prepared. The full-scale application was designed to meet the following objectives for remediation of a hypothetical site:

- Remove coarse and fine lead contaminants in the soil
  - A total lead concentration less than either 500 or 1000 mg/kg of soil, depending on land use, and
  - A TCLP concentration for lead less than 5 mg/L.
- Construct a modular, mobile, and cost-effective unit.

4.2 ENGINEERING DESIGN

4.2.1 Plant Design Criteria

Before designing a treatment plant for a firing range soil, representative bulk samples must be obtained and treatability studies completed on composite samples of the soil. The conceptual plant design was prepared for a hypothetical DOD firing range. The hypothetical site was derived by combining particle size and lead distribution data from the treatability studies conducted on soils from Fort Dix and Twentynine Palms. Table 4.1 summarizes the process treatability data that were used in the design. The table also lists the particle size distribution and lead contents for the hypothetical soil fractions. Details of the treatability data are given in Appendix Table I.1.

The full-scale treatment plant was designed to have an operating capacity of 13.60 t/hr (15 ton/hr) of −50.8 mm (−1 in.) soil. It was also assumed that the firing range contained 5.26 kt (5800 ton) of soil to be processed. The block diagram for the treatment system is shown in Figure 4.1. The complete flow sheet is provided in Appendix Figure I.1. Using the size distribution, treatability data, and process flow diagram, the mass balance in Appendix Table I.2 was calculated for the 13.60 t/hr treatment plant.

4.2.2 Conceptual Design

The full-scale plant was designed to be a mobile, modularized system. The technologies and equipment evaluated for the design were commonly available. The soil washing system included four principal unit operations: deagglomeration, particle sizing, gravity separation, and dewatering.

Table 4.1 Treatability Results for the Soil at the Hypothetical Firing Range

	Soil Before Treatment		Concentrate		Tailings/Middlings	
	Size Fraction Weight %	Total Lead mg/kg	Size Fraction Weight %	Total Lead mg/kg	Size Fraction Weight %	Total Lead mg/kg
Soil Sample Total	100.00	8495	10.45	78,440	89.55	330
Gravel Fraction Total	7.16	419	0.00	0	10.55	455
+25.00 mm (+1 in)	4.47	475	0.00	0	4.45	475
+19.00 mm (+3/4 in.)	2.69	325	0.00	0	2.70	325
Coarse Fraction Total	17.26	44,880	4.79	161,200	12.47	260
+9.50 mm (+3/8 in.)	7.78	72,180	3.98	140,530	3.80	525
+4.76 mm (+4 mesh)	3.73	31,185	0.56	211,700	3.17	75
+3.38 mm (+8 mesh)	5.75	16,415	0.24	382,710	5.51	180
Sand Fraction Total	67.33	751	4.56	9775	52.77	90
+1.19 mm (+16 mesh)	13.17	1015	0.56	20,960	12.61	130
+590 μm (+30 mesh)	14.11	1040	1.03	12,980	13.08	100
+297 μm (+50 mesh)	15.75	1070	1.15	13,535	14.60	85
+105 μm (+150 mesh)	24.30	235	1.82	2135	22.48	60
Fine (Slime) Fraction Total	8.25	2590	1.10	3110	7.15	2510
−105 μm (−150 mesh)	8.25	2590	1.10	3110	7.15	2510

Note: The particle sizes in millimeters or micrometers and mesh use the U.S. Standard Sieve series.

4.2.2.1 Deagglomeration

Deagglomeration is the process of separating particles from each other by mechanical means, allowing accurate particle sizing. Lead particles must be separated from the soil matrix in order to recover the lead particles. The selection of the proper equipment depended on the type of soil, and four types were considered: a trommel, a log washer, a high-pressure water spray on vibrating screen, and an attrition scrubber.

The trommel, log washer, and high-pressure water spray on a vibrating screen were all effective in separating lead particles from the soil matrix. For coarse sandy soils, spraying water on the soils while on a vibrating screen was sufficient for deagglomeration. For soils with either a high clay or high fines content, a trommel, a log washer, or a combination of high-pressure spray on a vibrating screen with a trommel or log washer should be used. Based on the hypothetical soil, a high-pressure water spray on a vibrating screen was used in the full-scale plant design.

The attrition scrubber was found to be less desirable than the other methods for deagglomeration of the lead contaminated soils. The antimony in the lead bullets reacted over time, making the older bullets brittle. The grinding action of the agitators could break the bullets into small fragments, which would deposit in the fines. Lead recovery from fines is more difficult than from coarse materials and may require leaching. Another negative factor in the use of the attrition scrubber for deagglomeration was the additional maintenance requirement caused by the coarse soil fractions wearing the scrubber liner.

4.2.2.2 Particle Sizing

In the conceptual design, particle sizing was performed four times within the process. The system required flexibility to allow specific particle size ranges to be separated and then removed from the process if they were considered clean (total lead less than 500 or 1000 mg/kg). While particle sizing has been utilized in many industries with a variety of applications and equipment, this design used equipment most commonly available.

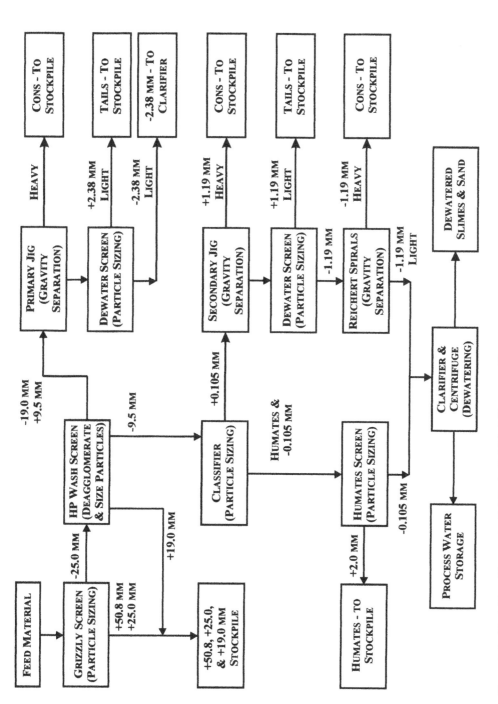

Figure 4.1 Block Flow Diagram for Full-Scale Soil Wash Treatment Plant.

The initial particle sizing of the feed material occurred on the grizzly-screen unit, which was a vibrating screen. The vibration provided some deagglomeration and minimized the spillage of the −50.8 mm (−2 in.) soil. The screen system should be either two single-deck units with screen openings at 50.8 mm (2 in.) and 25.0 mm (1 in.) or preferably a two-deck system. If it is a two-deck system, the screen openings should also be at 50.8 mm and 25.0 mm.

The second particle sizing operation utilized a two-deck or three-deck vibrating screen with a high-pressure spray immediately following the grizzly-screen unit. The sizes of the screen openings were determined from the treatability study, and took into account the particle size operating ranges for the gravity separation equipment.

The third particle sizing operation, known as desliming, utilized a spiral classifier for separation of the +150 mesh soil from the humates and the −150 mesh soil. This operation was the final deagglomeration step before the soil was washed and dewatered by the action of the spiral conveyor in the classifier's tank.

The fourth particle sizing operation used vibrating screens to remove water and undersize soil particles from the primary and secondary jig and tailings and water from the Reichert spiral concentrate. A screen was also used to remove coarse humates from the spiral classifier overflow slurry.

4.2.2.3 Gravity Separation

Lead, with the very high specific gravity of 11.4, is easily separated from soil components with lower specific gravities (2.6 to 3.0) through gravity separation processes. The conceptual design evaluated three types of gravity separation techniques for the coarse and sand fractions, +1.19 mm (+16 mesh) and −1.19 mm by +105 μm (−16 mesh by +150 mesh), respectively.

Shaking tables were considered for both the coarse and sand fractions but were eliminated from use in the full-scale plant. The tables lacked mobility, requiring construction of a base support. The table also had specific uniform feed requirements that would be difficult to achieve in a mobile field unit.

The use of mineral jigs and riffles were evaluated for separation of coarse soil particles. Duplex mineral jigs were selected to remove coarse lead in the conceptual design. Mineral jigs have been used in the mineral industry since the early 1900s and are easily obtained. A properly designed mineral jig collects the heavy particles in a concentrated stream while allowing the light particles to pass over the heavies. To prevent the mineral jig from overloading, a cup draw was included in the design. A mineral jig was found to produce the same effective lead removal as riffles with similar water requirements. However, riffles were labor intensive and produced a concentrate that was less concentrated than that produced by mineral jigs. Although not specific to the conceptual design assumption, riffles are also inefficient for soils with high concentrations of clay.

For the sand fraction, both Reichert spiral and Knelson bowl concentrators were evaluated. The pilot demonstration at NAS Miramar used both the Reichert spiral and Knelson bowl concentrators, and both gravity separation techniques successfully separated the lead from the soil. For the conceptual design, Reichert spiral concentrators were chosen because they are mobile, easy to install, have no mechanical parts, and are inexpensive to operate.

4.2.2.4 Dewatering

Dewatering is the critical operation for any soil washing system and can be very expensive. The dewatering cost is driven by both the total system flow and the concentration of slimes, 105 μm (−150 mesh) particles, in the soil. The conceptual design evaluated four dewatering techniques.

The least costly dewatering technique was to use multiple watertight roll-off boxes placed in series with addition of a flocculent to increase settling of the fines. However, this technique required a large surface area and, to be truly effective, must have a coarser soil than that in the conceptual design. This technique also required periodic shutdown of boxes to allow for removal of the settled soil. The settled material was 10 to 20% solids and was not stackable.

The second type of dewatering technique evaluated was the use of a large centrifuge to produce a stackable cake in the 40 to 60% solids range while handling the flow rate produced in the system. Flocculent would be injected at the centrifuge to facilitate settling of soil particles. Large centrifuges require upwards of 300 hp to operate, adding significantly to the operating cost of a system. This large centrifuge was considered impractical for the conceptual design.

The use of a clarifier, a dewatering technique which has been successfully applied for many years, was evaluated. A variety of clarifiers is available. The solids settle in the clarifier to produce a clear overflow. However, the underflow solids were 10 to 20% solids, which would also be unstackable.

The fourth type of dewatering technique evaluated was a filter press. The filter press produced a stackable cake at a reasonable cost. However, the presses are labor intensive to operate and to maintain and were not considered for the conceptual design.

The dewatering technique selected for the full-scale plant design was a combination of a clarifier and a small centrifuge. A plate clarifier was modified to fit on a lowboy for mobility. The centrifuge was sized based on the underflow from the clarifier, which was expected to be 10 to 20% solids. The centrifuge was to produce a stackable material of 40 to 60% solids.

4.2.2.5 Ancillary Operations

Ancillary operations included pumping, materials handling, power generation, process control, and process monitoring. As all of these operations have been applied in the minerals industry for an extended time, the equipment selected would produce predictable results. The conceptual design specified only standard equipment size ranges.

4.2.3 Equipment Selection and Design

The soil contained considerable amounts of sand particles, −1.19 mm to +105 µm (−16 mesh to +150 mesh), with some large pieces of gravel, +19.0 mm (+3/4 in.), before treatment (Table 4.1). The treatability results (Table 4.1) showed that the removal of large stones and gravel, +50.8 mm (+2 in.) and +25.0 mm (+1 in.) fractions, with a grizzly-screen unit produced a clean product. The remaining gravel fraction, +19.0 mm (+3/4 in.), was removed with the high-pressure (HP) wash screen. The largest coarse fraction, −19.0 mm to +9.5 mm (−3/4 in. to +3/8 in.), required treatment in the primary mineral jig and screening to produce a clean product. The −9.5 mm (−3/8 in.) fraction was then treated in the spiral classifier, which removed the −105 µm (−150 mesh) fine (slimes) fraction. The coarse and sand fractions, −9.5 mm to +105 µm, from the classifier's sand screw were then treated in the secondary mineral jig. The mineral jig, having a 1.19 mm (16 mesh) screen, produced a +1.19 mm concentrate. Using a 1.19-mm (16 mesh) dewatering screen, the tailings were separated into a +1.19 mm clean fraction and a −1.19 mm fraction containing lead particles. The −1.19 fraction was treated in the two Reichert spiral concentrators to produce a lead concentrate and a clean −1.19 mm tailings fraction. A 2.0 mm (10 mesh) screen removed the humates from the classifier overflow, the −105 µm fraction. The −105 µm and −1.19 mm fractions were combined in a lamella clarifier. The lamella overflow was recycled as process water, and the underflow was dewatered in a centrifuge. The dewatered slimes and sand particles were placed in a stockpile to be mixed with the coarser clean fractions. A block diagram of the unit operations for the treatment plant is shown in Table 4.1.

After the screening and gravity separation steps, about 83% of the soil was considered clean with a lead content of about 410 mg/kg. The combined concentrates, representing about 9% of the soil, contained 8.7% lead. The slimes fraction, −105 µm, was not treated in a gravity separator.

The lead content of this fraction was about 2590 mg/kg, which exceeded the treatment requirement of 1000 mg/kg. However, the lead content of the recombined soil fractions, representing 91% of the total, was about 594 mg/kg. This lead content was close to the treatment standards of 500 mg/kg.

A Knelson bowl concentrator can remove a portion of the lead from –105 μm fraction. Removal of about 50% of the lead would produce a recombined soil with a lead content below 500 mg/kg. The demonstration results (Table 2.12) showed that the Knelson bowl decreased the lead content in the hydrocyclone overflow from 5160 to 1740 mg/kg, about 65% removal of lead. At 65% removal, the lead content of the recombined soil in the full-scale plant design would decrease to 450 mg/kg.

Particle size and lead distributions and the treatability data (Table 4.1) were used to calculate a modified mass balance in Appendix Table J.1. For these calculations, it was assumed that the treatment plant would operate between 9.07 t/hr and 18.14 t/hr (10 ton/hr and 20 ton/hr) with a nominal capacity of 13.60 t/hr (15 ton/hr). Accordingly, a maximum capacity of 18.14 t/hr (20 ton/hr) was assumed to be the treatment rate for the modified mass balance. The mass and component data from this balance were used to size the equipment for the full-scale plant. The equipment size calculations are detailed in Appendix J. Table 4.2 lists the capacity requirements for the installed equipment. Finally, Table 4.3 lists the sizes of each piece of equipment determined in Appendix J.

Equipment layout drawings for a typical full-scale plant are shown in Appendix K:

- Figure K.1: Site Layout for Full-Scale Plant
- Figure K.2: Piping Layout for Full-Scale Plant
- Figure K.3: Skid #1, Primary Gravity Separation Units
- Figure K.4: Skid #2, Secondary and Tertiary Gravity Separation Units
- Figure K.5: High-pressure Water Pump Skid
- Figure K.6: Low-pressure Water Pump Skid
- Figure K.7: Main Electrical Diagram
- Figure K.8: Electrical Diagram for Skids #1 and #2
- Figure K.9: Electrical Diagram for Dewatering Trailer

Field conditions for a firing range requiring remediation will deviate in particle size and lead distributions from the assumptions made for this hypothetical site. The plant design is dependent on treatability studies on actual soil samples. Differences in particle sizes and treatability studies could affect the location, capacity, and size of each piece of equipment. Hence, the objective to design and construct a mobile, modularized treatment system was essential. The equipment in this system could easily be rearranged or replaced with larger or smaller equipment.

Table 4.2 Equipment Design Capacities for 13.60 t/hr (15 tons/hr) System

Equipment	Design Capacity Requirement kg/hr (Solid) or L/hr (Water)				Comments
	Mass Balance	Minimum	Maximum	Design	
Grizzly-Screen with Hopper	14,240	9495	18,990	31,400	Grizzly: −50.8 mm (−2 in.) Solid
	13,605	9070	18,140	31,400	Screen: −25.0 mm (−1 in.) Solid
Conveyor #1	13,605	9070	18,140	18,140	−25.0 mm (−1 in.) Solid
Double-Deck Wet Screen	13,240	8825	17,650	35,100	Top Deck: −19.0 mm (−3/4 in.) Solid
	12,180	8120	16,241	16,300	Bottom Deck: −9.5 mm (−3/8 in.) Solid
	25,205	16,805	33,610	33,600	L/hr HP Wash Water
Conveyor #2	365	245	490	500	+19.0 mm Solid
Primary Jig	1060	705	1410	1900	−19.0 mm +9.5 mm Solid
	1060	705	1415	2400	L/hr LP Wash Water
Dewatering Screen # 1	25	15	35	8000	−2.38 mm (−8 mesh) Solid
	285	190	385	>400	L/hr LP Wash Water
	1240	830	1656	>1700	kg/hr Undersize Slurry at 2.1% Solids
Conveyor #3	1060	705	1410	1400	−19.0 mm, +9.5 mm Solid
Spiral Classifier	11,000	7335	14,667	16,400	−9.5 mm +105 micron Solid in Sand Fraction
	1180	790	1575	18,700	−105 micron (−150 mesh) Solid in Overflow Slurry
Secondary Jig	11,000	7330	14,665	14,300	−9.5 mm +105 micron Solid
	11,015	7345	14,690	14,400	L/hr LP Wash Water
Dewatering Screen # 2	7620	5080	10,155	16,100	−1.19 mm (−16 mesh) Solid
	3065	2040	4085	6500	L/hr LP Wash Water
Conveyor # 4	25,820	17,215	34,430	53,600	kg/hr Undersize Slurry at 30% Solids
	3185	2125	4250	6500	+1.19 mm (+16 mesh) Solid
Primary Spiral Concentrator	9285	6190	12,380	13,900	Feed Solid in Slurry at 30% Solids
	0	0	0	0	L/hr LP Dilution Water
Secondary Spiral Concentrator	1680	1120	2240	4600	Feed Solid in Slurry at 30% Solids
	2115	1410	2820	5800	L/hr LP Dilution Water
Dewatering Screen # 3	0	0	0	0	−105 micron (−150 mesh) Solid
	0	0	0	0	L/hr LP Wash Water
Humates Screen	560	375	750	45,000	L/hr Undersize Slurry at 0.0% Solids
	1180	775	1575	<7100	−2.0 mm (−10 mesh) Solid
	655	440	880	>900	L/hr HP Wash Water
	22,230	14,820	29,640	>30,000	kg/hr Undersize Slurry at 5.5% Solids

continued

Table 4.2 (continued) Equipment Design Capacities for 13.60 t/hr (15 tons/hr) System

Equipment	Design Capacity Requirement kg/hr (Solid) or L/hr (Water)				Comments
	Mass Balance	Minimum	Maximum	Design	
Clarifier	8240	5490	10,895	21,600	Feed Solid in Slurry at 16.2% Solids
	41,190	27,460	54,925	94,700	kg/hr Underflow Slurry at 20.0% Solids
Centrifuge	41,190	27,460	54,925	68,000	L/hr Feed Slurry At 20.0% Solids
	13,730	9155	18,310	25,800	kg/hr Underflow Sludge At 60.0% Solids
P1: Hi-pressure Water Pump	25,865	17,240	34,485	41,380	L/hr (182 gpm); 0% Solids; 690 kPa (230 ft)
P2: Lo-pressure Pump	17,540	11,695	23,390	30,410	L/hr (135 gpm); 0% Solids; 206 kPa (70 ft)
P3: Classifier Slurry Pump	24,475	14,320	28,635	33,200	L/hr (146 gpm); 6% Solids; 206 kPa (70 ft)
P4: Screen #1 Underflow Pump	1215	810	1620	1950	L/hr (9 gpm); 2% Solids; 206 kPa (70 ft)
P5: Spiral Feed Pump	21,140	14,090	28,185	33,815	L/hr (149 gpm); 30% Solids; 206 kPa (70 ft)
P6: Spiral Mids Pump	2105	1400	2805	3365	L/hr (15 gpm); 40% Solids; 206 kPa (70 ft)
P7: Spiral Feed Pump	3135	2090	4180	5015	L/hr (22 gpm); 29% Solids; 206 kPa (70 ft)
P8: Spiral Mids Pump	1585	1055	2110	2530	L/hr (11gpm); 29% Solids; 206 kPa (70 ft)
P9: Spiral Tails Pump	2890	1925	3850	4620	L/hr (20 gpm); 17% Solids; 206 kPa (70 ft)
P10: Sands Pump	22,900	15,265	30,530	36,640	L/hr (161 gpm); 26% Solids; 206 kPa (70 ft)
P11: Centrifuge Feed Pump	36,120	24,080	48,160	57,790	L/hr (255 gpm); 20% Solids; 206 kPa (70 ft)
P12: Clarifier Overflow Pump	9555	6370	12,740	15,285	L/hr (68 gpm); 0% Solids; 206 kPa (70 ft)
Main Disconnect					3-phase 440 volts, 250 KVA

Table 4.3 Full-Scale Plant Equipment List and Specifications

Equipment Item	Quantity	Design Specifications
Grizzly and Vibrating Screen with Hopper and Belt Feeder	1	Grizzly with bar spacing at 50.8 mm (2 in.)
		Vibrating Screen: 0.91 m width by 1.83 m length (3 ft by 6 ft) with screen opening at 25.0 mm (1 in.)
		Hopper with belt feeder: minimum capacity of 4 to 6 m³ (5 to 7 yd³)
Conveyor #1	1	45 cm (18 in.) width by 10.7 m (35 ft) length
Double-Deck Wet Screen	1	Double-Deck Vibrating Screen: 0.91 m width by 1.83 m length (3 ft by 6 ft)
		Top Deck opening at 19.0 mm (3/4 in.)
		Bottom Deck opening at 9.5 mm (3/8 in.)
Conveyors #2, #3, #4	3	45 cm (18 in.) width by 9.1 m (30 ft) length
Primary Mineral Jig	1	Simplex 16 × 24 with a total jig bed area of 0.25 m² (2.67 ft²)
Dewatering Screen #1	1	Single-Deck Vibrating Screen: 0.91 m width by 1.83 m length (3 ft by 6 ft) with screen opening at 2.38 mm (8 mesh)
Spiral Classifier	1	1500 mm (60 in.) Simplex classifier with 150% spiral submergence and modified flare tank
Secondary Jig	1	Duplex 24 × 36 with a total jig bed area of 2.24 m² (24 ft²)
Dewatering Screen #2	1	Single-Deck Vibrating Screen: 0.91 m width by 2.44 m length (3 ft by 8 ft) with screen opening at 1.19 mm (16 mesh)
Reichert Spiral Concentrator	3	Primary Concentrator: MD LG7 with triple start (3 spirals per column)
Reichert Spiral Concentrator	1	Secondary Concentrator: MD LG7 with triple start (3 spirals per column)
Dewatering Screen #3	1	Rapped Sieve-bend Screen: 1625 mm surface radius, 2390 mm surface length, and 610 mm width with 0.100 mm bar spacing (~150 mesh).
Humates Screen	1	Single-Deck Vibrating Screen: 0.91 m width by 1.83 m length (3 ft by 6 ft) with screen opening at 2.0 mm (10 mesh)
Clarifier	1	Lamella LTS, Model 100 with clarification area of 80 m² and thickening area of 20 m²
Centrifuge	2	Sharples Super-D-Canter, Model P3400, with a capacity up to 570 Lpm (150 gpm)
Polymer Injection System	1	Mix tank, low-capacity metering pumps, and in-line mixers
P1: High-pressure Water Pump	1	3 × 4 × 6-3/4 in.; 3500 rpm; Gould Model 3656, 16 Al/BF; 18.50 kW
P2: Low-pressure Pump	1	2-1/2 × 3 × 8-5/8 in.; 1750 rpm; Gould Model 3656, 11 Al/BF; 3.75 kW
P3: Classifier Slurry Pump	1	2-1/2 × 2 × 10 in.; 1600 rpm; Denver Model SRL; 7.50 kW
P4: Screen Underflow Pump	1	1-1/2 × 1-1/4 × 10 in.; 1800 rpm; Denver Model SRL; 1.50 kW
P5: Spiral Feed Pump	1	2-1/2 × 2 × 10 in.; 1800 rpm; Denver Model SRL; 7.50 kW
P6: Spiral Mids Pump	1	1-1/2 × 1-1/4 × 10 in.; 1600 rpm; Denver Model SRL; 1.50 kW
P7: Spiral Feed Pump	1	1-1/2 × 1-1/4 × 10 in.; 1800 rpm; Denver Model SRL; 1.50 kW
P8: Spiral Mids Pump	1	1-1/2 × 1-1/4 × 10 in.;1800 rpm; Denver Model SRL; 1.50 kW
P9: Spiral Tails Pump	1	1-1/2 × 1-1/4 × 10 in.; 1800 rpm; Denver Model SRL; 1.50 kW
P10: Sands Pump	1	2-1/2 × 2 × 10 in.; 1800 rpm; Denver Model SRL; 7.50 kW
P11: Centrifuge Feed Pump	1	2-1/2 × 2 × 10 in.; 1800 rpm; Denver Model SRL; 7.50 kW
P12: Clarifier Overflow Pump	1	1-1/2 × 2-10 in.; 1750 rpm; Gould Model 3656, 11 Al/BF; 3.75 kW
Electrical Disconnect	1	3-phase, 440 volts, 250 KVA
Motor Starter Board	1	
Process Control Software	1	

CHAPTER 5

Economic Analysis

The economic analysis provides cost information to be used to evaluate the applicability of soil washing technology for small arms firing range lead removal projects. With a realistic knowledge of the contents of this section, it is possible to estimate the economics of operating similarly sized systems at other sites.

5.1 INTRODUCTION

In this section, the capital costs of equipment, fabrication costs of the skid-mounted system components, and operating costs for a full-scale commercial soil washing system were estimated. The estimates were based on the pilot demonstration conducted at NAS Miramar, other full-scale soil washing pilot demonstration projects, and best engineering judgment. Costs were defined as applicable to typical remediation activities at Superfund and RCRA sites, and were determined using outside sources, purchasing experience, and good engineering practices (Dataquest, 1997; Evans, 1997; Means, 1997). The order-of-magnitude costs in 1997 dollars are considered accurate within 30% below and 50% above actual costs.

Gravity separation combined with vat leaching was successfully applied for the remediation of lead contaminated soils at a small arms firing range in the NAS Miramar demonstration. A small percentage of fine material was produced with lead contents greater than the desired 500 or 1000 mg/kg, and these fines required additional treatment. Since the cost of multiple leaching processes for these materials could be high, leaching unit operations were eliminated in the design of the full-scale plant. In cases where the $-105\ \mu m$ (-150 mesh) materials exceeded 20% of the soil, vat and agitation leaching of the coarse, sand, and slimes fractions should be considered as a treatment technology.

Certain costs were not included in this analysis because they were site specific. Preliminary site preparation, permits and regulatory requirements, initiation of monitoring programs, waste disposal, sampling and analysis, and site cleanup and restoration were considered to be the responsible party's or site owner's obligations. The cost estimates represented the charges typically invoiced to the client by the vendor for the soil washing process as a separate technology and did not include any state or local taxes or profit. Wherever possible, applicable information was provided whenever site-specific calculations could be required.

5.2 ASSUMPTIONS

Several assumptions were used to develop the costs of the full-scale commercial soil washing system. In general, the assumptions were based on the NAS Miramar pilot demonstration, the treatability studies for Twentynine Palms and Fort Dix soils, and published data. Certain assumptions accounted for variable site and waste parameters. They will need to be refined for each remediation project to reflect site-specific conditions.

From information provided by the U.S. Navy and U.S. Army, the sizes of small arms firing ranges vary significantly from base to base. For purposes of this full-scale analysis, three range sizes were selected: small, less than 300 firing lines; medium, 300 to 1000 firing lines; and large, greater that 1000 firing lines. For these three classifications, typical berm heights ranged from 4.5 m to 6.1 m (15 ft to 20 ft) with an average of 5.3 m (17.5 ft). The lane width averaged 1.8 m (6.0 ft). The average depth of lead contamination for all three classifications was 2.0 m (6.5 ft). Based on these data, the following was calculated:

Small Ranges	152 lane average, 5.260 kt (5800 ton) for treatment
Medium Ranges	650 lane average, 22.310 kt (24,600 ton) for treatment
Large Ranges	1000 lanes, 34.380 kt (37,900 ton) for treatment

A commercial cleanup of a hypothetical small firing range berm was assumed. The volume of contaminated soil to be treated was approximately 2950 m³ (3870 yd³) with 10% (dwb) moisture. The soil weight on a dry basis was about 5.260 kt (5800 ton). Pretreatment of the feed soil was not required. The goal was to decrease the volume of lead-contaminated soil to be treated or disposed of to less than 20% by weight.

The site was assumed to be on a military facility within 1000 miles of the source of the skid fabrication site. Soil contamination at the site resulted from accumulation of lead bullets and lead particles at the small arms firing range and access roads. Adequate paved storage areas for treated and untreated soils and utility lines, such as electricity and telephone lines, were assumed to exist on site.

Oversized material, +19.0 mm (+3/4 in.), constituted about 7% of the feed soil and was disposed of as on-site backfill. Treated soil was used as backfill at the site. It was estimated that 15% of the wash water was considered lost due to soil retention and evaporation. Hence, 85% of the wash water was recycled until the project was complete. Makeup water was added to compensate for the water loss. The recycled and makeup water was stored in a fractionation tank and then disposed of offsite at the end of the project. Offsite disposal costs were site specific.

The on-line percentage takes into account periodic shutdowns to respond to maintenance or operational problems. For extended periods of operating time, 100% on-line conditions cannot be expected. However, it is common to achieve an on-line factor as high as 90 to 95% in the sand and gravel industry. An on-line factor of 90% was assumed.

The costs for the full-scale plant design were based on:

- A feed rate of 13.60 t/hr (15 ton/hr),
- Bulk density of excavated soil at 1780 kg/m³ (111 lb/ft³),
- An on-line percentage of 90%,
- An operating time of 10 hrs/day and 5 days/wk, and
- A total project time of about 3 months.

5.3 COST ANALYSIS FOR A HYPOTHETICAL SITE

The mobile commercial soil washing process design is applicable for removal of lead from soils at small arms firing ranges. A full-scale plant with a nominal capacity of 13.60 t/hr (15 ton/hr) was used for the remediation of a firing range containing 2950 m³ (3870 yd³) or 5.260 kt (5800 ton) of contaminated soil. Assuming 90% on-line operating efficiency, the total project cost to remediate the site was estimated to be $694,100 or $131.90/t ($119.60/ton).

The overall project schedule was:

- Two (2) weeks for site preparation and soil excavation,
- One (1) week for mobilization and equipment setup,
- One (1) week for treatment plant startup,
- Nine (9) weeks to process the soil, and
- One (1) week for demobilization

for a total of 14 weeks or about 3 1/4 months.

The total project costs are summarized in Table 5.1, with details found in Appendix L. The cost for the initiation of monitoring programs was not included in this analysis. Depending on the site and the location of the system, local authorities may impose specific guidelines for monitoring programs. The stringency and frequency of monitoring required may have significant impact on the project costs. Manifesting and shipping expenses and the expenses of RCRA landfill disposal were not included. Any decrease in cost due to recycling credits was not included.

5.3.1 Mobilization and Preparation Work

Preliminary site preparation was performed by the responsible party or site owners. Site preparation is site specific and includes site design and layout, surveys and site logistics, legal searches, obtaining access rights, installing access roads, installing utility connections, preparing support and decontamination facilities, and erecting auxiliary buildings. These site-specific costs were not included in the costs.

5.3.1.1 Preconstruction Submittals and Treatability Studies

Preconstruction submittals included development of a site-specific Work Plan and a Health and Safety Plan. A project manager, a project engineer, and clerical help were required to prepare these plans. The Health and Safety Plan was reviewed and approved by a Certified Industrial Hygienist (CIH). Treatability study costs included collecting samples, analyzing the results, and preparing design and operating criteria for the full-scale treatment plant.

5.3.1.2 Mobilization

Mobilization of the soil washing system consisted of transportation to the site, assembly, shakedown, on-line training of local personnel, and initiation of health and safety monitoring. It was anticipated that 5 working days at 10 hours per day would be required to accomplish these tasks.

The soil washing system was preassembled on skids, which could be transported by flatbed trucks and assembled at the site. For this analysis, it was assumed that the preassembled skids were located less than 1000 miles from the site. Five trucks were needed to transport the plant at a cost of $1.65 per mile per truck.

The costs for assembly at the site included:

- Installation of a plastic liner in and hay-bale berm around the equipment assembly area to collect any liquid spillage;
- Setup of a trailer for an on-site office and equipment storage;
- Installation of electrical power, water, and portable toilets; and
- Unloading and assembly of the soil washing equipment skids.

Table 5.1 Project Costs for 5.261 kt (5800 tons) of Soil By a 13.60 t/hr (15 tons/hr) Unit, Operating 10 hrs/day, 5 days/week with an On-Line Factor of 90%

	Work Breakdown Structure (WBS) Number and Cost Element	Item Cost	Total	Cost $/t	% of Total
Total Project Cost			**$694,100**	**$131.90**	**100.0**
Before Treatment Costs	**3.4.1 Mobilization and Preparatory Work**		**$88,900**	**$16.90**	**12.8**
	3.4.1.1 Preconstruction Submittals and Implementation Plan		$34,200	$6.50	4.9
	• Management Labor	$7200			
	• Monitoring, Sampling, Testing, and Analysis	$27,000			
	• Geotechnical Testing (Treatability Study)				
	3.4.1.2 Mobilization of Construction Equipment and Facilities		$41,400	$7.90	6.0
	• Transportation of Treatment Plant to Firing Range Site	$8300			
	• Construction Equipment	$2500			
	• Labor	$13,700			
	• Supplies	$16,900			
	3.4.1.3 Site Preparation (Soil Excavation)		$13,300	$2.50	1.9
	• Equipment	$9300			
	• Labor	$4000			
Treatment Costs	**3.4.2 Physical Treatment: Soil Washing**		**$564,800**	**$107.30**	**81.4**
	3.4.2.1 Equipment Costs		$244,800	$46.50	35.3
	• Capital Equipment	$194,800			
	• Rental Equipment	$19,100			
	• Maintenance	$19,300			
	• Repair Materials	$11,600			
	3.4.2.2 Labor and Administration		$254,200	$50.00	37.9
	• Labor Costs	$260,900			
	• Administration Costs	$2300			
	3.4.2.3 Consumables		$23,300	$4.40	3.4
	3.4.2.4 Utilities		$33,500	$6.40	4.8
	• Equipment and Consumables	$27,000			
	• Process Make-up Water	$6500			
	• Electric Power (Provided by Motor-Generator Set)	$0			
Post Treatment Costs	**3.4.3 Demobilization of Construction Equipment and Facilities**		**$40,400**	**$7.70**	**5.8**
	• Labor and Equipment	$16,200			
	• Transportation of Treatment Plant to Temporary Storage	$8300			
	• Post-construction Report	$15,900			

A 25-ton crane and a forklift were the only equipment required for installation. The cost to rent a 25-ton crane with an operator and a rigger was $1100 per day. The cost to rent an all-terrain forklift was $1150 per month, prorated for the duration of the project. The forklift was retained for the duration of the project to assist in moving heavy objects, while the crane was rented for two days.

Assembly and operation of the soil washing system required a lead process operator, two process operators, and a laborer. Training was conducted simultaneously with installation and shakedown. Since all personnel associated with the process were 40-hr OSHA-trained, no additional training was needed.

The only contamination concerns were noise pollution and contact with lead contaminated soils. A decibel meter was kept on site, and daily readings were monitored upon reaching full-scale operation. Contact with lead contaminated soil was minimal since all operations were automated and contained within the equipment. Therefore, the personal protection requirement was set at Level D consisting of Tyvek suits, safety boots, hardhats, ear plugs, safety glasses, and double gloves. An eyewash station and decontamination supplies were available on site.

5.3.1.3 Solids Collection

Solids collection activities, such as excavating and stockpiling soil from the firing range berm, are required at all sites. The costs of these activities were based on rental of heavy equipment and fuel. Using a rate of 65 t/hr (72 ton/hr) for the 5.621 kt (5800 ton) of soil, the excavation activities were completed in 2 weeks when working 8 hours per day, 5 days per week. Rental equipment required included a 1.15 m^3 (1.5 yd^3) excavator, a 2.29 m^3 (3.0 yd^3) front-end loader, and a 9.2 m^3 (12 yd^3) dump truck. The equipment consumed about 47 L/hr (15 gal/hr) of diesel fuel per day. One equipment operator and one dump truck driver were required. These personnel were local hires with 40-hr OSHA training.

5.3.2 Physical Treatment

The soil washing process must be modified for site-specific conditions. Contaminant level, cleanup criteria, soil mineralogy, and particle size distribution must be considered when designing a treatment system. The soil must first be characterized, and then a strategy developed to effect the separations necessary to achieve the volume reduction required to meet regulatory goals. These tasks are accomplished by conducting a treatability study including particle sizing, contamination distribution, gravity separation, and analysis. The equipment layout must then be optimized to produce the target cleanup goal. The number, size, and type of component units required to accomplish the necessary separations will impact the capital costs.

Construction of the treatment plant required 5 workdays (1 week) to complete. The startup required 5 workdays (1 week) at 10 hr per day. To treat 5.261 kt (5800 ton) of feed soil at 13.60 t/hr (15 ton/hr) would take about 40 workdays, working 10 hr per day, 5 days per week. To account for both scheduled maintenance and unscheduled shutdowns, a 10% downtime was included. Consequently, the actual treatment time required 45 workdays (9 weeks). Demobilization required 5 workdays (1 week). Therefore, the total on-site time for the treatment plant was 84 calendar days, 12 calendar weeks, or about 3 months.

Operating and maintenance schedules for the soil washing plant were based on processing 13.60 t/hr (15 ton/hr) for 5 days per week, 10 hr per day. Loading feed material from the stockpile with the front-end loader was scheduled for 10 hr per day, 5 days per week. In addition to the front-end loader, a forklift and a 75 m^3 (20,000 gal) fractionation tank for recycle water storage were included in the operation costs.

Maintenance labor and materials costs vary with the nature of the waste and performance of the equipment. For estimating purposes, total maintenance costs, labor, and materials were assumed to be 10% of the equipment costs on an annual basis. Materials costs were estimated at 60% of the total annual maintenance costs and were prorated for the time required for treatment. Scheduled maintenance was performed by a mechanic during the day shift. Costs for design adjustments, facility modifications, and equipment replacement were included in maintenance costs.

5.3.2.1 Capital Equipment

The capital equipment charge was prorated from the annualized equipment cost, and it was based on the time required for equipment depreciation, interest rate, total fabricated equipment cost, and equipment salvage value. In Appendix L, Work Breakdown Structure (WBS) # 3.4.2.1, the annualized equipment cost was derived using the following formula:

$$\text{Annualized Equipment Cost} = \frac{(V - V_S)(i)(1 + i)^n}{(1 + i)^n - 1}$$

where V = total equipment and fabrication cost = \$1,723,300
          V_S = equipment salvage value (10% of V) = \$172,300
          i = annual interest rate = 7%
          n = equipment depreciation time = 3 years

The annualized equipment cost was estimated to be \$591,000 per year. The total time for equipment setup, plant startup, soil treatment, and demobilization was about 12 weeks or about 3.0 months. After adding the 10% contingency and 7.5% insurance costs, the total capital charge for the project was \$194,800 or \$37.00/t (\$33.60/ton).

The cost to purchase the full-scale soil washing plant was estimated to be slightly over \$1 million (Table 5.2). The equipment list and quantities were developed for the flow diagram in Appendix Figure I.1. The equipment size calculations are detailed in Appendix J. The total capital cost to fabricate the equipment skids for the full-scale plant is approximately \$1.7 million (Table 5.3). An annual contingency cost of 10% of the equipment capital costs was assumed for any unforeseen or unpredictable cost conditions, such as strikes, weather, and price variations, and was prorated for the duration of the project.

The usual annual costs of insurance and of taxes were approximately 1% and 2 to 4% of the total capital cost, respectively. Insurance costs for hazardous waste projects can be several times more than normal insurance costs. Therefore, the annual insurance cost was assumed to be 7.5% of the equipment capital costs, and was prorated for the duration of the project. Since only DOD projects were considered, taxes were not included in the cost analysis.

5.3.2.2 Labor Costs

Labor costs were composed of salaries and living and travel expenses for non-local personnel. The staff for startup included one site manager, one lead process operator, two process operators, and one laborer. The staff for process operations included one site manager, one lead process operator, two process operators, one heavy-equipment operator, one laborer, and one sampling and monitoring technician. Each worked 40 hours plus 10 overtime hours per week. The lead process operator and the sampling and monitoring technician were non-local personnel. All other personnel were hired locally. The per-diem for each non-local employee was \$150. An allowance of \$500 per round trip airfare was included with four trips per person.

5.3.2.3 Supplies and Consumables

The costs for flocculent, startup sand, site preparation materials, and health and safety equipment were considered part of mobilization costs. Disposable PPE included Tyvek coveralls, gloves, and

Table 5.2 Purchased Equipment Costs for Full-Scale Plant

Equipment Item	Equipment Size	Quantity	Equipment Cost
Grizzly-Screen, Hopper, Belt Feeder	W × L: 0.91 m × 1.83 m	1	$37,000
Conveyor #1	W × L: 45 cm × 10.7 m	1	$19,500
Double-Deck Vibrating Wet Screen	W × L: 0.91 m × 1.83 m	1	$19,200
Conveyors #2, #3, #4	W × L: 45 cm × 9.1 m	3	$36,000
Primary Jig	Simplex 16 × 24	1	$28,700
Vibrating Dewatering Screen #1	W × L: 0.91 m × 1.83 m	1	$7800
Spiral Classifier	1500 mm Simplex 150 MF	1	$126,000
Secondary Jig	Duplex 24 × 36	1	$59,900
Vibrating Dewatering Screen #2	W × L: 0.91 m × 2.44 m	1	$8800
Reichert Spiral Concentrator	MD LG7, Triple Start	3	$30,000
Reichert Spiral Concentrator	MD LG7, Triple Start	1	$10,000
Sieve-Bend Dewatering Screen #3	W × L: 600 mm × 800 mm	1	$5100
Vibrating Humates Screen	W × L: 0.91 m × 1.83 m	1	$7800
Lamella Clarifier	Lamella LTS, Mod 100	1	$134,000
Sharples Centrifuge	Super-D-Canter, Mod 3400	2	$310,000
Polymer Injection System		1	$5500
P1: High-pressure Water Pump	Goulds Mod 3656, 16 AI/BF	1	$2000
P2: Low-pressure Water Pump	Goulds Mod 3656, 11 AI/BF	1	$2200
P3: Classifier Slurry Pump	Denver Model SLR Pump	1	$6900
P4: Screen Underflow Pump	Denver Model SLR Pump	1	$1000
P5: Spiral Feed Pump	Denver Model SLR Pump	1	$7000
P6: Spiral Mids Pump	Denver Model SLR Pump	1	$3800
P7: Spiral Feed Pump	Denver Model SLR Pump	1	$4200
P8: Spiral Mids Pump	Denver Model SLR Pump	1	$3500
P9: Spiral Tails Pump	Denver Model SLR Pump	1	$4000
P10: Sands Pump	Denver Model SLR Pump	1	$7200
P11: Centrifuge Feed Pump	Shanley Mod 550, Prog Cav	1	$3000
P12: Clarifier Overflow Pump	Goulds Mod 3656, 11 AI/BF	1	$1100
Electrical Main Disconnect	3-phase, 440 volts, 250 kVA	1	$10,000
Motor Starter Board		1	$50,000
Process Control Software		1	$50,000
Total Purchased Equipment Cost			**$1,006,300**

earplugs. The treatment system operators used PPE at all times and changed twice daily. The sampling and monitoring technician used PPE during sample collection. The heavy equipment operators were not required to wear PPE unless working close to excavated soil. PPE was needed for the personnel for the duration of the project. The cost of spare parts and office/general supplies were included under the Treatment Phase.

Fuel consumption for soil washing operations was limited to that used by the rented support equipment. The front-end loader utilized about 5 gallons of diesel fuel per hour, and the forklift required about 1 gallon per hour.

5.3.2.4 Utilities

Water and electricity had to be supplied to the soil washing plant and auxiliary equipment. The total amount of water required to treat 5.26 kt (5800 ton) of soil over the duration of the project was about 2470 m³ (653,000 gal). About 85% of the total water required for the soil washing process was recycled and reused, with the remainder lost to soil retention and evaporation and requiring replenishing. Water cost can vary by as much as 100% depending on the geographic location of the site, availability of water, and distance to the nearest water main. A 13.3 m³ (3500 gal) water truck was included in the cost of water for this analysis, and the estimated cost of water was calculated at an additional $0.01 per gallon. Upon project completion, the remaining water was placed in a storage tank prior to off-site disposal. Disposal cost was not included in the estimate.

Table 5.3 Equipment Assembly Costs

Description	Purchased Equipment Cost	Equipment Installation Factor 43.00%	Piping and Instrument Factor 15.00%	Electrical Factor 15.00%	Total Fabricated Cost
Soil Feed System					
Grizzly-Hopper-Feeder	$37,000.00				
Feed Conveyor	$19,500.00				
Feed System, Misc.	0.00				
Total Feed System	**$56,500.00**	$0.00	$0.00	$0.00	$0.00
Skid 1					
Wet Screen	$19,200.00				
Primary Jig	$28,900.00				
Dewatering Screen #1	$7800.00				
Conveyors (2 units)	$24,000.00				
Skid, Misc.	$11,459.00				
Total Skid 1	**$91,200.00**	$28,900.00	$10,100.00	$10,100.00	$140,300.00
Skid 2					
Secondary Jig	$59,900.00				
Dewatering Screen #2	$8800.00				
Spiral Concentrators (4 units)	$40,000.00				
Sieve-bend Screen	$5100.00				
Conveyors (1 unit)	$12,000.00				
Skid, Misc.	$29,600.00				
Total Skid 2	**$155,400.00**	$61,700.00	$21,500.00	$21,500.00	$260,100.00
Skid 3					
Spiral Classifier	$126,000.00				
Skid, Misc.	$0.00				
Total Skid 3	**$126,000.00**	$54,200.00	$18,900.00	$18,900.00	$218,000.00

Dewatering Trailer					
Centrifuge	$310,000.00				
Clarifier	$134,400.00				
Humates Screening	$7800.00				
Polymer Injection	$5500.00				
Trailer, Misc.	$33,938.00				
Total Dewatering Trailer	**$491,700.00**	**$211,400.00**	**$73,800.00**	**$73,800.00**	**$850,700.00**
Pump Units					
Pump Skid 1 (P1)	$2000.00	$900.00	$300.00	$300.00	$3500.00
Pump Skid 2 (P2)	$2200.00	$900.00	$300.00	$300.00	$3700.00
Classifier Overflow Sump/Pump (P3)	$6900.00	$3000.00	$1000.00	$1000.00	$11,900.00
Sump/Pump Skid 1 (P4)	$1000.00	$400.00	$200.00	$200.00	$1800.00
Sump/Pump Skid 2 (P5)	$7000.00	$3000.00	$1100.00	$1100.00	$12,200.00
Spiral Mids Sump/Pump (P6)	$3800.00	$1600.00	$600.00	$600.00	$6600.00
Spiral Cons Sump/Pump (P7)	$4200.00	$1800.00	$600.00	$600.00	$7200.00
Spiral Mids Sump/Pump (P8)	$3500.00	$1500.00	$500.00	$500.00	$6000.00
Spiral Tails Sump/Pump (P9)	$4000.00	$1700.00	$600.00	$600.00	$6900.00
Spiral Tails Sump/Pump (P10)	$7200.00	$3100.00	$1100.00	$1100.00	$12,500.00
Centrifuge Feed Pump Skid (P11)	$7700.00	$3300.00	$1200.00	$1200.00	$13,400.00
Clarifier Overflow Pump (P12)	$1100.00	$500.00	$200.00	$200.00	$2000.00
Total Pump Units	**$50,600.00**	**$21,700.00**	**$7700.00**	**$7700.00**	**$87,700.00**
Miscellaneous					
Electrical Disconnect	$10,000.00				
Motor Starter Board	$50,000.00				
Process Control Software	$50,000.00				
Total Miscellaneous	**$110,000.00**	**$0.00**	**$0.00**	**$0.00**	**$110,000.00**
Total Item Cost	**$1,081,400.00**	**$377,900.00**	**$132,000.00**	**$132,000.00**	**$1,723,300.00**

Note: Assembly was not required for the Grizzly-Hopper-Feeder, Conveyors, and Fractionation Tank. Required piping and electrical connections were included in the costs of the nearest skid.

A 440-volt 3-phase power source was required for the operation of the soil washing equipment. Since most firing range sites are remote, a 250 KVA (200 kW) generator was included in the plant design. The generator used about 13 gal per hour of diesel fuel. Long-term operations (lasting longer than 4 months) would justify the use of electricity from the local power company. The full-scale system required about 189 kW of power.

5.3.3 Demobilization

Following the treatment process, equipment had to be decontaminated, the plant disassembled, the site restored, the construction equipment returned to suppliers, the plant transported away from the site, and the post-construction report prepared. Demobilization required one heavy equipment operator, one lead processor operator, and two laborers working 10 hr per day, 5 days per week.

The soil washing system was preassembled on skids to be transported by flatbed trucks. It was assumed that the site was less than 1000 miles from the plant storage site and that five trucks were needed to transport the full-scale plant. The transportation cost was $1.65 per mile per truck.

5.3.4 Effluent Treatment and Disposal Cost

The effluent streams generated by the soil washing system should consist of clean material. Solid lead particles may be recycled or disposed of at a RCRA-permitted facility. Lead contaminated fines could be further treated by active leaching, encapsulation, or disposed in a landfill if the lead concentrations meet the minimum requirements. The wastewater could be discharged at a POTW if it contained less than 1.0 mg/L of total lead. The responsible party or site owner must obtain a discharge permit from the local municipality. The costs of treatment and disposal were not included in this analysis because the quantities and distances to the appropriate facilities were site specific.

5.3.5 Residual Waste Shipping and Handling

Shipping, handling, manifesting, and waste profile analysis were considered to be the obligation of the responsible party and, therefore, were not included in the cost estimate. Disposal costs for contaminated health and safety gear, protective plastic sheeting, and other residuals were also assumed to be the obligation of the responsible party, and were not included in the estimate.

5.3.6 Permitting and Regulatory Requirements

Permitting and regulatory costs depend on whether the site is a Superfund or a RCRA corrective action site and on how treated effluent and solid wastes are disposed of. Superfund sites require remedial actions to be consistent with applicable or relevant and appropriate requirements (ARAR) of environmental laws, ordinances, regulations, and statutes including federal, state, and local standards and criteria. In general, the ARAR must be determined on a site-specific basis. RCRA corrective action sites require additional monitoring records and sampling protocols, which can increase permitting and regulatory costs by an additional 5%.

Permitting and regulatory costs are generally the obligation of the responsible party or site owner. These costs may include actual permit costs, system monitoring, and analytical protocols. No permitting costs were included in this analysis.

5.3.7 Analytical Services

The analysis assumed that the untreated firing range berm soils at the site were well character-ized. Analytical costs for environmental monitoring were not included in this analysis as such programs are agency and site specific. State and local agencies could impose specific sampling and monitoring criteria, and analytical requirements could add significantly to the cost of a specific project. Additionally, the client could elect to initiate an independent sampling and analytical program at extra expense. The cost of laboratory analysis, including sample analysis, data reduction and tabulation, quality assurance and quality control (QA/QC), and reporting of results, was not included in the cost analysis.

5.4 MAXIMIZING TREATMENT RATE

For performing the cost analysis, the remediation of a hypothetical small firing range berm at a DOD site was assumed. The soil mass on a dry basis was about 5.260 kt (5800 ton) or about 2950 m³ (3870 yd³). The nominal soil feed rate to the treatment plant was 13.6 t/hr (15 ton/hr). Using hypothetical soil characteristics, all unit operations were designed to handle a soil feed rate that could vary from 9.1 t/hr (10 ton/hr) to 18.1 t/hr (20 ton/hr). The plant operated 5 days per week, 10 hours per day.

The treatment rate was maximized by one or more of the following modifications to operational procedures:

- Two 10 hr shifts per day for 5 days per week,
- Three 8 hr shifts per day for 5 days per week,
- Three 8 hr shifts per day for 7 days per week, and
- Operation at near maximum design capacity of 18.1 t/hr.

Any combination of these changes in the operation of the plant increased the amount of soil treated per month and decreased the unit cost.

5.5 POTENTIAL EFFECTS ON COST OR PERFORMANCE

The unit treatment costs and percentages of the total cost in Table 5.1 can be summarized:

- **Total Cost: $131.90/t (100%)**
- Preconstruction submittals and implementation plan: $6.50/t (5%)
- Mobilization of construction equipment and facilities: $7.90/t (6%)
- Site preparation: $2.50/t (2%)
- Physical treatment: $107.30/t (81%)
  - Equipment costs: $46.50/t (35%)
  - Labor and administrative: $50.00/t (38%)
  - Consumables: $4.40/t (3%)
  - Utilities: $6.40/t (5%)
- Demobilization of construction equipment and facilities: $7.70/t (6%)

The physical treatment cost was the major contributor to the unit cost. Equipment, labor, and administrative costs were the major components of the expense of physical treatment. Variations in capital depreciation, operational parameters, and amount of soil treated impacted the unit cost of physical treatment. However, the other costs remained about the same independent of the project.

The unit treatment cost was decreased by the increasing treatment rate, adding more shifts per day, and increasing the work week from 5 to 7 days. Other factors that decreased the unit cost were:

- Increasing the depreciation time for the capital investment from 3 to 10 years,
- Increasing the amount of soil treated per project, and
- Purchasing refurbished (used) equipment at a lower cost per unit.

A few factors that increased the unit cost were:

- Operating three shifts per day, 7 days per week increased the downtime from 10 to 15% or more.
- Use of vat or agitation leaching if the amounts of the contaminated sand and slimes fractions increased after particle sizing and gravity separation.

The effects on the unit cost for the nonleaching factors are summarized in Table 5.4. These costs were derived by modifying the calculations in Appendix L and Table 5.1 for each operating factor. In Case 1, for example, Appendix L, Section 3.4.2.1 was changed to reflect the increased time of 10 years for capital depreciation. The unit cost in Table 5.1 was recalculated using the revised capital equipment charge. Consequently, the unit cost decreased from $131.90/t ($119.60/ton) to $74.00/t ($67.00/ton) by making the following changes to the operation and capital expenditure for the hypothetical treatment plant:

- Replaced one 10-hr shift with two 10-hr shifts per day,
- Increased treatment rate from 13.60 t/hr (15 ton/hr) to 18.14 t/hr (20 ton/hr),
- Increased time for the depreciation of capital from 3 to 10 years, and
- Increased amount of treated soil from 5261 t (5800 ton) to 7030 t (7750 ton).

Table 5.4 Operational Parameter Effects on Unit Cost

		Unit Cost	
	Description	$/t	$/ton
Base	Set Treatment Rate at 13.60 t/hr (15 ton/hr) Depreciate Capital over 3 years Treat 5.261 kt (5800 ton) of Soil per Project	131.90	119.60
Case 1:	Increase Capital Depreciation from 3 years to 10 years Treatment Rate at 13.60 t/hr (15 ton/hr) Treat 5.261 kt (5800 ton) of Soil per Project	112.60	102.10
Case 2:	Maximize Treatment Rate at 18.14 t/hr (20 ton/hr) Depreciate Capital over 3 years Treat 5.261 kt (5800 ton) of Soil per Project	104.10	94.40
Case 3:	Use Two 10-hour Shifts per Day Set Treatment Rate at 13.60 t/hr (15 ton/hr) Depreciate Capital over 3 years Treat 5.261 kt (5800 ton) of Soil per Project	110.40	100.10
Case 4:	Treat 7.030 kt (7750 ton) of Soil per Project Maximize Treatment Rate at 18.14 t/hr (20 ton/hr) Depreciate Capital over 3 years	99.90	90.60
Case 5:	Maximize Treatment Rate at 18.14 t/hr (20 ton/hr) Increase Capital Depreciation from 3 years to 10 years Treat 7.030 kt (7750 ton) of Soil per Project	85.40	77.50
Case 6:	Use Two 10-hour Shifts per Day Maximize Treatment Rate at 18.14 t/hr (20 ton/hr) Increase Capital Depreciation from 3 years to 10 years Treat 7.030 kt (7750 ton) of Soil per Project	74.00	67.00

The use of refurbished equipment in the fabricated plant had a major effect on the capital depreciation charge. Purchased equipment costs accounted for 58% of the fabricated plant cost of $1.7 million (Table 5.3). Replacing the two new centrifuges with refurbished equipment decreased the purchased equipment cost by about $50,000 or about 5% (Vendor Quotes, Appendix L).

Vat leaching (Section 2.3.5) can be used on combinations of coarse soil fractions and coarse and sand fractions. The leaching step usually can be carried out in modified roll-off boxes. The soil can be easily loaded and removed using a front-end loader. After percolating leach solution through the soil and washing with fresh water, gravity drainage of the water is sufficient to remove residual water from the soil. Filtration or centrifugation of the tailings is not required. Reagent cost can also be quite low. The −1.19 mm by +297 μm (−16 mesh by +50 mesh) fraction from the

Twentynine Palms soil represented 40% of the dry weight, and required 8.9 kg/t of nitric acid to decrease the lead content from 780 mg/kg to 30 mg/kg. At $0.25/kg, the reagent cost would be $2.20/t or $0.90/t of feed to the gravity separation plant. Adding capitalized leaching equipment (e.g., $250,000), a front-end loader to handle the leach feed and tailings, two process operators, and other consumables could bring the vat leaching cost to about $23.60/t ($21.40/ton) of feed.

Before using an agitation leach on the slimes fraction, -105 μm (-150 mesh), an attempt should be made to remove lead particles with a hydrocyclone and Knelson bowl concentrator. The hydrocyclone would remove the clay particles from the slimes fraction producing a clay-free underflow. The Knelson concentrator would remove the lead from the underflow stream. If the concentrator does not produce a clean fraction, it can be mixed with coarse and sand fractions and treated in the vat leach plant. This mode of operation can decrease the amount of material added to the agitation leach. A similar process was proposed for the treatment of the slimes fraction in the NAS Miramar demonstration. The amount of the -105 μm fraction in the NAS Miramar soil was about 40%. The hydrocyclone treatment decreased the amount of the slimes fraction by about 35% (Sections 2.3.1 and 2.3.3). When the hydrocyclone underflow was treated by the Knelson concentrator, the amount of soil fed to the agitation leach was about 25% of the plant feed. Agitation leach could cost as much as $65 to $110 per tonne ($59 to $100/ton) of gravity plant feed (M&E, 1998). Therefore, pretreatment of the slimes could decrease agitation leach cost by $20 to $40 per tonne ($18 to $36/ton).

CHAPTER 6

Performance and Potential Application

The major goal of this evaluation was to determine general applicability of soil washing technology to remove heavy metals from all sizes and types of small arms ranges within Department of Defense (DOD) control. For the remediation of small arms ranges, three principles govern the remedial work. First, the amount of clean reusable soil must be maximized. Second, particles of coarse lead must be produced through gravity separations. Third, the process must be cost effective.

Soil washing is an *ex situ* water-based process that separates contaminants from the clean soil matrix. To achieve separation and physical sizing, gravity separation and attrition scrubbing are used. All process water is recycled through a closed-loop treatment system, minimizing water consumption and disposal requirements. Treated soil can be reused on-site, undergo further treatment, or be stabilized, if required, for use as foundation material for subsequent berm reconstruction. The concentrated lead streams recovered during the process are recycled, with the salvage value offsetting a portion of the cleanup cost. Since soil washing is a highly efficient mineral separation process, little, if any, soil ends up in the concentrated lead streams, resulting in premium quality salvageable metals.

6.1 OPERABILITY OF THE TECHNOLOGY

The pilot treatment system operated during the NAS Miramar demonstration consisted of gravity separation and leaching equipment (Section 1.0). The system, a specifically designed mobile unit, was operated by former BOM personnel. The system experienced few operational problems during the demonstration. These consisted of failed pumps, failure of a new sampling device, and failure of the leaching equipment. None of the failures adversely affected the gravity separation portion of the operation, and enough data was generated prior to the failure of the leaching system to properly evaluate performance. Throughput was the primary operating parameter that influenced the performance of the system. The equipment furnished by the former BOM was capable of a limited throughput of about 0.907 kt/hr (2000 lb/hr).

The placement of mining equipment on mobile equipment skids provided an economical and functional separation system for the full-scale remediation of a hypothetical contaminated site. This system was a considerable improvement over the design and operation of the gravity pilot plant used in the demonstration. The transportable equipment skids could be easily modified or rearranged to fit the treatment scheme specified by a treatability study of the contaminated soils.

6.2 APPLICABLE WASTES

The soil washing technology is applicable for the treatment of many heavy metal contaminated wastes. Generally, the process is tolerant of normal swings in soil particle size. Potential problems arise if the soil exhibits wide swings in the coarse material content. If the maximum concentration of coarse material is known at the design phase, it can be properly accounted for. Treatability studies should be undertaken with any candidate soil.

Leaching, stabilization, or disposal is the treatment of choice if the lead contamination is concentrated in the slimes fraction. Using one of these methods, the waste can be economically processed. For soil washing to be applicable, the fine fraction should not represent a large percentage of the overall soil. Ideally, the slimes fraction should not exceed 30 to 35% by weight of the overall soil feed. The lower the fine soil concentration, the more favorable will be the potential economics of the soil washing process.

The soil washing process is not tolerant of free oil and grease. A separate pretreatment is necessary if soil washing is to be applied on a soil with free oil and grease. Containerization of the concentrated wastes, including lead bullets and particles, fines, the humic fraction, and possibly the liquid waste, should be considered along with transport of the residues and effluents for disposal.

6.3 ADVANTAGES OF THE DESIGNED SOIL WASHING SYSTEM

Soil washing technology provides an economic method of removing lead bullets and particles from firing range soils, and thereby decreasing the quantity of material requiring further treatment or disposal off-site. The flexibility of skid-mounted components that can be interchanged or modified to perform the maximum separation of the materials is a major advantage of the full-scale system designed in this project over other soil washing systems.

The full-scale soil washing gravity separation and leaching units and support equipment are mounted on skids for loading on flatbed trailers and are easily transported. Once on-site, the full-scale treatment system can be in operation within 5 days if all necessary facilities, utilities, and supplies are available. On-site assembly and maintenance requirements are minimal.

Demobilization activities include decontaminating the on-site equipment, disconnecting utilities, and disassembling the equipment skids. Demobilization requires only 5 days also.

6.4 SITE CHARACTERISTICS SUITABLE FOR THE TECHNOLOGY

6.4.1 Site Selection

The applicability of the soil washing technology is not restricted by the geological features of the site. The equipment might be erected within the confines of the contaminated area or placed off-site with the waste transported to the unit. Usually, operation is most cost-effective if the system is located on-site. Major geological constraints that can render a site unsuitable for location of the soil washing plant include landslide potential, volcanic activity, and fragile geological formations that may be disturbed by heavy loads or vibrational stress. Whether on or off the contaminated site, the soil washing plant must be located at a site suitable for construction with appropriate access.

6.4.2 Surface, Subsurface, and Clearance Requirements

The site area required for the skid arrangement of the full-scale treatment system is about 36.5 m by 30.5 m (120 ft by 100 ft). Additional space must be made available for storage of stockpiled feed soils and any waste generated during the treatment process. All equipment should be located in a manner to facilitate convenient access for handling feed soil, processed lead materials, and treated soil fractions. The site must be cleared to allow assembly and operational activities to take place. The extent of the clearing depends on the operational configuration selected and must allow construction of and access to the facility. This is not an issue if the equipment is to be operated off-site. In any case, a cleared treatment area is required for stockpiling, storage, and loading and unloading activities. Asphalt roads may be necessary to provide support for oversize and heavy equipment.

Surface requirements for the operation of the soil washing system include a level, graded area capable of supporting the equipment. In most cases, subsurface preparation is not required since all treatment activities take place above the soil surface. If the feed material is to be excavated and then treated on-site, all subsurface obstacles (i.e., underground piping, cables, etc.) must be removed prior to excavation.

6.4.3   Utility Requirements

The utility requirements for the soil washing system include electricity and water. The full-scale electrical system required one 300 ampere, 480 volt, triple-phase electrical circuit and can be operated using a 250 KVA generator. However, a generator is normally noisy, and, at sites with nearby residential communities, an alternate source of electricity must be found. If available, a 3-phase, 200 kW power supply from the local electric company was desirable.

An abundant water supply must be readily available and easily accessible. Water is needed to operate the wash unit and to decontaminate equipment. Approximately 57 m³ (15,000 gal) of water must be supplied initially and about 6400 L/hr (28 gpm) supplied during operation. Although the treatment system requires quantities of water greater than these specified parameters, a large quantity of water is recycled through the system, thereby decreasing water usage. To eliminate the cost of purchasing water from the local utility company, the water may be obtained from a nearby source such as a river or lake, if feasible, or a well might be drilled. The water need not be potable, but must be free of debris. Water streams containing debris may be used if a basket filter is added to the pumping system.

6.4.4   Site Support Facilities and Equipment

Support facilities include a contaminated soil staging area and a treated soil storage area. Contaminated soil is generally stockpiled in manageable piles with easy access to the system soil feed module. Recovered lead can be stored in roll-off boxes or soil piles. Treated soils can be stockpiled for reuse as backfill. In addition, a tank area to store wastewater might be required at some sites. These support facilities must be contained to control run-on and runoff.

Support equipment for the system includes earth-moving equipment, containers for recovered lead and containers for wastewater. Earth-moving equipment, including backhoes, front-end loaders, and, at some sites, dump trucks, are needed to excavate and move soils to the soil washing unit. An all-terrain forklift is also required to move heavy objects and containers.

6.4.5   Climate

Moderate temperatures and low wind conditions are desirable for operation of a soil washing plant. Below-freezing temperatures can profoundly affect operation since the soil washing system utilizes large amounts of water. Windy conditions may be detrimental to the conveyor belts. Severe storms may result in hazardous operating conditions as the system is fully exposed to the weather. To diminish the effect of many climatic attributes, the system might be erected in an enclosure, such as a fixed structure or a tent. Steam heating might be used to maintain an acceptable operating temperature within the enclosure and to alleviate problems associated with freezing temperatures.

6.5 MATERIALS HANDLING REQUIREMENTS

Contaminated soil will be excavated, staged, transported, and loaded into the soil washing plant. Soil should be kept wet to minimize airborne particulates. At the conclusion of washing, treated soil will be placed in containers or on the ground. The treated soil may contain an appreciable amount of water and require runoff control measures. Arrangements for storage of the recovered lead and subsequent transport to a recycling facility will be required.

6.6 LIMITATIONS OF THE TECHNOLOGY

In general, soil washing only decreases contaminant volume. A greater reduction than expected in contaminant mass and volume was achieved in the NAS Miramar soil washing demonstration because both soil washing and leaching were performed on the lead contaminated soil. Typically, leaching is not performed due to the small quantity of contaminated material requiring leaching and the high cost of the leach process.

Contaminants in silty or clayey soils are usually strongly sorbed and difficult to remove. For these types of soils, soil washing is generally physically and economically ineffective.

6.7 OBJECTIVES AND PERFORMANCE AS COMPARED TO ARAR

This subsection discusses specific environmental regulations pertinent to the operation of the soil washing system, including the transport, treatment, storage, and disposal of wastes and treatment residuals, and analyzes these regulations in view of the demonstration results. State and local regulatory requirements, which may be more stringent, will also have to be addressed by remedial managers.

Section 121 of the Comprehensive Environmental Response, Compensation, and Liability Act (CERCLA) requires that subject to specified exceptions, remedial actions must be undertaken in compliance with Applicable or Relevant and Appropriate Requirements (ARAR), federal laws, and more stringent state laws, as necessary to protect human health and the environment. ARAR include:

- The Comprehensive Environmental Response, Compensation, and Liability Act (CERCLA),
- The Resource Conservation and Recovery Act (RCRA),
- The Clean Air Act (CAA),
- The Safe Drinking Water Act (SDWA),
- The Toxic Substances Control Act (TSCA), and
- The Occupational Safety and Health Administration (OSHA) regulations.

These six general ARAR are discussed below; specific ARAR must be identified by remedial managers for each site. Some specific ARAR, which may be applicable to soil washing technology, are identified and discussed in Table 6.1.

6.7.1 Comprehensive Environmental Response, Compensation and Liability Act (CERCLA) and Superfund Amendments and Reauthorization Act (SARA)

CERCLA authorizes the federal government to respond to releases or potential releases of any hazardous substance into the environment, as well as the releases of pollutants or contaminants that may present an imminent or significant danger to public health and welfare or the environment. As part of the requirements of CERCLA, the Environmental Protection Agency (EPA) has prepared the National Contingency Plan (NCP) for hazardous substance response. The NCP is codified in

Title 40 of Federal Regulations (CFR) Part 300, and delineates the methods and criteria used to determine the appropriate extent of removal and cleanup for hazardous waste contamination.

The Superfund Amendments and Reauthorization Act (SARA) amended CERCLA and directed the EPA to do the following:

- Use remedial alternatives that permanently and significantly decrease the volume, toxicity, or mobility of hazardous substances, pollutants, or contaminants.
- Select remedial actions that protect human health and the environment, are cost-effective, and involve permanent solutions and alternative treatment or resource recovery technologies to the maximum extent possible.
- Avoid off-site transport and disposal of untreated hazardous substances or contaminated materials when practicable treatment technologies exist (Section 121(b)).

In general, two types of responses are possible under CERCLA: removal and remedial actions. Soil washing technology is likely to be considered a part of a CERCLA remedial action. Between 1986 and 1993, various soil washing technologies have been selected as source control remedies at eight Superfund sites (Warminsky, unpublished).

Remedial actions are governed by the SARA amendments to CERCLA. As stated above, these amendments promote remedies that permanently reduce the volume, toxicity, and mobility of hazardous substances, pollutants, or contaminants. In general, soil washing technologies only transfer contaminants from one media to another or decrease the volume of contaminated material. However, the volume of contaminant to be transferred is significantly decreased, and raw lead material can be recycled as a permanent treatment.

On-site remedial actions must comply with federal and more stringent state ARAR. These ARAR are determined on a site-by-site basis and may be waived under the following six conditions:

- The action is an interim measure and the ARAR will be met at completion;
- Compliance with the ARAR would pose a greater risk to health and the environment than noncompliance;
- It is technically impracticable to meet the ARAR;
- The standard of performance of an ARAR can be met by an equivalent method;
- A state ARAR has not been consistently applied elsewhere; and
- ARAR compliance would not provide a balance between the protection achieved at a particular site and demands on the Superfund for other sites.

These waiver options apply only to Superfund actions taken on site, and justification for the waiver must be clearly demonstrated.

6.7.2 Resource Conservation and Recovery Act (RCRA)

RCRA, an amendment to the Solid Waste Disposal Act (SWDA), was passed in 1976 to address the problem of how to safely dispose the enormous volume of municipal and industrial solid waste generated annually. RCRA specifically addressed the identification and management of hazardous wastes. The Hazardous and Solid Waste Amendments of 1984 (HSWA) greatly expanded the scope and requirements of RCRA.

The presence of RCRA-defined hazardous waste determines whether RCRA regulations apply to the soil washing technology. If soils are determined to be hazardous according to RCRA, all RCRA requirements regarding the management and disposal of hazardous wastes will need to be addressed. RCRA regulations define hazardous wastes and regulate their transport, treatment, storage, and disposal. Wastes defined as hazardous under RCRA include characteristic and listed wastes. Criteria for identifying characteristic hazardous wastes are included in 40 CFR, Part 261,

Table 6.1 Federal and State ARAR for Soil Washing Technology

Process Activity	ARAR	Description	Basis	Response
Waste characterization (untreated waste)	RCRA 40 CFR, Part 261 or state equivalent	Identifying and characterizing the waste as treated	A requirement of RCRA prior to managing and handling the waste.	Chemical and physical analyses must be performed.
	TSCA 40 CFR, Part 761 or state equivalent	Standards that apply to the treatment and disposal of wastes containing PCBs	During waste characterization, PCBs may be identified in contaminated soils, and are therefore subject to TSCA regulations.	Chemical and physical analyses must be performed; if PCBs are identified, soils will be managed according to TSCA regulations.
Soil excavation	Clean Air Act 40 CFR 50.6, and 40 CFR 52, Subpart K or state equivalent	Management of toxic pollutants and particulate matter in the air	Fugitive air emissions may occur during excavation and material handling and transport.	If necessary, the waste material should be watered down or covered to eliminate or minimize dust generation.
	RCRA 40 CFR, Section 262 or state equivalent	Standards that apply to generators of hazardous waste	The soils are excavated for treatment.	If possible, soils should be fed directly into the wash unit for treatment.
Storage prior to processing	RCRA 40 CFR, Part 264 or state equivalent	Standards applicable to the storage of hazardous waste	Excavation may generate a hazardous waste that must be stored in a waste pile.	If in a waste pile, the material should be placed on and covered with plastic and tied down to minimize fugitive air emissions and volatilization; the time between excavation and treatment should be kept to a minimum.
Waste processing	RCRA 40 CFR, Parts 264 and 265 or state equivalent	Standards applicable to the treatment of hazardous waste at permitted and interim status facilities	Treatment of hazardous waste must be conducted in a manner that meets the operating and monitoring requirements; the treatment process occurs in a tank.	Equipment must be operated and maintained daily. Tank integrity must be monitored and maintained to prevent leakage or failure; the tank must be decontaminated when processing is complete. Air emissions must be characterized by continuous emissions monitoring.
Storage after processing	RCRA 40 CFR, Part 264 or state equivalent	Standards that apply to the storage of hazardous waste in containers	The treated soil will be placed in tanks prior to a decision on final disposition.	The treated soils must be stored in containers that are well maintained; container storage area must be constructed to control run-on and runoff.
Waste characterization (treated waste)	RCRA 40 CFR, Part 261 or state equivalent	Standards that apply to waste characteristics	A requirement of RCRA prior to managing and handling the waste; it must be determined if treated soil is RCRA hazardous waste.	Chemical and physical tests must be performed on treated soils prior to disposal.

Activity	Regulation	Standards	Concern	Action required
	TSCA 40 CFR, Part 761 or state equivalent	Standards that apply to the treatment and disposal of wastes containing PCBs	Soils treated may still contain PCBs.	Chemical and physical tests must be performed on treated soils. If PCBs are identified, a proper disposal method will be selected.
On-site/off-site disposal	RCRA 40 CFR, Part 264 or state equivalent	Standards that apply to landfilling hazardous waste	Treated soils may still contain contaminants in levels above required cleanup action levels and therefore be subject to LDRs.	Treated wastes must be disposed of at a RCRA-permitted hazardous waste facility, or approval must be obtained from EPA to dispose of the wastes on site.
	TSCA 40, Part 761 or state equivalent	Standards that restrict the placement of PCBs in or on the ground	Treated soils containing less than 5 mg/kg PCBs may be landfilled or incinerated.	If untreated soil contained PCBs, then treated soil should be analyzed for PCB concentration. Approved PCB landfills or incinerators must be used for disposal.
	RCRA 40 CFR, Part 268 or state equivalent	Standards that restrict the placement of certain wastes in or on the ground	The nature of the waste may be subject to the LDRs.	The waste must be characterized to determine if the LDRs apply; treated wastes must be tested and results compared.
	SARA Section 121(d)(3)	Requirements for the off-site disposal of wastes from a Superfund site	The waste is being generated from a response action authorized under SARA.	Wastes must be disposed of at a RCRA-permitted hazardous waste facility.
	RCRA 40 CFR, Part 262 or state equivalent	Manifest requirements and packaging and labeling requirements prior to transporting	The treated soil may need to be manifested and managed as a hazardous waste.	An identification (ID) number must be obtained from EPA.
Transportation for off-site disposal	RCRA 40 CFR, Part 263 or state equivalent	Transportation standards	Treated soil may need to be transported as a hazardous waste.	A transporter licensed by EPA must be used to transport the hazardous waste according to EPA regulations.
Wastewater discharge	Clean Water Act 40 CFR, Parts 301, 304, 306, 307, 308, 402, and 403	Standards that apply to discharge of wastewater into sewage treatment plant or surface water bodies	The wastewater may be a hazardous waste.	Determine if wastewater could be directly discharged into a sewage treatment plant or surface water body. If not, the wastewater may need to be further treated to meet discharge requirements by conventional processes.
	Safe Drinking Water Act 40 CFR, Parts 144 and 145	Standards that apply to the disposal of contaminated water in underground injection wells	Wastewater may require disposal in underground injection wells.	If underground injection is selected as a disposal means for contaminated wastewater, permission must be obtained from EPA to use existing permitted underground injection wells, or to construct and operate new wells.

Subpart C. Listed wastes from nonspecific and specific industrial sources, off-specification products, spill cleanups, and other industrial sources are itemized in 40 CFR, Part 261, Subpart D.

Once contaminated soils are processed by the soil washing system, the processed soils may still contain hazardous constituents at levels above required cleanup action levels. Such soils need to be managed as hazardous waste and are subject to land disposal restrictions (LDRs) under both RCRA and CERCLA. Applicable RCRA requirements could include a Uniform Hazardous Waste Manifest if the treated soils are transported, restrictions on placing the treated soils in land disposal units, time limits on accumulating treated soils, and permits for storing treated soils.

Requirements for corrective action at RCRA-regulated facilities are provided in 40 CFR, Part 264, Subpart F and Subpart S. These subparts also generally apply to remediation at Superfund sites. Subparts F and S include requirements for initiating and conducting RCRA corrective actions, remediating ground water, and ensuring that corrective actions comply with other environmental regulations. Subpart S also details conditions under which particular RCRA requirements may be waived for temporary treatment units operating at corrective action sites. Thus, RCRA mandates requirements similar to CERCLA, and as proposed, allows treatment units such at the soil washing system to operate without full permits.

6.7.3 Clean Air Act (CAA)

The CAA requires that treatment, storage, and disposal facilities comply with primary and secondary ambient air quality standards. During the excavation, transportation, and treatment of soils, fugitive emissions are possible. Steps must be taken to prevent or minimize the impact from fugitive emissions, such as watering down the soils, or covering them with industrial-strength plastic prior to treatment. State air quality standards may require additional measures to prevent fugitive emissions.

6.7.4 Safe Drinking Water Act (SDWA)

The SWDA of 1974, as most recently amended by the Safe Drinking Water Amendments of 1986, requires the EPA to establish regulations to protect human health from contaminants in drinking water. The legislation authorizes national drinking water standards and a joint federal-state system for ensuring compliance with these standards. The National Primary Drinking Water Standards are found in 40 CFR, Parts 141 through 149. Since process water is recycled through a closed-loop system, on-site disposal is not an issue.

6.7.5 Toxic Substances Control Act (TSCA)

The TSCA grants the EPA the authority to prohibit or control the manufacturing, importing, processing, use, and disposal of any chemical substance that presents an unreasonable risk of injury to human health or the environment. These regulations may be found in 40 CFR 761. With respect to hazardous waste regulation, TSCA focuses on the use, management, disposal, and cleanup of PCBs. Materials with less than 50 ppm PCB are classified as normal PCBs. Since PCBs are not a contaminant of concern on firing ranges, requirements for TSCA compliance are not an issue.

6.7.6 Occupational Safety and Health Administration Requirements (OSHA)

CERCLA remedial actions and RCRA corrective actions must be performed in accordance with OSHA requirements detailed in 20 CFR, Parts 1900 through 1926, especially Part 1910.120, which provides for the health and safety of workers at hazardous waste sites. On-site construction activities at Superfund or RCRA corrective actions sites must be performed in accordance with Part 1926 of OSHA, which provides safety and health regulations for construction sites. State OSHA requirements, which may be significantly stricter than federal standards, must also be met.

All technicians operating the soil washing system are required to have completed an OSHA training course and must be familiar with all OSHA requirements relevant to hazardous waste sites. For most sites, minimum personal protective equipment (PPE) for technicians will include gloves, hard hats, steel toe boots, and coveralls. Depending on contaminant types and concentrations, additional PPE may be required. Noise levels should be monitored to ensure that workers are not exposed to noise levels above a time-weighted average of 85 decibels over an 8-hour day. If operation of the system causes noise levels to increase above this limit, then workers will be required to wear ear protection.

6.8 PERSONNEL REQUIREMENTS

The labor requirements for handling the materials for the soil washing unit are one heavy equipment operator on a full-time basis. The operator must move soil, load the pretreatment screening process and handle solid products from the plant.

The gravity separation plant is fully automated and incorporates modem computer monitoring and control. Direct contact by the operator with the process equipment or material is not often necessary. The process requires one lead and two full-time operators and a laborer for routine operation as well as to initiate repairs. A technician to perform sampling procedures is also required. The process operates continuously, and runs well once equilibrium conditions are established. Therefore, operators are required for 120 hours per week of operation.

6.9 TRAINING

Operator training specific to the soil washing gravity separation system is required, and personnel must be trained in the operation and maintenance of each unit within the system. This is necessary to develop a safe and effective operating technique. This training is conducted by the lead operator during mobilization and startup.

Personnel involved with sampling or other activities close to the unit are required to wear Level D protection. The 40-hr OSHA training on PPE applications, health and safety, emergency response procedures, and quality assurance-quality control are required. Additional training to address site activities, procedures, and monitoring associated with the technology is recommended. Personnel should also be briefed when new operations are planned, work practices changed, or site conditions changed.

6.10 HEALTH AND SAFETY

The U.S. Department of Labor (USDOL) specifies occupational health and safety standards for general industry in 29 CFR 1910. Within these regulations are detailed sections applying to occupation protection of employees at hazardous waste treatment sites. Health and safety regulations for construction are specified by the USDOL in 29 CFR 1926. The health and safety issues involved in using the soil washing include those presented in the above-mentioned regulations. Additional issues are generally the same as those that apply to all hazardous waste treatment facilities as detailed in 40 CFR 264, Subparts B through G, and Subpart X. If the operations are to take place at a hazardous waste treatment site, USEPA (40 CFR) and USDOL (26 CFR) also apply.

The principal occupational hazard of the soil washing system is probably not the health risk from exposure to toxic substances (except in extreme cases), but rather the risk of injury. Although it poses no significant threat of explosion or fire, the system is comprised of equipment with heights, moving machinery, high voltage, pinch points, conveyor belts, and other safety hazards. Therefore, the physical hazards may be greater than the chemical hazards, depending on the waste.

When appropriate, workers should have routine medical exams to monitor for exposure to lead. Health and safety monitoring and incident reports must be routinely filed; records of occupational illnesses and injuries (OSHA Forms 101 and 200) must be maintained. Audits ensuring compliance with the health and safety plan should be performed.

Proper PPE should be available for use by on-site personnel. Different levels of personal protection will be required, based on the potential hazards associated with the site and the work activities.

Site monitoring should be conducted to identify the extent of hazards and to document exposures at the site. The monitoring results should be maintained and posted.

CMRI Treatability Study Results

CONTENTS

Colorado Minerals Research Institute (CMRI) completed the treatability studies in late May 1996 and issued their final report on November 12, 1996. The report was prepared by CMRI for Metcalf & Eddy, Inc. under the M&E project number 962017. The results of this study were used to design the equipment and prepare the layout of the pilot plant for the demonstration at NAS Miramar.

A.1 INTRODUCTION, SUMMARY, AND CONCLUSIONS

CMRI was authorized to perform treatability studies on berm samples from three military firing ranges. The samples were from Naval Air Station (NAS) Miramar, Twentynine Palms Marine Base, and Fort Dix Army Base. The initial treatability was to deagglomerate and size the samples. The size fractions then were gravity-separated. The products of gravity separation were concentrate, middlings, and tailings.

From 79 to 95% of the lead was associated with the concentrate fractions of the soil. These gravity fractions contained 9 to 17% of the total sample weights.

About one half of the middlings and tailings fractions were below the target lead content of 500 mg/kg. The lead content of the combined tailing and middling fractions from the Fort Dix samples averaged less than 500 mg/kg and therefore did not require further study.

Attrition scrubbing of the –4.76 mm by +1.19 mm (–4 mesh by +16 mesh) tailing fractions of the Miramar samples was not effective in producing fractions of significant weight that was less than 500 mg/kg. However, the scrubbing action did break down agglomerated soil particles.

The coarse fractions of the Twentynine Palms samples, coarser than 105 μm (150 mesh), containing more than 500 mg Pb/kg showed good amenability to percolation leaching with nitric acid at 2 g/L. The residues after three days leaching contained 30 to 150 mg Pb/kg.

The lead content for the –105 μm (–150 mesh) fractions from the NAS Miramar and Twentynine Palms samples averaged 3870 mg/kg and 19,900 mg/kg, respectively. The –105 μm fraction from Miramar A sample was quite refractory to leaching, whereas the lead concentrations in both Twentynine Palms fractions were very high. Single-stage leaching was not efficient at decreasing the lead content of the refractory –105 μm Miramar fraction or both Twentynine Palms samples to less than 500 mg Pb/kg. However, two-stage leaching decreased the lead to less than 500 mg/kg. The most effective lixiviant was a mixture of nitric acid, hydrochloric acid, and hydrogen peroxide. Achieving good leaching of the Miramar A sample required heating the leach solution to 40 or 60°C.

A.2 SAMPLES

Six 5-gal pails of berm samples were received in late May 1996 for treatability studies. The samples consisted of two pails each from NAS Miramar, Twentynine Palms, and Fort Dix.

The Miramar samples were not labeled as different samples. CMRI labeled the samples as "A" and "B" and treated them separately throughout the study. The Twentynine Palms samples were identified as LR-23 and LR-30. The Fort Dix samples were identified as BK-1 and BK-3.

A.3 SIZING - GRAVITY SEPARATION

Each of the six pails was subjected to a treatability study consisting of an initial deagglomeration by wet screening on a 9.50 mm (3/8 in.) sieve. No significant agglomeration was noted in any of the samples. The balance of the material then was screened through the following sieves:

- 4.76 mm (4 mesh)
- 1.19 mm (16 mesh)
- 297 μm (50 mesh)
- 105 μm (150 mesh)

The fractions generated by screening were separated by gravity methods as follows:

- Humates — skimmed from +3/8 in. tailings
- +3/8 in. — Jig producing a tailings and concentrate
- –3/8 in. by +4 mesh — Jig producing a tailings and concentrate
- –4 mesh by +16 mesh — Jig producing a tailings and concentrate
- –16 mesh by +50 mesh — Shaking table producing a tailings, middlings, and concentrate
- –50 mesh by +150 mesh — Shaking table producing a tailings, middlings, and concentrate
- –150 mesh (First Sample) — Shaking table producing a tailings, middlings, and concentrate
- –150 mesh (Second Sample) — Knelson bowl concentrator producing a tailings and concentrate

Each product was analyzed for lead, antimony, and copper. The results of particle sizing and gravity separation tests on NAS Miramar, Twentynine Palms, and Fort Dix samples are summarized in Tables A.1, A.2, and A.3 with details given in Appendix A.1. A visual description was made of

Table A.1 Treatability Results for Particle Sizing and Gravity Separation – NAS Miramar Samples

	Screen Analyses			Lead Analyses		
Size	Sample A weight %	Sample B weight %	Averages weight %	Sample A mg/kg	Sample B mg/kg	Averages mg/kg
Humates						
+9.50 mm (+3/8 in.)	0.16%	0.35%	0.25%	5902	2563	3610
Tailings						
+9.50 mm (+3/8 in.)	3.50%	2.90%	3.20%	114	72	95
+4.76 mm (+4 mesh)	3.20%	1.70%	2.45%	229	130	195
+1.19 mm (+16 mesh)	4.59%	3.90%	4.25%	2330	3887	3045
+297 μm (+50 mesh)	4.00%	9.90%	6.95%	603	192	310
+105 μm (+150 mesh)	8.09%	12.89%	10.49%	747	237	434
−105 μm (−150 mesh)	37.26%	42.48%	39.87%	5580	2370	3870
Middlings						
+297 μm (+50 mesh)	16.68%	6.60%	11.64%	503	184	413
+105 μm (+150 mesh)	9.49%	10.79%	10.14%	655	184	404
Concentrate						
+9.50 mm (+3/8 in.)	1.10%	0.90%	1.00%	400,700	242,800	329,622
+4.76 mm (+4 mesh)	1.10%	0.50%	0.80%	349,000	180,200	296,228
+1.19 mm (+16 mesh)	2.70%	1.80%	2.25%	145,100	53,600	108,487
+297 μm (+50 mesh)	3.90%	1.50%	2.70%	36,800	18,100	31,603
+105 μm (+150 mesh)	2.20%	2.90%	2.55%	37,800	10,200	22,102
−105 μm (−150 mesh)	2.05%	0.90%	1.47%	33,200	23,400	30,209
Total	100.00%	100.00%	100.00%	17,533	6080	11,806
Samples	20,241 kg	24,273 kg	22,257 kg	17,644	5966	11,805

Gravity Treatment

Humates: skimmed from +3/8 in. tailings

+9.50 mm (+3/8 in.): Jig tailings and concentrate

+4.76 mm (+4 mesh): Jig tailings and concentrate

+1.19 mm (+16 mesh): Jig tailings and concentrate

+297 μm (+50 mesh): Shaking table tailings, middlings, and concentrate

+105 μm (+150 mesh): Shaking table tailings, middlings, and concentrate

−105 μm (−150 mesh) Sample A: Shaking table tailings, middlings, and concentrate

−105 μm (−150 mesh) Sample B: Knelson bowl concentrator tailings and concentrate

−105 μm (−150 mesh) middlings, Sample A: weight and lead % added to −105 μm tailings

each of the size/gravity fractions. The descriptions are also given in Appendix A.1. A brief discussion of the distributions and analyses follows:

Miramar "A" split

- Average grade for total sample was 17,600 mg Pb/kg, 640 mg Sb/kg, and 8100 mg Cu/kg.
- The humates, 0.16% of the weight, contained 5900 mg Pb/kg.
- The tailings coarser than 105 μm (150 mesh), 23.3% of the weight, contained an average of 915 mg Pb/kg. The majority of the lead content was in the −4.76 mm by +1.19 mm (−4 mesh by +16 mesh) fraction.
- The −105 μm (−150 mesh) fraction was separated using a shaking table.
- The −105 μm tailings and middlings, 37.3% of the weight, contained 5600 mg Pb/kg.
- The total concentrate, 13.0% of the weight, contained an average of 117,000 mg Pb/kg.
- The concentrates that were coarser than 4.76 mm (4 mesh) contained about 30% copper.

Only two fractions, the +9.50 mm (+3/8 in.) and the −9.50 mm by +4.76 mm (−3/8 in. by +4 mesh) gravity tails, contained less than 500 mg Pb/kg. These two fractions represent only 6.7% of the weight. The other tailings and middlings contained from 500 to 6000 mg Pb/kg. The highest lead

Table A.2 Treatability Results for Particle Sizing and Gravity Separation – Twentynine Palms Samples

	Screen Analyses			Lead Analyses		
Size	LR-23 weight %	LR-30 weight %	Averages weight %	LR-23 mg/kg	LR-30 mg/kg	Averages mg/kg
Humates						
+9.50 mm (+3/8 in.)	0.00%	0.00%	0.00%	0	0	0
Tailings						
+9.50 mm (+3/8 in.)	8.51%	6.99%	7.75%	46	85	64
+4.76 mm (+4 mesh)	5.81%	4.60%	5.20%	255	386	313
+1.19 mm (+16 mesh)	13.01%	12.19%	12.60%	314	820	559
+297 µm (+50 mesh)	28.83%	33.97%	31.40%	674	599	633
+105 µm (+150 mesh)	11.21%	7.29%	9.25%	418	525	460
−105 µm (−150 mesh)	8.21%	8.89%	8.55%	15,121	24,300	19,894
Middlings						
+297 µm (+50 mesh)	5.51%	5.09%	5.30%	1318	1483	1397
+105 µm (+150 mesh)	5.31%	4.00%	4.65%	768	941	842
Concentrate						
+9.50 mm (+3/8 in.)	0.20%	0.60%	0.40%	149,400	54,600	78,336
+4.76 mm (+4 mesh)	1.50%	2.80%	2.15%	92,000	77,800	82,760
+1.19 mm (+16 mesh)	3.70%	6.19%	4.95%	59,500	57,500	58,248
+297 µm (+50 mesh)	3.10%	3.80%	3.45%	64,400	69,800	67,371
+105 µm (+150 mesh)	3.80%	2.40%	3.10%	53,000	81,700	64,096
−105 µm (−150 mesh)	1.30%	1.20%	1.25%	54,500	59,900	57,089
Total	100.00%	100.00%	100.00%	10,263	14,031	12,147
Samples	31,501 kg	30,281 kg	30,891 kg	10,309	13,989	12,113

Gravity Treatment

> Humates: skimmed from +3/8 in. tailings
>
> +9.50 mm (+3/8 in.): Jig tailings and concentrate
>
> +4.76 mm (+4 mesh): Jig tailings and concentrate
>
> +1.19 mm (+16 mesh): Jig tailings and concentrate
>
> +297 µm (+50 mesh): Shaking table tailings, middlings, and concentrate
>
> +105 µm (+150 mesh): Shaking table tailings, middlings, and concentrate
>
> −105 µm (−150 mesh) Sample LR-23: Shaking table tailings, middlings, and concentrate
>
> −105 µm (−150 mesh) Sample LR-30: Knelson bowl concentrator tailings and concentrate
>
> −105 µm (−150 mesh) middlings, Sample LR-23: weight and lead % added to −105 µm tailings

values were in the −105 µm (−150 mesh) tailings and middlings. However, the −4.76 mm by +1.19 mm (−4 mesh by 16 mesh) tailings also ranked high in lead at 2330 mg/kg.

Miramar "B" split

- Average grade for total sample was 6000 mg Pb/kg, 270 mg Sb/kg, and 3900 mg Cu/kg.
- The humates, 0.35% of the weight, contained 2600 mg Pb/kg.
- The tailings coarser than 105 µm (150 mesh), 32% of the weight, contained an average of 680 mg Pb/kg.
- The −105 µm (−150) was separated using the Knelson concentrator.
- The −105 µm tailings, 42.5% of the weight, contained 2400 mg Pb/kg.
- The total concentrate, 8.5% of the weight, contained an average of 55,000 mg Pb/kg.
- The concentrates that were coarser than 4.76 mm (4 mesh) contained about 27% copper.

The "B" sample generated only two tailings fractions that exceeded 500 mg Pb/kg. These were the Knelson tailings at 2370 mg Pb/kg and the −4.76 mm by +1.19 mm (−4 mesh by +16 mesh) gravity tailings at 3887 mg Pb/kg. The fractions that were less than 500 mg Pb/kg represented about 45% of the weight.

Table A.3 Treatability Results for Particle Sizing and Gravity Separation – Fort Dix Samples

Size	Screen Analyses			Lead Analyses		
	BK-1 weight %	BK-3 weight %	Averages weight %	BK-1 mg/kg	BK-3 mg/kg	Averages mg/kg
Humates						
+9.50 mm (+3/8 in.)	0.02%	0.02%	0.02%	13,700	4400	9050
Tailings						
+9.50 mm (+3/8 in.)	0.20%	0.10%	0.15%	742	76	520
+4.76 mm (+4 mesh)	1.00%	0.70%	0.85%	58	95	73
+1.19 mm (+16 mesh)	21.80%	25.57%	23.69%	231	136	180
+297 μm (+50 mesh)	13.30%	18.78%	16.04%	98	63	78
+105 μm (+150 mesh)	25.00%	19.28%	22.14%	75	96	84
−105 μm (−150 mesh)	8.90%	6.09%	7.50%	2967	1840	2509
Middlings						
+297 μm (+50 mesh)	6.80%	10.49%	8.64%	95	95	95
+105 μm (+150 mesh)	14.10%	9.99%	12.05%	67	78	72
Concentrate						
+9.50 mm (+3/8 in.)	1.50%	1.50%	1.50%	104,100	177,000	140,532
+4.76 mm (+4 mesh)	0.10%	0.20%	0.15%	147,100	244,100	211,745
+1.19 mm (+16 mesh)	1.30%	0.80%	1.05%	19,900	19,700	19,824
+297 μm (+50 mesh)	1.20%	2.70%	1.95%	14,100	13,300	13,546
+105 μm (+150 mesh)	3.70%	2.60%	3.15%	1700	2200	1906
−105 μm (−150 mesh)	1.10%	1.20%	1.15%	4000	2300	3113
Total	100.00%	100.00%	100.00%	2608	3937	3272
Samples	26,074 kg	24,835 kg	25,455 kg	2669	3875	3257

Gravity Treatment

Humates: skimmed from +3/8 in. tailings

+9.50 mm (+3/8 in.): Jig tailings and concentrate

+4.76 mm (+4 mesh): Jig tailings and concentrate

+1.19 mm (+16 mesh): Jig tailings and concentrate

+297 μm (+50 mesh): Shaking table tailings, middlings, and concentrate

+105 μm (+150 mesh): Shaking table tailings, middlings, and concentrate

−105 μm (−150 mesh) Sample BK-1: Shaking table tailings, middlings, and concentrate

−105 μm (−150 mesh) Sample BK-3 Knelson bowl concentrator tailings and concentrate

−105 μm (−150 mesh) middlings, Sample BK-1: weight and lead % added to −105 μm tailings

Twentynine Palms LR-23

- Average grade for total sample was 10,300 Pb, 290 Sb, and 8100 Cu (mg/kg).
- The tailings coarser than 105 μm (150 mesh), 67% of the weight, contained an average of 469 mg Pb/kg.
- The −105 μm (−150 mesh) was separated using a shaking table.
- The −105 μm tailings and middlings, 8.2% of the weight, contained 15,100 mg Pb/kg.
- The total concentrates, 13.7% of the weight, contained an average of 63,000 mg Pb/kg.
- The concentrates that were coarser than 4.76 mm (4 mesh) contained almost 40% copper.

Four gravity tailings fractions contained less than 500 mg lead/kg; the +9.50 mm (+3/8 in.), −9.50 mm by +4.76 mm (−3/8 in. by +4 mesh), −4.76 mm by +1.19 mm (−4 mesh by +16 mesh), and −297 μm by + 105 μm (−50 mesh by +150 mesh). These fractions represented almost 40% of the weight. The other tailings and all middlings fractions exceeded 500 mg Pb/kg, containing from 674 to 1300 mg Pb/kg. The highest lead values were in the −105 μm (−150 mesh) tailings and middlings. However, the −1.19 mm by +297 μm (−16 mesh by +50 mesh) tailings and middlings also were high at 674 mg Pb/kg.

Twentynine Palms LR-30

- Average grade for total sample was 14,000 mg Pb/kg, 780 mg Sb/kg, and 18,000 mg Cu/kg.
- The tailings coarser than 105 μm (150 mesh), 65% of the weight, contained an average of 560 mg Pb/kg.
- The –105 μm (150 mesh) was separated using the Knelson concentrator.
- The –105 μm (–150 mesh) tailings, 8.9% of the weight, contained 24,300 mg Pb/kg.
- The combined concentrates, 16.9% of the weight, contained an average of 67,000 mg Pb/kg.
- The concentrates that were coarser than 4.76 mm (4 mesh) contained over 30% copper.

Only two fractions, the +9.50 mm (+3/8 in.) and –9.50 mm by +4.76 mm (–3/8 in. by –4 mesh) gravity tailings, contained less than 500 mg Pb/kg. These two fractions represent only 11.6% of the weight. The other tailings and middling fractions contained from 525 to 24,300 mg Pb/kg. The highest lead value was in the –105 μm (–150 mesh) tailings. The highest lead content of the other fractions was 1483 mg/kg in the –4.76 mm by +297 μm (–16 mesh by +50 mesh) middlings.

Fort Dix BK-1

- Average grade for total sample was 2700 mg Pb/kg, 66 mg Sb/kg, and 6800 mg Cu/kg.
- The humates, 0.02% of the weight, contained 13,700 mg Pb/kg.
- The tailings coarser than 105 μm (150 mesh), 61% of the weight, contained an average of 140 mg Pb/kg.
- The –105 μm (–150 mesh) was separated using a shaking table.
- The –105 μm tailings and middlings, 8.9% of the weight, contained 3000 mg Pb/kg.
- The combined concentrates, 8.9% of the weight, contained an average of 25,900 mg Pb/kg.
- The concentrates that were coarser than 4.76 mm (4 mesh) contained over 40% copper.

Only two fractions, the +9.50 mm (+3/8 in.) and the –105 μm (–150 mesh) gravity tailings, contained more than 500 mg lead/kg. These fractions represented about 9% of the total weight. The –105 μm (–150 mesh) tailings fraction contained 3000 mg Pb/kg. The other tailings and middlings fractions containing less than 500 mg Pb/kg represented 85% of the weight. If all of the tailing and middling fractions were combined, the average lead content was about 400 mg/kg.

Fort Dix BK-3

- Average grade for total sample was 3900 mg Pb/kg, 90 mg Sb/kg, and 8000 mg Cu/kg.
- The humates, 0.02% of the weight, contained 4400 mg Pb/kg.
- The tailings coarser than 105 μm (150 mesh), 65% of the weight, contained an average of 103 mg Pb/kg.
- The –105 μm (–150 mesh) fraction was separated using the Knelson concentrator.
- The –105 μm (–150 mesh) tailings, 6.1% of the weight, contained 1800 mg Pb/kg.
- The combined concentrates, 9.0% of the weight, contained an average of 41,000 mg Pb/kg.
- The concentrates that were coarser than 4.76 mm (4 mesh) contained over 45% copper.

Only one of the tailings and middlings fractions, the –105 μm (–150 mesh) tailings, contained more than 500 mg Pb/kg (1840 mg Pb/kg). This fraction represented only 6.1% of the total weight. The other tailings and middlings fractions containing less than 500 mg Pb/kg, represented 85% of the weight. If all of the tailings and middling fractions were combined, the average lead content was about 215 mg/kg.

A.4 ATTRITION SCRUBBING OF MIRAMAR COARSE FRACTION

The –4.76 mm by +1.19 mm (–4 mesh by +16 mesh) portions of the Miramar A and B samples were elevated in lead content, 2330 and 3887 mg/kg, respectively. Portions of both fractions were treated in an attrition scrubber and then treated in a jig to determine whether the lead contamination could be concentrated. The products were jig tailings, jig concentrates, and -0.59 mm (–30 mesh) fractions. The tailings and concentrates were also screened at 1.19 mm (16 mesh) such that five fractions were generated in total. Data sheets for these tests are included in Appendix A.2.

The lead contents of the +1.19 mm (+16 mesh) tailings and concentrate for Miramar A were decreased to less than 500 mg/kg and represented about 70% of the weight. The lead content of the –1.19 mm (–16 mesh) tailings was 658 mg/kg. Both the –1.19 mm concentrate and the -0.59 mm (30 mesh) fraction were high in lead, 2920 mg/kg and 2592 mg/kg, respectively.

The lead content of +1.19 mm (+16 mesh) tailings fraction for Miramar B was decreased to less than 171 mg/kg and represented about 39% of the weight. The other fractions were similar in lead content ranging from 675 mg/kg for the –1.19 mm (–16 mesh) tailings to 1653 mg/kg for the –1.19 mm (–16 mesh) concentrate.

However, the sized products, without gravity separation, showed some concentration of lead in the two –1.19 mm (–16 mesh) fractions. The average results for samples A and B are as follows:

NAS Miramar Sample	Weight		Lead		
	Grams	Distn	mg/kg	mg	Distn
Averaged Calculated Feed	496.0	100.0%	766	379.8	100.0%
Averaged +1.19 mm (+16 mesh) Total	337.0	67.9%	505	168.5	43.8%
Averaged –1.19 mm (–16 mesh) Total	117.1	23.6%	1145	133.3	35.4%
Averaged -0.59 mm (–30 mesh) Total	42.0	8.5%	1952	78.0	20.8%
Distn: weight or lead distribution					

The lead content of +1.19 mm (+16 mesh) fraction for Miramar samples was about 500 mg/kg, representing about 44% of the weight. The –1.19 mm (–16 mesh) and –0.59 mm (–30 mesh) fractions were higher in lead content, 1145 mg/kg and 1952 mg/kg, respectively. Hence, attrition scrubbing seemed to break apart the coarse soil particles releasing fine lead particles.

A.5 HYDROCYCLONE STUDIES

The –105 μm (–150 mesh) tailings from Miramar "A" and "B" samples were evaluated for amount of size and gravity separation using a hydrocyclone. The hydrocyclone was a Krebs D4B-12 unit with a 9.5 mm (3/8 in.) apex and a (1 1/4 in.) vortex finder. Because of the limited amount of feed, a 5 to 10% solids slurry was prepared and tested at three or four pressures. To conserve sample, the underflow and overflow were sampled for different lengths of time. The sampling time was not recorded as the overflow was a partial split of the stream. The sampling method caused the feed for subsequent tests to be biased with higher fines (not noticeable from the analyses). The overflow was wet-screened at 37 μm (400 mesh) and the underflow at 53 μm (270 mesh), 44 μm (325 mesh), and 37 μm (400 mesh). The two overflow fractions were combined for analysis as were the +53 μm (270 mesh) and –53 μm by +44 μm (–270 mesh by +325 mesh) fractions of the underflow.

Table A.4 lists the averaged results for samples A and B. The results include the weight and lead distributions and the lead contents within the overflow and underflow for the tests carried out at the operating pressure of 69 kPa gauge (10 psig). The complete test results for each sample at several operating pressures and at 69 kPa gauge (10 psig) are given in Appendix A.3. In the appendix

Table A.4 Hydrocyclone Data for NAS Miramar Samples A & B at 69 kPa gauge (10 psig)

Overflow	Weight %	Accumulative Weight %	Pb mg/kg	Pb Distribution	Accumulative Pb Distribution
+37 µm (+400 mesh)	0.71%	0.71%	3890	0.74%	0.74%
−37 µm (−400 mesh)	35.11%	35.82%	4946	46.91%	47.65%
Overflow Total	35.82%		4925	47.65%	
Underflow					
+53 µm (+270 mesh)	24.81%	60.63%	964	6.46%	54.11%
+44 µm (+325 mesh)	3.13%	63.76%	1452	1.23%	55.34%
+37 µm (+400 mesh)	9.20%	72.96%	2185	5.43%	60.77%
−37 µm (−400 mesh)	27.04%	100.00%	5370	39.23%	100.00%
Underflow Total	64.18%		3019	52.35%	
Total	100.00%		3702		
Feed Analysis			3870		

tables, an estimate of the weight split was made by adjusting the percentage split so that the head analysis calculated from the test products matched the feed analysis.

From Table A.4, the average lead analyses for each fraction are summarized as follows:

Overflow	Miramar A & B Lead Content mg/kg
−37 µm (−400 mesh)	4950
Underflow	
+53 µm (+270 mesh)	970
+44 µm (+325 mesh)	1450
+37 µm (+400 mesh)	2180
−37 µm (−400 mesh)	5370
Total Underflow	3019

The data indicated no general trend in lead content since most differences are within the precision of the analysis. The most significant result was that the −37 µm (−400 mesh) fraction in the underflow was considerably upgraded in lead, and the +53 µm (+270 mesh) fractions were not below levels of concern. The inference is that with the Miramar soils treating the −105 µm (−150 mesh) fraction with a hydrocyclone will not concentrate the lead into either the overflow or underflow fractions. If the cut can be made closer to 44 µm (325 mesh), then a marginally clean fraction can be generated from the Miramar B sample. However, this fraction, −105 µm (−150 mesh) by +37 µm (+325 mesh), represented about 6% of the total soil weight, and 36% of the soil weight still required chemical treatment. There were no fractions from the −105 µm (−150 mesh) material for the Miramar A sample that can avoid chemical treatment.

A.6 PERCOLATION LEACHING

Only two coarse tailings and middlings fractions of the Twentynine Palms samples exceeded 500 mg Pb/kg. These fractions were the −1.19 mm by +297 µm (−16 mesh by +50 mesh) fraction of LR-23 and the −4.76 mm by +297 µm (−4 mesh by +50 mesh) fractions of LR-30. Tests to evaluate the potential requirements for percolation leaching were performed on these two samples. The detailed test results are included in Appendix A.4.

The initial tests on these two fractions were designed to evaluate the response to a nitric acid agitation leach. One hundred grams of each material was added to water containing 5 ml reagent nitric acid (HNO_3) to form a 33% solids slurry. The slurry was agitated on rolls for 24 hours. The lead in the LR-23 sample was decreased from 780 to 30 mg Pb/kg and the lead in the LR-30 sample was decreased from 710 to 30 mg Pb/kg. The acid additions for both tests were excessive as the final pH values were 0.5 or less.

Two percolation tests then were performed on 400-gram portions of each material. In each case, the material was leached for 72 hours by a dilute solution of percolating nitric acid. The initial contacts used 5 g HNO_3/L and decreased the residual lead to 46 and 53 mg/L, respectively, for LR-23 and LR-30. The HNO_3 additions were about 22 kg/t (44 lb/ton) of the soil fraction leached.

The second set of percolation leaches used a 2 g HNO_3/L solution. The total acid additions were about 10 kg/t (19 lb/ton) of material. The residue from the LR-23 leach contained 30 mg Pb/kg, whereas the residue from the LR-30 leach contained 150 mg Pb/kg.

A.7 AGITATION LEACHING

The −105 μm (−150 mesh) fractions of the NAS Miramar and Twentynine Palms samples all had high lead content. The list below summarizes the lead and antimony analyses for these fractions:

Miramar A	5580 mg Pb/kg	73 mg Sb/kg	Sb = 1.3% of Pb
Miramar B	2370 mg Pb/kg	42 mg Sb/kg	Sb = 1.8% of Pb
LR-23	15,100 mg Pb/kg	190 mg Sb/kg	Sb = 1.3% of Pb
LR-30	24,300 mg Pb/kg	430 mg Sb/kg	Sb = 1.8% of Pb

Agitation leaching tests were performed on these samples. The detailed leaching conditions and results are included in the test data sheets in Appendix A.5. Initial leaching tests on the −105 μm (−150 mesh) fractions were single-stage leach with nitric acid (HNO_3). In these tests, the pH was controlled by incremental additions of nitric acid over a 4-hour period. The target pH values were 2.0, 1.5, and 1.2 pH. None of the samples was leached successfully by this technique. A summary of the results is shown below:

Sample	Target pH	Final pH	HNO_3 Addition kg/t	Soluble %Pb	Residue Lead mg/kg
Miramar A	2.0	2.04	22	32	3850
Miramar A	1.5	1.56	23	42	3230
Miramar A	1.2	1.26	21	29	3940
Miramar B	2.0	2.15	25	38	1510
Miramar B	1.5	1.53	30	52	1150
Miramar B	1.2	1.24	35	59	970
LR-23	2.0	2.47	60	76	3838
LR-23	1.5	1.57	65	84	2651
LR-23	1.2	1.25	70	88	2092
LR-30	2.0	2.05	60	87	3314
LR-30	1.5	1.51	64	90	2601
LR-30	1.2	1.12	67	92	2204

The Miramar samples were very refractory to the nitric acid leaching. The Twentynine Palms samples showed high levels of dissolution, but because of the very high lead content of the material, the residues still contained more than 2000 mg/kg.

The −105 μm fraction form Miramar B was subjected to a number of leach chemistries to determine what conditions were necessary to satisfactorily solubilize the lead. Adequate corrosion of the lead metal probably would require more aggressive conditions, i.e., more acidic and oxidizing. This series of tests evaluated the following reagents and combination of reagents using single-stage leaching:

- Acetic acid (CH_3COOH),
- Citric acid ($HOC(CH_2COOH)_2COOH$),
- Nitric acid (HNO_3) with hydrogen peroxide (H_2O_2), and
- Nitric acid (HNO_3) and hydrochloric acid (HCl) with hydrogen peroxide (H_2O_2).

The results show that aggressive conditions were necessary:

- Acetic acid was ineffective: 1160 mg Pb/kg in the residue.
- Citric acid was ineffective: 1030 mg Pb/kg in the residue.
- Nitric acid plus hydrogen peroxide: 704 mg Pb/kg in the residue.
- Nitric and hydrochloric acids plus peroxide: 389 mg Pb/kg in the residue.

A number of leach tests were then performed on the Miramar B fraction to evaluate mixtures of nitric and hydrochloric acids plus peroxide or other additives. These tests, B-9 through B-17, show that single stage leaching with approximately 50 kg/t each of nitric and hydrochloric acids decreased the residual lead to less than 500 mg/kg. In addition, a two-stage leach at a slightly lower acid addition decreased the residual lead to 313 mg/kg.

Since the Miramar A sample appeared to be the most refractory to leaching, only two-stage leaching studies were performed on it. It was found that excessive acid additions at ambient temperature were not capable of decreasing the residual lead to less than 500 mg/kg. Increasing the leaching temperature to 40°C and 60°C decreased the residual lead to 473 and 327 mg Pb/kg, respectively.

Two-stage leaching appeared to be necessary for leaching the Twentynine Palms samples so that the final residue would not be in equilibrium with a solution having a high lead content.

A pair of two-stage tests on the LR-23 material showed that the residual lead decreased to 170 and 360 mg/kg. Two-stage leach tests on the LR-30 material decreased the residual lead to 176 to 250 mg/kg.

A.8 DISCLAIMER

This work was performed for Metcalf & Eddy, Inc. based on studies conducted on samples provided by Metcalf & Eddy. The results and data reported herein pertain to samples provided to CMRI for study and CMRI makes no claim as to their application to any other samples or their representatives of the material requiring remediation.

In conducting this study CMRI has applied engineering and/or scientific judgment and used a level of effort consistent with accepted standards of care and professional practice. Except as set forth herein, CMRI makes no warranty, expressed or implied, in fact or by law, whether of merchantability, fitness for a particular purpose or otherwise, concerning any of the materials or services which have been furnished by CMRI under this project.

APPENDIX A.1 SIZE AND GRAVITY SEPARATION DATA SHEETS

Table A.5 Treatability Results for NAS Miramar Sample A

Fraction	Particle Size	grams (dwb)	Wt% (dwb)	Retain %	Pass %	Pb mg/kg	Pb %	Retain %	Pass %	Sb mg/kg	Sb %	Retain %	Pass %	Cu mg/kg	Cu %	Retain %	Pass %
Humates																	
+3/8 in.	9.50 mm	32	0.2%	0.2%	99.8%	5902	0.1%	0.1%	99.9%	103	0.0%	0.0%	100.0%	256	0.0%	0.0%	100.0%
Tailings																	
+3/8in.	9.50 mm	705	3.5%	3.6%	96.4%	114	0.0%	0.1%	99.9%	20	0.1%	0.1%	99.9%	15	0.0%	0.0%	100.0%
+4 mesh	4.76 mm	639	3.2%	6.8%	93.2%	229	0.0%	0.1%	99.9%	20	0.1%	0.2%	99.8%	13	0.0%	0.0%	100.0%
+16 mesh	1.19 mm	938	4.6%	11.4%	88.6%	2330	0.6%	0.7%	99.3%	62	0.4%	0.7%	99.3%	353	0.2%	0.2%	99.8%
+50 mesh	297 μm	808	4.0%	15.4%	84.6%	603	0.1%	0.9%	99.1%	26	0.2%	0.8%	99.2%	20	0.0%	0.2%	99.8%
+150 mesh	105 μm	1641	8.1%	23.5%	76.5%	747	0.3%	1.2%	98.8%	30	0.4%	1.2%	98.8%	50	0.1%	0.3%	99.7%
-150 mesh	105 μm	5994	29.6%	53.1%	46.9%	6030	10.1%	11.3%	88.7%	79	3.7%	4.9%	95.1%	270	1.0%	1.3%	98.7%
Middlings																	
+50 mesh	297 μm	3375	16.7%	69.8%	30.2%	503	0.5%	11.8%	88.2%	21	0.5%	5.4%	94.6%	168	0.3%	1.6%	98.4%
+150 mesh	105 μm	1916	9.5%	79.3%	20.7%	655	0.4%	12.2%	87.8%	25	0.4%	5.8%	94.2%	73	0.1%	1.7%	98.3%
-150 mesh	105 μm	1554	7.7%	87.0%	13.0%	3850	1.7%	13.8%	86.2%	49	0.6%	6.4%	93.6%	200	0.2%	1.9%	98.1%
Concentrates																	
+3/8 in.	9.50 mm	222	1.1%	88.1%	11.9%	400,700	24.9%	38.7%	61.3%	10,000	17.2%	23.5%	76.5%	316,800	42.9%	44.8%	55.2%
+4 mesh	4.76 mm	232	1.1%	89.2%	10.8%	349,000	22.7%	61.5%	38.5%	9400	16.9%	40.4%	59.6%	275,000	38.9%	83.7%	16.3%
+16 mesh	1.19 mm	538	2.7%	91.9%	8.1%	145,100	21.9%	83.3%	16.7%	12,000	49.9%	90.3%	9.7%	37,700	12.4%	96.1%	3.9%
+50 mesh	297 μm	786	3.9%	95.7%	4.3%	36,800	8.1%	91.4%	8.6%	800	4.9%	95.1%	4.9%	6500	3.1%	99.2%	0.8%
+150 mesh	105 μm	447	2.2%	98.0%	2.0%	37,800	4.7%	96.2%	3.8%	1000	3.4%	98.6%	1.4%	2000	0.5%	99.7%	0.3%
-150 mesh	105 μm	414	2.0%	100.0%	0.0%	33,200	3.8%	100.0%	0.0%	440	1.4%	100.0%	0.0%	1070	0.3%	100.0%	0.0%
Total		20,241				17,644	100.0%			640	100.0%			8106	100.0%		

Table A.6 Treatability Results For NAS Miramar Sample B

Fraction / Particle Size	grams (dwb)	Wt% (dwb)	Retain %	Pass %	Pb mg/kg	Pb %	Retain %	Pass %	Sb mg/kg	Sb %	Retain %	Pass %	Cu mg/kg	Cu %	Retain %	Pass %
Humates																
+3/8 in. 9.50 mm	85	0.35%	0.35%	99.6%	2563	0.2%	0.2%	99.8%	59	0.1%	0.1%	99.9%	262	0.0%	0.0%	100.0%
Tailings																
+3/8in. 9.50 mm	714	2.9%	3.3%	96.7%	72	0.0%	0.2%	99.8%	36	0.4%	0.5%	99.5%	25	0.0%	0.0%	100.0%
+4 mesh 4.76 mm	418	1.7%	5.0%	95.0%	130	0.0%	0.2%	99.8%	24	0.2%	0.6%	99.4%	28	0.0%	0.1%	99.9%
+16 mesh 1.19 mm	950	3.9%	8.9%	91.1%	3887	2.6%	2.8%	97.2%	129	1.9%	2.5%	97.5%	75	0.1%	0.1%	99.9%
+50 mesh 297 μm	2400	9.9%	18.8%	81.2%	192	0.3%	3.1%	96.9%	19	0.7%	3.2%	96.8%	13	0.0%	0.2%	99.8%
+150 mesh 105 μm	3130	12.9%	31.7%	68.3%	237	0.5%	3.6%	96.4%	29	1.4%	4.6%	95.4%	23	0.1%	0.2%	99.8%
-150 mesh 105 μm	10,311	42.5%	74.2%	25.8%	2370	16.9%	20.5%	79.5%	42	6.7%	11.4%	88.6%	160	1.8%	2.0%	98.0%
Middlings																
+50 mesh 297 μm	1594	6.6%	80.8%	19.2%	184	0.20%	20.7%	79.3%	18	0.4%	11.8%	88.2%	13	0.0%	2.0%	98.0%
+150 mesh 105 μm	2618	10.8%	91.5%	8.5%	194	0.35%	21.0%	79.0%	19	0.8%	12.6%	87.4%	23	0.1%	2.1%	97.9%
-150 mesh 105 μm	0	0.0%	91.5%	8.5%		0.00%	21.0%	79.0%		0.0%	12.6%	87.4%		0.0%	2.1%	97.9%
Concentrates																
+3/8 in. 9.50 mm	217	0.9%	92.4%	7.6%	242,800	36.3%	57.3%	42.7%	13,400	45.0%	57.6%	42.4%	309,800	71.8%	73.9%	26.1%
+4 mesh 4.76 mm	110	0.5%	92.9%	7.1%	180,200	13.7%	71.0%	29.0%	8900	15.2%	72.7%	27.3%	193,400	22.7%	96.6%	3.4%
+16 mesh 1.19 mm	427	1.8%	94.6%	5.4%	53,600	15.8%	86.8%	13.2%	2900	19.2%	91.9%	8.1%	5500	2.5%	99.1%	0.9%
+50 mesh 297 μm	368	1.5%	96.2%	3.8%	18,100	4.6%	91.4%	8.6%	600	3.4%	95.4%	4.6%	400	0.2%	99.3%	0.7%
+150 mesh 105 μm	712	2.9%	99.1%	0.9%	10,200	5.0%	96.5%	3.5%	300	3.3%	98.7%	1.3%	700	0.5%	99.8%	0.2%
-150 mesh 105 μm	219	0.90%	100.0%	0.0%	23,400	3.5%	100.0%	0.0%	390	1.3%	100.0%	0.0%	690	0.2%	100.0%	0.0%
Total	24,273				5966	100.0%			266	100.0%			3850	100.0%		

Table A.7 Treatability Results For Twentynine Palms Sample LR-23

Fraction	Particle Size	grams (dwb)	Wt% (dwb)	Retain %	Pass %	Pb mg/kg	Pb %	Retain %	Pass %	Sb mg/kg	Sb %	Retain %	Pass %	Cu mg/kg	Cu %	Retain %	Pass %
Humates																	
+3/8 in.	9.50 mm	0	0.0%	0.0%	100.0%	0	0.0%	0.0%	100.0%	0	0.0%	0.0%	100.0%	0	0.0%	0.0%	100.0%
Tailings																	
+3/8in.	9.50 mm	2689	8.5%	8.5%	91.5%	46	0.0%	0.0%	100.0%	19	0.6%	0.6%	99.4%	9	0.0%	0.0%	100.0%
+4 mesh	4.76 mm	1840	5.8%	14.4%	85.6%	255	0.1%	0.2%	99.8%	24	0.5%	1.1%	98.9%	15	0.0%	0.0%	100.0%
+16 mesh	1.19 mm	4100	13.0%	27.4%	72.6%	314	0.4%	0.6%	99.4%	23	1.0%	2.1%	97.9%	26	0.0%	0.1%	99.9%
+50 mesh	297 µm	9070	28.8%	56.2%	43.8%	674	1.9%	2.5%	97.5%	26	2.6%	4.7%	95.3%	21	0.1%	0.1%	99.9%
+150 mesh	105 µm	3513	11.2%	67.3%	32.7%	418	0.5%	2.9%	97.1%	27	1.1%	5.8%	94.2%	56	0.1%	0.2%	99.8%
-150 mesh	105 µm	1404	4.5%	71.8%	28.2%	20,800	9.0%	11.9%	88.1%	270	4.2%	10.0%	90.0%	940	0.5%	0.7%	99.3%
Middlings																	
+50 mesh	297 µm	1732	5.5%	77.3%	22.7%	1318	0.7%	12.6%	87.4%	41	0.8%	10.8%	89.2%	45	0.0%	0.8%	99.2%
+150 mesh	105 µm	1673	5.3%	82.6%	17.4%	768	0.4%	13.0%	87.0%	27	0.5%	11.3%	88.7%	87	0.1%	0.8%	99.2%
-150 mesh	105 µm	1165	3.7%	86.3%	13.7%	8200	2.9%	15.9%	84.1%	95	1.2%	12.5%	87.5%	420	0.2%	1.0%	99.0%
Concentrates																	
+3/8 in.	9.50 mm	67	0.2%	86.5%	13.5%	149,400	3.1%	19.0%	81.0%	4300	3.2%	15.7%	84.3%	343,300	9.0%	10.0%	90.0%
+4 mesh	4.76 mm	464	1.5%	88.0%	12.0%	92,000	13.1%	32.2%	67.8%	6800	35.1%	50.8%	49.2%	400,700	72.5%	82.5%	17.5%
+16 mesh	1.19 mm	1177	3.7%	91.7%	8.3%	59,500	21.6%	53.8%	46.2%	1100	14.4%	65.2%	34.8%	23,100	10.6%	93.1%	6.9%
+50 mesh	297 µm	990	3.1%	94.9%	5.1%	64,400	19.6%	73.4%	26.6%	1500	16.5%	81.7%	18.3%	12,400	4.8%	97.9%	2.1%
+150 mesh	105 µm	1209	3.8%	98.7%	1.3%	53,000	19.7%	93.1%	6.9%	900	12.1%	93.7%	6.3%	3500	1.6%	99.5%	0.5%
-150 mesh	105 µm	411	1.3%	100.0%	0.0%	54,500	6.9%	100.0%	0.0%	1370	6.3%	100.0%	0.0%	2960	0.5%	100.0%	0.0%
Total		31,501				10,309	100.0%			286	100.0%			8145	100.0%		

Table A.8 Treatability Results For Twentynine Palms Sample LR-30

Fraction / Particle Size	grams (dwb)	Wt% (dwb)	Retain %	Pass %	Pb mg/kg	Pb %	Retain %	Pass %	Sb mg/kg	Sb %	Retain %	Pass %	Cu mg/kg	Cu %	Retain %	Pass %
Humates																
+3/8 in.	0	0.0%	0.0%	100.0%	0	0.0%	0.0%	100.0%	0	0.0%	0.0%	100.0%	0	0.0%	0.0%	100.0%
Tailings																
+3/8in.	2109	7.0%	7.0%	93.0%	85	0.0%	0.0%	100.0%	27	0.2%	0.2%	99.8%	8	0.0%	0.0%	100.0%
+4 mesh	1401	4.6%	11.6%	88.4%	386	0.1%	0.2%	99.8%	24	0.1%	0.4%	99.6%	17	0.0%	0.0%	100.0%
+16 mesh	3704	12.2%	23.8%	76.2%	820	0.7%	0.9%	99.1%	30	0.5%	0.8%	99.2%	126	0.1%	0.1%	99.9%
+50 mesh	10,281	34.0%	57.8%	42.2%	599	1.5%	2.3%	97.7%	23	1.0%	1.8%	98.2%	38	0.1%	0.2%	99.8%
+150 mesh	2205	7.3%	65.1%	34.9%	525	0.3%	2.6%	97.4%	29	0.3%	2.1%	97.9%	62	0.0%	0.2%	99.8%
-150 mesh	2693	8.9%	74.0%	26.0%	24,300	15.4%	18.1%	81.9%	430	4.9%	7.0%	93.0%	1090	0.5%	0.7%	99.3%
Middlings																
+50 mesh	1541	5.1%	79.0%	21.0%	1483	0.5%	18.6%	81.4%	42	0.3%	7.3%	92.7%	48	0.0%	0.7%	99.3%
+150 mesh	1221	4.0%	83.1%	16.9%	941	0.3%	18.9%	81.1%	43	0.2%	7.5%	92.5%	101	0.0%	0.8%	99.2%
-150 mesh	0	0.0%	83.1%	16.9%		0.0%	18.9%	81.1%		0.0%	7.5%	92.5%		0.0%	0.8%	99.2%
Concentrates																
+3/8 in.	178	0.6%	83.7%	16.3%	54,600	2.3%	21.2%	78.8%	1100	0.8%	8.3%	91.7%	297,900	9.7%	10.4%	89.6%
+4 mesh	848	2.8%	86.5%	13.5%	77,800	15.6%	36.7%	63.3%	9000	32.1%	40.4%	59.6%	385,000	59.4%	69.8%	30.2%
+16 mesh	1878	6.2%	92.7%	7.3%	57,500	25.5%	62.2%	37.8%	3700	29.2%	69.6%	30.4%	63,300	21.6%	91.4%	8.6%
+50 mesh	1142	3.8%	96.4%	3.6%	69,800	18.8%	81.1%	18.9%	3700	17.8%	87.4%	12.6%	34,500	7.2%	98.6%	1.4%
+150 mesh	714	2.4%	98.8%	1.2%	81,700	13.8%	94.8%	5.2%	2200	6.6%	94.0%	6.0%	7700	1.0%	99.6%	0.4%
-150 mesh	366	1.2%	100.0%	0.0%	59,900	5.2%	100.0%	0.0%	3900	6.0%	100.0%	0.0%	5750	0.4%	100.0%	0.0%
Total	30,281				13,989	100.0%			785	100.0%			18,145	100.0%		

Table A.9 Treatability Results for Fort Dix Army Base Sample BK-1

Fraction / Particle Size	grams (dwb)	Wt% (dwb)	Retain %	Pass %	Pb mg/kg	Pb %	Retain %	Pass %	Sb mg/kg	Sb %	Retain %	Pass %	Cu mg/kg	Cu %	Retain %	Pass %
Humates																
+3/8 in.	5	0.0%	0.0%	100.0%	13,700	0.1%	0.1%	99.9%	88	0.0%	0.0%	100.0%	676	0.0%	0.0%	100.0%
Tailings																
+3/8in.	55	0.2%	0.2%	99.8%	742	0.1%	0.2%	99.8%	22	0.1%	0.1%	99.9%	24	0.0%	0.0%	100.0%
+4 mesh	265	1.0%	1.2%	98.8%	58	0.0%	0.2%	99.8%	17	0.3%	0.4%	99.6%	9	0.0%	0.0%	100.0%
+16 mesh	5679	21.8%	23.0%	77.0%	231	1.9%	2.1%	97.9%	21	6.9%	7.2%	92.8%	20	0.1%	0.1%	99.9%
+50 mesh	3468	13.3%	36.3%	63.7%	98	0.5%	2.6%	97.4%	16	3.2%	10.5%	89.5%	8	0.0%	0.1%	99.9%
+150 mesh	6508	25.0%	61.3%	38.7%	75	0.7%	3.3%	96.7%	21	7.9%	18.3%	81.7%	6	0.0%	0.1%	99.9%
-150 mesh	1396	5.4%	66.6%	33.4%	4240	8.5%	11.8%	88.2%	50	4.0%	22.4%	77.6%	370	0.3%	0.4%	99.6%
Middlings																
+50 mesh	1780	6.8%	73.5%	26.5%	95	0.2%	12.0%	88.0%	12	1.2%	23.6%	76.4%	5	0.0%	0.4%	99.6%
+150 mesh	3678	14.1%	87.6%	12.4%	67	0.4%	12.4%	87.6%	14	3.0%	26.6%	73.4%	4	0.0%	0.4%	99.6%
-150 mesh	925	3.5%	91.1%	8.9%	1000	1.3%	13.7%	86.3%	20	1.1%	27.7%	72.3%	50	0.0%	0.4%	99.6%
Concentrates																
+3/8 in.	393	1.5%	92.6%	7.4%	104,100	58.8%	72.5%	27.5%	1600	36.4%	64.0%	36.0%	429,200	95.3%	95.7%	4.3%
+4 mesh	37	0.1%	92.8%	7.2%	147,100	7.8%	80.3%	19.7%	6000	12.8%	76.8%	23.2%	167,000	3.5%	99.2%	0.8%
+16 mesh	334	1.3%	94.0%	6.0%	19,900	9.5%	89.8%	10.2%	900	17.4%	94.1%	5.9%	3900	0.7%	99.9%	0.1%
+50 mesh	305	1.2%	95.2%	4.8%	14,100	6.2%	96.0%	4.0%	300	5.3%	99.4%	0.6%	200	0.0%	99.9%	0.1%
+150 mesh	959	3.7%	98.9%	1.1%	1700	2.3%	98.3%	1.7%	0	0.0%	99.4%	0.6%	100	0.1%	100.0%	0.0%
-150 mesh	288	1.1%	100.0%	0.0%	4000	1.7%	100.0%	0.0%	35	0.6%	100.0%	0.0%	105	0.0%	100.0%	0.0%
Total	26,074				2669	100.0%			66	100.0%			6796	100.0%		

Table A.10 Treatability Results for Fort Dix Army Base Sample BK-3

Fraction / Particle Size	grams (dwb)	Wt% (dwb)	Retain %	Pass %	Pb mg/kg	Pb %	Retain %	Pass %	Sb mg/kg	Sb %	Retain %	Pass %	Cu mg/kg	Cu %	Retain %	Pass %
Humates																
+3/8 in. 9.50 mm	4	0.0%	0.0%	100.0%	4430	0.0%	0.0%	100.0%	50	0.0%	0.0%	100.0%	260	0.0%	0.0%	100.0%
Tailings																
+3/8in. 9.50 mm	33	0.1%	0.1%	99.9%	76	0.0%	0.0%	100.0%	21	0.0%	0.0%	100.0%	7	0.0%	0.0%	100.0%
+4 mesh 4.76 mm	168	0.7%	0.8%	99.2%	95	0.0%	0.0%	100.0%	15	0.1%	0.2%	99.8%	5	0.0%	0.0%	100.0%
+16 mesh 1.19 mm	6358	25.6%	26.4%	73.6%	136	0.9%	0.9%	99.1%	15	4.3%	4.4%	95.6%	20	0.1%	0.1%	99.9%
+50 mesh 297 μm	4660	18.8%	45.2%	54.8%	63	0.3%	1.2%	98.8%	17	3.5%	8.0%	92.0%	5	0.0%	0.1%	99.9%
+150 mesh 105 μm	4788	19.3%	64.5%	35.5%	96	0.5%	1.7%	98.3%	20	4.3%	12.2%	87.8%	7	0.0%	0.1%	99.9%
-150 mesh 105 μm	1517	6.1%	70.6%	29.4%	1840	2.9%	4.6%	95.4%	26	1.8%	14.0%	86.0%	116	0.1%	0.2%	99.8%
Middlings																
+50 mesh 297 μm	2596	10.5%	81.0%	19.0%	95	0.3%	4.9%	95.1%	12	1.4%	15.4%	84.6%	4	0.0%	0.2%	99.8%
+150 mesh 105 μm	2477	10.0%	91.0%	9.0%	78	0.2%	5.1%	94.9%	14	1.5%	16.9%	83.1%	5	0.0%	0.2%	99.8%
-150 mesh 105 μm	0	0.0%	91.0%	9.0%		0.0%	5.1%	94.9%		0.0%	16.9%	83.1%		0.0%	0.2%	99.8%
Concentrates																
+3/8 in. 9.50 mm	378	1.5%	92.5%	7.5%	177,000	69.6%	74.6%	25.4%	2800	47.3%	64.3%	35.7%	483,300	92.2%	92.4%	7.6%
+4 mesh 4.76 mm	38	0.2%	92.7%	7.3%	244,100	9.7%	84.4%	15.6%	12,200	20.9%	85.2%	14.8%	242,900	4.7%	97.1%	2.9%
+16 mesh 1.19 mm	208	0.8%	93.5%	6.5%	19,700	4.3%	88.6%	11.4%	600	5.6%	90.8%	9.2%	26,000	2.7%	99.9%	0.1%
+50 mesh 297 μm	663	2.7%	96.2%	3.8%	13,300	9.2%	97.8%	2.2%	200	5.9%	96.7%	3.3%	300	0.1%	100.0%	0.0%
+150 mesh 105 μm	649	2.6%	98.8%	1.2%	2200	1.5%	99.3%	0.7%	100	2.9%	99.6%	0.4%	100	0.0%	100.0%	0.0%
-150 mesh 105 μm	298	1.2%	100.0%	0.0%	2300	0.7%	100.0%	0.0%	29	0.4%	100.0%	0.0%	63	0.0%	100.0%	0.0%
Total	24,835				3875	100.0%			90	100.0%			7980	100.0%		

Table A.11 Description of Size Fractions from NAS Miramar A

Humates/light

> 3/8 in. Humates	Grass and plant stalks
Tails	
> 3/8 in. light	Rock fragments
> 4 mesh light	Rock fragments
> 16 mesh light	Mainly rock fragments with a high percentage of quartz, wide range of color
> 50 mesh light	99% quartz
> 150 mesh light	99% gravel
Middlings	
> 50 mesh mid	Mainly light color gravel, some black rock
> 150 mesh mid	Mainly light color river sands, some black rock
< 150 mesh mid	Multi-color river sands
Heavies	
> 3/8 in. heavy	Lead, copper jacketed lead, trace of humates
> 4 mesh heavy	All lead and copper
> 16 mesh heavy	Multi-color rock fragments with numerous lead pieces
> 50 mesh heavy	Mainly quartz, small amounts of red and black color rock, trace of lead
> 150 mesh heavy	Over 90% quartz, less than 10% other type of rocks, no visible lead
< 150 mesh heavy	Over 90% quartz, less than 10% other type of rocks, no visible lead

Table A.12 Description of Size Fractions from NAS Miramar B

Humates/light

> 3/8 in. humates	Grass and plant roots
Tails	
> 3/8 in. light	Multi-color rock fragments
> 4 mesh light	Multi-color rock fragments
> 16 mesh light	Multi-color rock fragments, trace of lead
> 50 mesh light	Mainly quartz, trace of humates and mica
> 150 mesh light	Mainly quartz, trace of mica
< 150 mesh light	
Middlings	
> 50 mesh mid	Over 90% quartz, some black rock
> 150 mesh mid	Over 90% quartz, some black rock, trace of mica
Heavies	
> 3/8 in. heavy	All lead and copper, copper in excess, one whole bullet
> 4 mesh heavy	All lead and copper, lead in excess
> 16 mesh heavy	Multi-color rock fragments with some lead pieces
> 50 mesh heavy	Over 95% quartz, one small piece of lead
> 150 mesh heavy	Over 90% quartz, some black rocks, trace of mica
< 150 mesh heavy	Over 90% quartz, some black rocks, trace of mica

Table A.13 Description of Size Fractions from Twentynine Palms LR-23

Humates/light	
> 3/8 in. humates	Large piece of gravel
Tails	
> 3/8 in. light	All rock fragments
> 4 mesh light	99% light color rock fragments, trace of iron
> 16 mesh light	Rock fragments with light color
> 50 mesh light	90% quartz, trace of mica
> 150 mesh light	90% quartz, trace of mica
< 150 mesh light	99% clay
Middlings	
> 50 mesh mid	Quartz, trace of black rock
> 150 mesh mid	Quartz, more black rock
Heavies	
> 3/8 in. heavy	Copper and lead with copper in excess
> 4 mesh heavy	Mainly lead and copper, a few iron tips
> 16 mesh heavy	Rock fragments with numerous copper and lead pieces, a few iron tips
> 50 mesh heavy	Quartz with black rock, some lead, copper, and iron
> 150 mesh heavy	Mainly quartz and black rocks, a few microscopic lead pieces
< 150 mesh heavy	

Table A.14 Description of Size Fractions from Twentynine Palms LR-30

Tails	
> 3/8 in. light	Light color rock fragments
> 4 mesh light	Light color rock fragments
> 16 mesh light	95% quartz, black rock, trace of iron
> 50 mesh light	Quartz with a trace of black colored rock fragments
> 150 mesh light	Quartz, trace of black rock, trace of mica
< 150 mesh light	Mainly clay, some quartz
Middlings	
> 50 mesh mid	Mainly quartz, some black rock
> 150 mesh mid	Mainly quartz, some black rock, trace of mica
Heavies	
> 3/8 in. heavy	Mainly copper jacketed bullets, some copper pieces
> 4 mesh heavy	Light colored rock fragments
> 16 mesh heavy	Quartz with some black rock, trace of iron
> 50 mesh heavy	Quartz with some black rock
> 150 mesh heavy	Quartz with same black rock, trace of mica
< 150 mesh heavy	Mainly clay, some quartz

Table A.15 Description of Size Fractions from Fort Dix BK-1

Humates/light

> 3/8 in. humates	Fine humates

Tails

> 3/8 in. light	Gravel, some with clay coating, one piece of carbon
> 4 mesh light	99% gravel, trace of humates, one piece of carbon
> 16 mesh light	Quartz, few pieces of carbon
> 50 mesh light	95% quartz, some carbon, trace of humates
> 150 mesh light	Quartz, some mica, trace of carbon
< 150 mesh light	Mainly clay, quartz, some mica and trace of carbon

Middlings

> 50 mesh mid	Quartz, some black rock, trace of mica
> 150 mesh mid	Quartz, some black rock, trace of mica

Heavies

> 3/8 in. heavy	Copper jacketed bullets, trace of humates
> 4 mesh heavy	Lead and copper, trace of iron
> 16 mesh heavy	90% quartz, some lead pieces, a few copper pieces
> 50 mesh heavy	Quartz, some black rock, trace of iron and lead
> 150 mesh heavy	Quartz, some black rock
< 150 mesh heavy	50% quartz, 50% black rock

Table A.16 Description of Size Fractions from Fort Dix BK-3

Humates/light

> 3/8 in. humates	Fine humates

Tails

> 3/8 in. light	Mainly gravel, some agglomerated clay, a few pieces of pumice and brick, one piece of pottery chip
> 4 mesh light	99% gravel, one piece of brick
> 16 mesh light	99% quartz, several carbon pieces
> 50 mesh light	99% quartz, several carbon pieces
> 150 mesh light	99% quartz, several carbon pieces
< 150 mesh light	Clay

Middlings

> 50 mesh mid	Quartz, some carbon pieces
> 150 mesh mid	Quartz, some carbon pieces

Heavies

> 3/8 in. heavy	Bullets, one piece of copper
> 4 mesh heavy	Lead, copper balls, copper jacketed lead
> 16 mesh heavy	Mainly quartz, a few pieces of copper
> 50 mesh heavy	Mainly quartz, some black rock
> 150 mesh heavy	Quartz and black rock
< 150 mesh heavy	Quartz and black rock

APPENDIX A.2 ATTRITION SCRUBBER DATA SHEETS

Sample: **Miramar "A" -4.76 mm by +1.19 mm (-4 mesh by +16-mesh) Tailings**

Lead Content: 2330 mg/kg

Instructions: Attrition scrub 500 grams at 67% solids for 5 minutes
                     Hand jig to generate clean tails sample
                     Screen tails (lights) on 16-mesh
                     Screen concentrates (heavies) on 16-mesh

Data

Sample	Weight		Lead		
	Grams	Distn	mg/kg	mg	Distn
"A" +1.19 mm (+16 mesh) Tailings	202.9	40.9%	290	58.8	16.3%
"A" -1.19 mm (-16 mesh) Tailings	79.3	16.0%	658	52.2	14.4%
"A" +1.19 mm (+16 mesh) Concentrate	144.3	29.1%	417	60.2	16.7%
"A" -1.19 mm (-16 mesh) Concentrate	33.2	6.7%	2920	96.9	26.8%
"A" -0.59 mm (-30 mesh)	35.9	7.2%	2592	93.1	25.8%
Calculated "A" +1.19 mm (+16 mesh) Total	347.2	70.1%	343	119.0	33.0%
Calculated "A" -1.19 mm (-16 mesh) Total	112.5	22.7%	1326	149.1	41.3%
Calculated "A" -0.59 mm (-30 mesh) Total	35.9	7.2%	2592	93.1	25.8%
Calculated Feed	495.6	100.0%	729	361.2	100.0%
Analytical Feed	500.0	100.0%	2330	1165.0	100.0%

Sample: **Miramar "B" -4.76 mm by +1.19 mm (-4 mesh by +16-mesh) Tailings**

Lead Content: 3887 mg/kg

Instructions: Attrition scrub 500 grams at 67% solids for 5 minutes
                     Hand jig to generate clean tails sample
                     Screen tails (lights) on 16-mesh
                     Screen concentrates (heavies) on 16-mesh

Data

Sample	Weight		Lead		
	Grams	Distn	mg/kg	mg	Distn
"B" +1.19 mm (+16 mesh) Tailings	191.8	38.6%	171	32.8	8.2%
"B" -1.19 mm (-16 mesh) Tailings	85.5	17.2%	675	57.7	14.5%
"B" +1.19 mm (+16 mesh) Concentrate	135.0	27.2%	1372	185.2	46.5%
"B" -1.19 mm (-16 mesh) Concentrate	36.1	7.3%	1653	59.7	15.0%
"B" -0.59 mm (-30 mesh)	48.0	9.7%	1312	63.0	15.8%
Calculated "B" +1.19 mm (+16 mesh) Total	326.8	65.8%	667	218.0	54.7%
Calculated "B" -1.19 mm (-16 mesh) Total	121.6	24.5%	965	117.4	29.5%
Calculated "B" -0.59 mm (-30 mesh) Total	48.0	9.7%	1312	63.0	15.8%
Calculated Feed	496.4	100.0%	803	398.4	100.0%
Analytical Feed	500.0	100.0%	3887	1943.5	100.0%

Summary of Samples A & B

Sample: Miramar "A" and "B" -4.76 mm by +1.19 mm (-4 mesh by +16-mesh) Tailings

Lead Content: 3109 mg/kg

Sample	Weight		Lead		
	Grams	Distn	mg/kg	mg	Distn
Averaged +1.19 mm (+16 mesh) Tailings	197.4	39.8%	231	45.8	12.3%
Averaged -1.19 mm (-16 mesh) Tailings	82.4	16.6%	667	54.9	14.5%
Averaged +1.19 mm (+16 mesh) Concentrate	139.7	28.2%	895	122.7	31.6%
Averaged -1.19 mm (-16 mesh) Concentrate	34.7	7.0%	2287	78.3	20.9%
Averaged -0.59 mm (-30 mesh)	42.0	8.5%	1952	78.0	20.8%
Averaged +1.19 mm (+16 mesh) Total	337.0	67.9%	505	168.5	43.8%
Averaged -1.19 mm (-16 mesh) Total	117.1	23.6%	1145	133.3	35.4%
Averaged -0.59 mm (-30 mesh) Total	42.0	8.5%	1952	78.0	20.8%
Averaged Calculated Feed	496.0	100.0%	766	379.8	100.0%
Averaged Analytical Feed	500.0	100.0%	3109	1554.3	100.0%

APPENDIX A.3 HYDROCYCLONE DATA SHEETS

Hydrocyclone Test Results for NAS Miramar; −150 mesh Tails from Sample A; 6028 mg Pb/kg

kPa (psig)	Stream	Grams	Distribution	Lead mg/kg	Split
21 (3)	Overflow				
	+37 μm (+400 mesh)	4.56	3.90%	6163	
	−37 μm (−400 mesh)	113.09	96.10%	6163	87.00%
	Underflow				
	+53 μm (+270 mesh)	31.46	21.40%	3049	
	+44 μm (+325 mesh)	9.38	6.40%	3049	
	+37 μm (+400 mesh)	27.57	18.70%	2824	
	−37 μm (−400 mesh)	78.73	53.50%	7009	13.00%
	Average Lead, mg/kg			**5126**	**6028**
35 (5)	Overflow				
	+37 μm (+400 mesh)	2.33	1.60%	6215	
	−37 μm (−400 mesh)	145.21	98.40%	6215	85.00%
	Underflow				
	+53 μm (+270 mesh)	27.29	18.40%	3493	
	+44 μm (+325 mesh)	8.44	5.70%	3493	
	+37 μm (+400 mesh)	23.58	15.90%	2685	
	−37 μm (−400 mesh)	89.36	60.10%	6179	15.00%
	Average Lead, mg/kg			**4979**	**6030**
69 (10)	Overflow				
	+37 μm (+400 mesh)	1.73	1.00%	6360	
	−37 μm (−400 mesh)	163.56	99.00%	6360	71.00%
	Underflow				
	+53 μm (+270 mesh)	33.68	15.00%	2892	
	+44 μm (+325 mesh)	9.21	4.10%	2892	
	+37 μm (+400 mesh)	39.13	17.40%	2918	
	−37 μm (−400 mesh)	142.83	63.50%	6545	29.00%
	Average Lead, mg/kg			**5217**	**6029**
104 (15)	Overflow				
	+37 μm (+400 mesh)	1.73	1.00%	6120	
	−37 μm (−400 mesh)	163.56	99.00%	6120	93.00%
	Underflow				
	+53 μm (+270 mesh)	33.68	15.00%	2754	
	+44 μm (+325 mesh)	9.21	4.10%	2754	
	+37 μm (+400 mesh)	39.13	17.40%	2921	
	−37 μm (−400 mesh)	142.83	63.50%	6067	7.00%
	Average Lead, mg/kg			**4888**	**6030**

Hydrocyclone Test Results for NAS Miramar; –105 micron (150 mesh) from Sample B; 2370 mg Pb/kg

Pressure kPa (psig)	Stream	Grams	Distribution	Lead mg/kg	Split
35 (5)	Overflow				
	+ 37 μm (+400 mesh)	3.67	3.40%	2725	
	−37 μm (−400 mesh)	104.14	96.60%	2725	78.00%
	Underflow				
	+53 μm (+270 mesh)	158.51	59.20%	546	
	+44 μm (+325 mesh)	14.63	5.50%	546	
	+37 μm (+400 mesh)	31.04	11.60%	1261	
	−37 μm (−400 mesh)	63.36	23.70%	2740	23.00%
	Average Lead, mg/kg			**1149**	**2370**
69 (10)	Overflow				
	+37 μm (+400 mesh)	1.83	2.20%	2768	
	−37 μm (−400 mesh)	82.71	97.80%	2768	72.00%
	Underflow				
	+53 μm (+270 mesh)	247.18	59.20	664	
	+44 μm (+325 mesh)	20.08	4.80	664	
	+37 μm (+400 mesh)	40.34	9.70	1497	
	−37 μm (−400 mesh)	109.70	26.30	2962	28.00%
	Average Lead, mg/kg			**1349**	**2370**
104 (15)	Overflow				
	+37 μm (+400 mesh)	1.39	1.70%	2709	
	−37 μm (−400 mesh)	79.62	98.30%	2709	76.00%
	Underflow				
	+53 μm (+270 mesh)	192.91	50.10%	515	
	+44 μm (+325 mesh)	19.12	5.00%	515	
	+37 μm (+400 mesh)	46.57	12.90%	1243	
	−37 μm (−400 mesh)	123.03	32.00%	2702	24.00%
	Average Lead, mg/kg			**1308**	**2370**

Hydrocyclone Data for NAS Miramar Samples A & B at 69 kPa gauge (10 psig)

Miramar A Sample Results

Overflow	Grams	Distn.	Accum Distn.	Pb mg/kg	Pb mg	Pb Distn.	Accum. Pb Distn.
+37 micron (+400 mesh)	1.73	0.44%	0.44%	6360	11.00	0.49%	0.49%
−37 micron (−400 mesh)	163.56	41.92%	42.37%	6360	1040.24	46.77%	47.26%
Overflow Total	165.29	42.37%		6360	1051.24	47.26%	
Underflow							
+53 micron (+270 mesh)	33.68	8.63%	51.00%	2892	97.40	4.38%	51.64%
+44 micron (+325 mesh)	9.21	2.36%	53.36%	2892	26.64	1.20%	52.84%
+37 micron (+400 mesh)	39.13	10.03%	63.39%	2918	114.18	5.13%	57.97%
−37 micron (−400 mesh)	142.83	36.61%	100.00%	6545	934.82	42.03%	100.00%
Underflow Total	224.85	57.63%		5217	1173.04	52.74%	
Total	390.14	100.00%		5701	2224.29		
Feed	390.14			5580	2176.98		

Miramar B Sample Results

Overflow	Grams	Distn.	Accum Distn.	Pb mg/kg	Pb mg	Pb Distn.	Accum. Pb Distn.
+37 micron (+400 mesh)	3.67	0.99%	0.99%	2725	10.00	1.67%	1.67%
−37 micron (−400 mesh)	104.14	27.97%	28.95%	2725	283.78	47.44%	49.11%
Overflow Total	107.81	28.95%		2725	293.78	49.11%	
Underflow							
+53 micron (+270 mesh)	155.51	41.76%	70.72%	546	84.91	14.19%	63.31%
+44 micron (+325 mesh)	14.63	3.93%	74.65%	546	7.99	1.34%	64.64%
+37 micron (+400 mesh)	31.04	8.34%	82.98%	1261	39.14	6.54%	71.19%
−37 micron (−400 mesh)	63.36	17.02%	100.00%	2720	172.34	28.81%	100.00%
Underflow Total	264.54	71.05%		1151	304.38	50.89%	
Total	372.35	100.00%		1606	598.16		
Feed	372.35			2370	882.47		

Miramar Combined Sample Results

Overflow	Grams	Distn.	Accum Distn.	Pb mg/kg	Pb mg	Pb Distn.	Accum. Pb Distn.
+37 micron (+400 mesh)	5.4	0.71%	0.71%	3890	21.00	0.74%	0.74%
−37 micron (−400 mesh)	267.7	35.11%	35.82%	4946	1324.02	46.91%	47.65%
Overflow Total	273.1	35.82%		4925	1345.03	47.65%	
Underflow							
+53 micron (+270 mesh)	189.19	24.81%	60.63%	964	182.31	6.46%	54.11%
+44 micron (+325 mesh)	23.84	3.13%	63.76%	1452	34.62	1.23%	55.34%
+37 micron (+400 mesh)	70.17	9.20%	72.96%	2185	153.32	5.43%	60.77%
−37 micron (−400 mesh)	206.19	27.04%	100.00%	5370	1107.16	39.23%	100.00%
Underflow Total	489.39	64.18%		3019	1477.42	52.35%	
Total	762.49	100.00%		3702	2822.45		
Feed	762.49			3870	2950.84		

APPENDIX A.4 PERCOLATION LEACH DATA SHEETS

Example Calculations

Reference Tests LR23-1 and LR 23-3

Grams / Liter Nitric Acid

1. Test LR 23-1
2. Calculate nitric acid solution concentration in g/L
3. Volume Reagent HNO_3 Added = 5.0 mL
4. Density Reagent HNO_3 = 1.42 g/mL
5. Reagent HNO_3 = 70.0%
6. Volume Leach Solution = 200 mL
7. Solution Concentration = ((5.0 mL)(1.42 g/mL)(70%/100%)(1000 mL/L))/(200 mL) = 24.9 g/L

Grams Nitric Acid (HNO₃) Added (Bottle Tests)

1. Test LR 23-1
2. Calculate amount of reagent nitric acid added
3. Volume Reagent HNO_3 Added = 5.0 mL
4. Density Reagent HNO_3 = 1.42 g/mL
5. Reagent HNO_3 Added = (5.0 mL)(1.42 g/mL) = 7.10 g

Grams Nitric Acid (HNO₃) Added (Percolation Column Tests)

1. Test LR 23-3
2. Calculate amount of reagent nitric acid added
3. Volume Time Sample "LR23P-3-1: 24 hr filtrate" = 780 mL
4. Reagent HNO_3 Added = ((780 mL)(2.0 g/L))/(1000 mL/L) = 1.56 g
5. Volume Final Sample "LR23P-3-4: 72 hr filtrate + wash" = 1310 mL
6. Volume Final Wash Water = 100 mL
7. Nitric Acid Concentration "LR23P-3" = 2.0 g/L
8. Reagent HNO_3 Added = ((1310 mL – 100 mL)(2.0 g/L))/(1000 mL/L) = 2.42 g

HNO₃, milliequivalents / gram soil

1. Test LR 23-1
2. Calculate acid milliequivalents of nitric acid added per gram of soil
3. Dry Wt. Soil = 100 g
4. Reagent HNO_3 Added = 7.1 g
5. Reagent HNO_3 = 70.0%
6. SUM[Wt. 100% HNO_3] = (7.1 g)(70%/100%) = 4.97 g
7. Milliequivalents Per gram Soil = ((4.97 g)(1000 meq/eq))/((63.01 g/eq)(100 grams/soil)) = 0.788 meq/g

HNO₃ Consumed

1. Test LR 23-1
2. Calculate the amount of 100% nitric acid consumed per 1000 kg (t) of soil
3. Dry Wt. Soil = 100 g
4. SUM["Grams Nitric Acid (HNO_3) Added"] = 7.10 g
5. Reagent HNO_3 = 70.0%
6. Nitric Acid Consumed = 49.7 ((7.1 g)(70%/100%)(1000 g/kg))/(100 g soil) = 49.7 g/kg or kg/1000 kg

16 × 50 mesh tailings + middlings

1. Test LR 23-1
2. Calculate the amount of lead in "tailings + middlings" in mg
3. Dry Wt. "tailings + middlings" = 100 g
4. Lead Analysis of "tailings + middlings" = 780 mg/kg
5. Calculated Lead Content = ((100 g)(780 mg/kg))/(1000 g/kg) = 78 mg Pb

"LR23P"-1-1 24 hr filtrate + wash

1. Calculate the amount of lead in "24 hr filtrate + wash" in mg
2. Volume "24 hr filtrate + wash" = 267 mL
3. Lead Analysis "24 hr filtrate + wash" = 292 mg/L
4. Calculated Lead Content = ((267 mL)(292 mg/L))/(1000 ml/L) = 78 mg Pb

"LR23P"-1-1 24 hr residue

1. Calculate the amount of lead in "24 hr residue" in mg
2. Dry Wt. "24 hr residue" = 98 g
3. Lead Analysis "24 hr residue" = 30 mg/kg
4. Calculated Lead Content = ((98 g)(30 mg/kg))/(1000 g/kg) = 3 mg Pb

Calculated Feed Analysis

1. Calculated Feed Analysis in mg/kg
2. Dry Wt. Soil = 100 grams
3. Calculated Feed = ((78 mg Pb + 3 mg Pb)(1000 g/kg))/(100 g) = 809 mg/kg

Calculated Leach Recovery

1. Calculated Leach Recovery in Wt.%
2. Calculated Recovery = ((78 mg Pb)(100))/(78 mg Pb + 3 mg Pb) = 96%

PERCOLATION LEACH DATA SHEET

Sample: 84 grams Twentynine Palms LR-23, 16 x 50 mesh tailings
16 grams Twentynine Palms LR-23, 16 x 50 mesh middlings
100 grams Twentynine Palms tailings + middlings

Instructions: Add the 100 grams of sample into 200 mL water in an 8 ounce bottle
Add 5.0 mL reagent nitric acid 24.9 g/L nitric acid
Place bottle on rolls and roll for 24 hours
Measure solution pH and readjust to <2.0 if necessary at 3 and 6 hours.
At 24 hours measure pH, flocculate, filter and wash with 100 mL water
Combine filtrate and wash, measure volume submit for lead analysis
Get wet and dry weight of filter cake, submit for lead analysis

Data:

Leach Time, hours	0	4.5		25		
Clock Time	11:00	15:30		12:00		
Solution Temperature, °C	25	29		27		
Solution pH		0.74		0.50		
Grams Nitric Acid (HNO_3) Added	7.10					
HNO_3, milliequivalents/gram soil	0.789	0.789	0.789	0.789		
HNO_3 Consumed	100% HNO_3	49.7	kg/1000 kg dry soil			

Test No.:	"LR23P"-1	Dry Wt.	Solid or Solution Analysis or Mass			
		or Volume	mg/kg, mg/L, or mg			
Sample ID	Description	g or mL	Pb			
16x50 mesh tailings + middlings		100	780			
			78	0	0	0	0
"LR23P"-1-1 24 hr filtrate + wash		267	292			
			78	0	0	0	0
"LR23P"-1-2 24 hr residue		98	30			
			3	0	0	0	0
Calculated Feed Analysis		–2%	809	0	0	0	0
Calculated Leach Recovery			96%			

PERCOLATION LEACH DATA SHEET

Sample: 336 grams Twentynine Palms LR-23, 16 x 50 mesh tailings
                           64 grams Twentynine Palms LR-23, 16 x 50 mesh middlings
                           400 grams Twentynine Palms tailings + middlings

Instructions: Charge 400 grams coarse soil to percolation column apparatus
                           Pump 5.0 g/L HNO_3 to top of apparatus at 0.5 mL/min
                           Flow lixiviant for 72 hours
                           At 72 hours wash residue with 100 mL water
                           Combine filtrate and wash, measure volume submit for lead analysis
                           Get wet and dry weight of filter cake, submit for lead analysis

Data:

Leach Time, hours	0					
Clock Time	15:30					
Solution Temperature, °C	25					
Solution pH						
Grams Nitric Acid (HNO_3) Added	8.75					
HNO_3, milliequivalents/gram soil	0.972	0.972	0.972	0.972		
HNO_3 Consumed	100% HNO_3	15.3	kg/1000 kg dry soil			

Test No.:	"LR23P"-2	Dry Wt. or Volume	Solid or Solution Analysis or Mass mg/kg, mg/L, or mg				
Sample ID	**Description**	**g or mL**	**Pb**				
16x50 mesh tailings + middlings		400	780				
			312	0	0	0	0
"LR23P"-2-1 72 hr filtrate + wash		1850	128				
			237	0	0	0	0
"LR23P"-2-2 72 hr residue		386	46				
			18	0	0	0	0
Calculated Feed Analysis		−3%	**636**	**0**	**0**	**0**	**0**
Calculated Leach Recovery			**93%**				

PERCOLATION LEACH DATA SHEET

Sample: 336 grams Twentynine Palms LR-23, 16 x 50 mesh tailings
64 grams Twentynine Palms LR-23, 16 x 50 mesh middlings
400 grams Twentynine Palms tailings + middlings

Instructions: Charge 400 grams coarse soil to percolation column apparatus
Pump 2.0 g/L HNO_3 to top of apparatus at 0.5 mL/min
Flow lixiviant for 72 hours, sample discharge at 24 and 48 hours
At 72 hours wash residue with 100 mL water
Combine filtrate and wash, measure volume submit for lead analysis
Get wet and dry weight of filter cake, submit for lead analysis

Data:

Leach Time, hours	0	24	48	72		
Date	16-Jul	17-Jul	18-Jul	19-Jul		
Clock Time	16:40	16:40	16:40	16:40		
Solution Temperature, °C	25					
Solution pH						
Grams Nitric Acid (HNO_3) Added	1.6	1.1	2.42			
HNO_3, milliequivalents/gram soil	0.173	0.296	0.564	0.564		
HNO_3 Consumed	100% HNO_3	8.9	kg/1000 kg dry soil			

Test No.: "LR23P"-3		Dry Wt. or Volume	Solid or Solution Analysis or Mass mg/kg, mg/L, or mg				
Sample ID	Description	g or mL	Pb				
16x50 mesh tailings + middlings		400	780				
			312	0	0	0	0
"LR23P"-3-1 24 hr filtrate + wash		780	144				
			112	0	0	0	0
"LR23P"-3-2 48 hr residue		550	133				
			73	0	0	0	0
"LR23P"-3-3 72 hr filtrate + wash		1310	35				
			46	0	0	0	0
"LR23P"-3-4 72 hr residue		394	30				
			12	0	0	0	0
Calculated Feed Analysis		−1%	**608**	0	0	0	0
Calculated Leach Recovery			**95%**				

PERCOLATION LEACH DATA SHEET

Sample: 8 grams Twentynine Palms LR-30, 3/8-in. x 4 mesh tails
                         22 grams Twentynine Palms LR-30, 4 x 16 mesh tails
                         61 grams Twentynine Palms LR-30, 16 x 50 mesh tails
                         9 grams Twentynine Palms LR-30, 16 x 50 mesh mids
                         100 grams Twentynine Palms tailings + middlings

Instructions: Add the 100 grams of sample into 200 mL water in an 8 ounce bottle
                         Add 5.0 mL reagent nitric acid 24.9 g/L nitric acid
                         Place bottle on rolls and roll for 24 hours
                         Measure solution pH and readjust to <2.0 if necessary at 3 and 6 hours.
                         At 24 hours measure pH, flocculate, filter and wash with 100 mL water
                         Combine filtrate and wash, measure volume submit for lead analysis
                         Get wet and dry weight of filter cake, submit for lead analysis

Data:

Leach Time, hours	0	4.5		25		
Clock Time	11:00	15:30		12:00		
Solution Temperature, °C	25	29		27		
Solution pH		0.66		0.43		
Grams Nitric Acid (HNO_3) Added	7.10					
HNO_3, milliequivalents / gram soil	0.789	0.789	0.789	0.789		
HNO_3 Consumed	100% HNO_3	49.7	kg/1000 kg soil added			

Test No:	"LR30P"-1	Dry Wt.	Solid or Solution Analysis or Mass			
		or Volume	mg/kg, mg/L, or mg			
Sample ID	Description	g or mL	Pb			
3/8x50 mesh tailings + middlings		100	710			
			71	0	0	0	0
"LR30P"-1-1 24 hr filtrate + wash		275	310			
			85	0	0	0	0
"LR30P"-1-2 24 hr residue		98	30			
			3	0	0	0	0
Calculated Feed Analysis		–2%	**882**	**0**	**0**	**0**	**0**
Calculated Leach Recovery			**97%**			

PERCOLATION LEACH DATA SHEET

Sample: 32 grams Twentynine Palms LR-30, 3/8-in. x 4-mesh tails
88 grams Twentynine Palms LR-30, 4- x 16-mesh tails
244 grams Twentynine Palms LR-30, 16- x 50-mesh tails
36 grams Twentynine Palms LR-30, 16- x 50-mesh mids
400 grams Twentynine Palms tailings + middlings

Instructions: Charge 400 grams coarse soil to percolation column apparatus
Pump 5.0 g/L HNO_3 to top of apparatus at 0.5 mL/min
Flow lixiviant for 72 hours
At 72 hours wash residue with 200 mL water
Combine filtrate and wash, measure volume submit for lead analysis
Get wet and dry weight of filter cake, submit for lead analysis

Data:

Leach Time, hours	0					
Clock Time	15:30					
Solution Temperature, °C	25					
Solution pH						
Grams Nitric Acid (HNO_3) Added	8.80					
HNO_3, milliequivalents / gram soil	0.978	0.978	0.978	0.978		
HNO_3 Consumed	100% HNO_3	15.4	kg/1000 kg soil added			

Test No:	"LR30P"-2	Dry Wt. or Volume	Solid or Solution Analysis or Mass mg/kg, mg/L, or mg				
Sample ID	Description	g or mL	Pb				
3/8x50 mesh tailings + middlings		400	710				
			284	0	0	0	0
"LR30P"-2-1 72 hr filtrate + wash		1960	193				
			378	0	0	0	0
"LR30P"-2-2 72 hr residue		396	53				
			21	0	0	0	0
Calculated Feed Analysis		-1%	**998**	**0**	**0**	**0**	**0**
Calculated Leach Recovery			**95%**				

PERCOLATION LEACH DATA SHEET

Sample: 32 grams Twentynine Palms LR-30, 3/8-in. x 4-mesh tails
                      88 grams Twentynine Palms LR-30, 4- x 16-mesh tails
                      244 grams Twentynine Palms LR-30, 16- x 50-mesh tails
                      36 grams Twentynine Palms LR-30, 16- x 50-mesh mids
                      400 grams Twentynine Palms tailings + middlings

Instructions: Charge 400 grams coarse soil to percolation column apparatus
                      Pump 2.0 g/L HNO_3 to top of apparatus at 0.5 mL/min
                      Flow lixiviant for 72 hours, sample discharge at 24 and 48 hours
                      At 72 hours wash residue with 100 mL water
                      Combine filtrate and wash, measure volume submit for lead analysis
                      Get wet and dry weight of filter cake, submit for lead analysis

Data:

Leach Time, hours	0	24	48	72		
Date	16-Jul	17-Jul	18-Jul	19-Jul		
Clock Time	16:30	16:30	16:30			
Solution Temperature, °C	30	30	30			
Solution pH						
Grams Nitric Acid (HNO_3) Added	1.56	1.10	2.44			
HNO_3, milliequivalents / gram soil	0.173	0.296	0.567	0.567		
HNO_3 Consumed	100% HNO_3	8.9	kg/1000 kg soil added			

Test No "LR30P"-3		Dry Wt.	Solid or Solution Analysis or Mass				
		or Volume	mg/kg, mg/L, or mg				
Sample ID	Description	g or mL	Pb				
3/8x50 mesh tailings + middlings		400	710				
			284	0	0	0	0
"LR30P"-3-1 24 hr filtrate		780	206				
			161	0	0	0	0
"LR30P"-3-2 48 hr filtrate		550	271				
			149	0	0	0	0
"LR30P"-3-3 72 hr filtrate/w		1320	109				
			144	0	0	0	0
"LR30P"-3-4 72 hr residue		399	150				
			60	0	0	0	0
Calculated Feed Analysis		0%	**1284**	**0**	**0**	**0**	**0**
Calculated Leach Recovery			**88%**				

APPENDIX A.5 AGITATION LEACH DATA SHEETS

Example Calculations

Reference Test "A"-1

HNO_3, milliequivalents / gram soil

1. Calculate acid milliequivalents of nitric acid added per gram of soil
2. Dry Wt. Soil = 100 g
3. At 0 minutes: Reagent HNO_3 Added = 1.43 g
4. At 60 minutes: Reagent HNO_3 Added = 1.25 g
5. Reagent HNO_3 = 70.0%
6. SUM[Wt. 100% HNO_3] = (1.43 + 1.25)(70%/100%) = 1.88 g
7. Milliequivalents Per Gram Soil = ((1.88 g)(1000 meq/eq))/((63.01 g/eq)(100 grams soil)) = 0.298 meq/g

HNO_3 Consumed

1. Calculate the amount of 100% nitric acid consumed per 1000 kg (1.0 t) of soil
2. Dry Wt. Soil = 100 g
3. At 0 minutes: Grams HNO_3 Added = 1.43 g
4. At 60 minutes: Grams HNO_3 Added = 1.25 g
5. At 120 minutes: Grams HNO_3 Added = 0.20 g
6. At 180 minutes: Grams HNO_3 Added = 0.20 g
7. SUM["Grams Nitric Acid (HNO_3) Added"] = (1.43 + 1.25 + 0.20 + 0.20) = 3.08 g
8. Reagent HNO_3 = 70.0%
9. Nitric Acid Consumed = ((3.08 g)(70%/100%)(1000 g/kg))/(100 g soil) = 21.6 g/kg or kg/1000 kg

<150 mesh tailings + middlings

1. Calculate the amount of lead in "tailings + middlings" in mg
2. Dry Wt. "tailings + middlings" = 100 g
3. Lead Analysis of "tailings + middlings" = 5570 mg/kg
4. Calculated Lead Content = ((100 g)(5570 mg/kg))/(1000 g/kg) = 557 mg Pb

"A"-1-1 4 hr filtrate + wash

1. Calculate the amount of lead in "4 hr filtrate + wash" in mg
2. Volume "4 hr filtrate + wash" = 293 mL
3. Lead Analysis "4 hr filtrate + wash" = 615 mg/L
4. Calculated Lead Content = ((293 mL)(615 mg/L))/(1000 mL/L) = 180 mg Pb

"A"-1-1 4 hr residue

1. Calculate the amount of lead in "4 hr residue" in mg
2. Dry Wt. "4 hr residue" = 98.1 g
3. Lead Analysis "4 hr residue" = 3850 mg/kg
4. Calculated Lead Content = ((98.1 g)(3850 mg/kg))/(1000 g/kg) = 378 mg Pb

Calculated Feed Analysis

1. Calculated Feed Analysis in mg/kg
2. Dry Wt. Soil = 100 grams
3. Calculated Feed = ((180 mg Pb + 378 mg Pb)(1000 g/kg))/(100 g) = 5580 mg/kg
4. Calculated Feed Error = ((98.1 g – 100 g/100 g))(100) = –1.9%.

Calculated Leach Recovery

1. Calculated Leach Recovery in Wt.%
2. Calculated Recovery = ((180 mg Pb)(100))/(180 mg Pb + 378 mg Pb) = 32%

Reference Test A-5

HCl Consumed

1. Calculate the amount of 100% hydrochloric acid consumed per 1000 kg (t) of soil
2. Dry Wt. Soil = 100 g
3. At 0 minutes: Grams HCl Added = 8.00 g
4. At 30 minutes: Grams HCl Added = 0.00 g
5. At 60 minutes: Grams HCl Added = 0.00 g
6. At 120 minutes: Grams HCl Added = 0.00 g
7. SUM["Grams Hydrochloric Acid (HCl) Added"] = (8.00 + 0.00 + 0.00 + 0.00) = 8.00 g
8. Reagent HCl = 37%
9. HCl Consumed = ((8.00 g)(37%/100%)(1000 g/kg))/(100 g soil) = 29.6 g/kg or kg/1000 kg

H_2O_2 Consumed

1. Calculate the amount of 100% hydrogen peroxide consumed per 1000 kg (t) of soil
2. Dry Wt. Soil = 100 g
3. At 0 minutes: Grams H_2O_2 Added = 0.00 g
4. At 30 minutes: Grams H_2O_2 Added = 1.00 g
5. At 60 minutes: Grams H_2O_2 Added = 1.00 g
6. At 120 minutes: Grams H_2O_2 Added = 0.00 g
7. SUM["Grams Hydrogen Peroxide (H_2O_2) Added"] = (0.00 + 1.00 + 1.00 + 0.00) = 2.00 g
8. Reagent H_2O_2 = 30%
9. H_2O_2 Consumed = ((2.00 g)(30%/100%)(1000 g/kg))/(100 g soil) = 6.0 g/kg or kg/1000 kg

AGITATION LEACH DATA SHEET

Sample: 79 grams Miramar "A," <150 mesh tails
21 grams Miramar "A," <150 mesh mids
100 grams Miramar "A" tailings + middlings

Instructions: Add 100 grams of sample in 200 mL water
Use 400 mL beaker and propeller stirrer
Adjust pH to 2.0 target with reagent nitric acid
Measure pH and readjust as necessary at 60, 120, and 180 minutes
At 240 minutes, flocculate, filter, and wash residue with 100 mL water
Combine filtrate and wash, measure volume, submit for lead analysis
Get wet and dry weight of filter cake, submit for lead analysis

Data:

Leach Time, minutes	0	60	120	180	240	
Clock Time	9:30	10:30	11:30	12:30	13:30	
Solution Temperature, °C	25	25	25	25	25	
Solution pH	7.00	4.74	1.80	2.13	2.04	
Grams Nitric Acid (HNO_3) Added	1.43	1.25	0.20	0.20		
HNO_3, milliequivalents / gram soil	0.159	0.298	0.320	0.342		
HNO_3 Consumed	100% HNO_3	21.6	kg/1000 kg soil added			

Test No:	"A"-1	Dry Wt. or Volume	Solid or Solution Analysis or Mass mg/kg, mg/L, or mg				
Sample ID	Description	g or mL	Pb				
<150 mesh tailings + middlings		100	5570				
			557	0	0	0	0
"A"-1-1 4 hr filtrate + wash		293	615				
			180	0	0	0	0
"A"-1-2 4 hr residue		98.1	3850				
			378	0	0	0	0
Calculated Feed Analysis		-2%	**5579**	**0**	**0**	**0**	**0**
Calculated Leach Recovery			**32%**				

AGITATION LEACH DATA SHEET

Sample: 79 grams Miramar "A," <150 mesh tails
                     21 grams Miramar "A," <150 mesh mids
                     100 grams Miramar "A" tailings + middlings

Instructions: Add 100 grams of sample in 200 mL water
                     Use 400 mL beaker and propeller stirrer
                     Adjust pH to 1.5 target with reagent nitric acid
                     Measure pH and readjust as necessary at 60, 120, and 180 minutes
                     At 240 minutes, flocculate, filter, and wash residue with 100 mL water
                     Combine filtrate and wash, measure volume, submit for lead analysis
                     Get wet and dry weight of filter cake, submit for lead analysis

Data:

Leach Time, minutes			0	60	120	180	240	
Clock Time			9:40	10:40	11:33	12:36	13:40	
Solution Temperature, °C			24	25	25	25	26	
Solution pH			8.42	4.94	1.54	1.51	1.56	
Grams Nitric Acid (HNO_3) Added			1.21	1.81	0.18	0.10		
HNO_3, milliequivalents / gram soil			0.134	0.336	0.355	0.367		
HNO_3 Consumed			100% HNO_3	23.1	kg/1000 kg soil added			

Test No:	"A"-2	Dry Wt.	Solid or Solution Analysis or Mass				
		or Volume	mg/kg, mg/L, or mg				
Sample ID	Description	g or mL	Pb				
<150 mesh tailings + middlings		100	5570				
			557	0	0	0	0
"A"-2-1 4 hr filtrate + wash		291	796				
			232	0	0	0	0
"A"-2-2 4 hr residue		98	3230				
			318	0	0	0	0
Calculated Feed Analysis		-2%	**5495**	**0**	**0**	**0**	**0**
Calculated Leach Recovery			**42%**				

AGITATION LEACH DATA SHEET

Sample: 79 grams Miramar "A," <150 mesh tails
21 grams Miramar "A," <150 mesh mids
100 grams Miramar "A" tailings + middlings

Instructions: Add 100 grams of sample in 200 mL water
Use 400 mL beaker and propeller stirrer
Adjust pH to 1.2 target with reagent nitric acid
Measure pH and readjust as necessary at 60, 120, and 180 minutes
At 240 minutes, flocculate, filter, and wash residue with 100 mL water
Combine filtrate and wash, measure volume, submit for lead analysis
Get wet and dry weight of filter cake, submit for lead analysis

Data:

Leach Time, minutes		0	60	120	180	240	
Clock Time		9:40	10:40	11:36	12:43	13:42	
Solution Temperature, °C		24	25	25	25	25	
Solution pH		8.44	1.62	1.19	1.23	1.26	
Grams Nitric Acid (HNO_3) Added		1.77	0.83	0.17	0.17		
HNO_3, milliequivalents / gram soil		0.197	0.289	0.308	0.327		
HNO_3 Consumed		100% HNO_3	20.6	kg/1000 kg soil added			

Test No:	"A"-3	Dry Wt.	Solid or Solution Analysis or Mass			
		or Volume	mg/kg, mg/L, or mg			
Sample ID	Description	g or mL	Pb			
<150 mesh tailings + middlings		100	5570			
			557	0	0	0	0
"A"-3-1 4 hr filtrate + wash		306	512			
			157	0	0	0	0
"A"-3-2 4 hr residue		99	3940			
			389	0	0	0	0
Calculated Feed Analysis		-1%	**5459**	0	0	0	0
Calculated Leach Recovery			**29%**			

AGITATION LEACH DATA SHEET

Sample: 79 grams Miramar "A," <150 mesh tails
                         21 grams Miramar "A," <150 mesh mids
                         100 grams Miramar "A" tailings + middlings

Instructions: Add 100 grams of sample in 200 mL water
                         Use 400 mL beaker and propeller stirrer
                         Add 10 grams $Fe_2(SO_4)_3xH_2O$
                         Adjust pH to 1.5 target with reagent nitric acid
                         Measure pH and readjust as necessary at 30, 60, 120, and 180 minutes
                         At 240 minutes, flocculate, filter, and wash residue with 100 mL water
                         Combine filtrate and wash, measure volume, submit for lead analysis
                         Get wet and dry weight of filter cake, submit for lead analysis

Data:

Leach Time, minutes			0	30	60	120	180	240
Clock Time			12:45	13:15	13:45	14:45	15:45	16:45
Solution Temperature, °C			24	25	25	25	25	
Solution pH			2.35	1.52	1.52	1.49	1.51	
Grams Nitric Acid (HNO_3) Added			3.88	0.00	0.00	0.19		
HNO_3, milliequivalents / gram soil			0.431	0.431	0.431	0.452		
HNO_3 Consumed			100% HNO_3	28.5	kg/1000 kg soil added			

Test No:	"A"-4	Dry Wt.	Solid or Solution Analysis or Mass				
		or Volume	mg/kg, mg/L, or mg				
Sample ID	Description	g or mL	Pb				
<150 mesh tailings + middlings		100	5570				
			557	0	0	0	0
"A"-4-1 4 hr filtrate + wash		900	6				
			5	0	0	0	0
"A"-4-2 4 hr residue		98	4640				
			454	0	0	0	0
Calculated Feed Analysis		-2%	4597	0	0	0	0
Calculated Leach Recovery			1%				

AGITATION LEACH DATA SHEET

Sample:　　　　79 grams Miramar "A," <150 mesh tails
　　　　　　　　21 grams Miramar "A," <150 mesh mids
　　　　　　　　100 grams Miramar "A" tailings + middlings

Instructions:　　Add 100 grams of sample in 400 mL water
　　　　　　　　Use 600 mL beaker and propeller stirrer
　　　　　　　　Add 8 grams HNO_3
　　　　　　　　Add 8 grams HCl
　　　　　　　　Add 1 + 1 grams H_2O_2
　　　　　　　　Measure pH and emf at 30, 60, 120, 180, and 240 minutes
　　　　　　　　At 240 minutes, flocculate, filter, and wash residue with 100 mL water
　　　　　　　　Combine filtrate and wash, measure volume, submit for lead analysis
　　　　　　　　Get wet and dry weight of filter cake, submit for lead analysis

Data:

Leach Time, minutes		0	30	60	120	180	240
Clock Time		10:15	10:45	11:15	12:15	13:15	14:15
Solution Temperature, °C		25	26	26	25	25	25
Solution pH		7.46	1.10	1.14	1.20	1.12	1.22
Solution emf, mv		202	460	530	547	536	541
Grams HCl Added		8.00					
Grams H_2O_2 Added			1.00	1.00			
Grams HNO_3 Added		8.00					
HNO_3, milliequivalents / gram soil		0.889	0.889	0.889	0.889		
HNO_3 Consumed	100% HNO_3	56	kg/1000 kg soil added				
HCl Consumed	100% HCl	30	kg/1000 kg soil added				
H_2O_2 Consumed	100% H_2O_2	6	kg/1000 kg soil added				

Test No:	"A"-5	Dry Wt.	Solid or Solution Analysis or Mass				
		or Volume	mg/kg, mg/L, or mg				
Sample ID	Description	g or mL	Pb				
<150 mesh tailings + middlings		100	5570				
			557	0	0	0	0
"A"-5-1 4 hr filtrate + wash		503	843				
			424	0	0	0	0
"A"-5-2 4 hr residue		94	1566				
			147	0	0	0	0
Calculated Feed Analysis		-6%	**5712**	**0**	**0**	**0**	**0**
Calculated Leach Recovery			**74%**				

AGITATION LEACH DATA SHEET

Sample: Residue from Test Miramar "A"-5 36.7 grams

Instructions: Add 36.7 grams of sample in 400 mL water
Use 600 mL beaker and propellor stirrer
Add 8 grams HNO_3
Add 8 grams HCl
Add 1 + 1 grams H_2O_2
Measure pH and emf at 30, 60, 120, 180, and 240 minutes
At 240 minutes, flocculate, filter, and wash residue with 100 mL water
Combine filtrate and wash, measure volume, submit for lead analysis
Get wet and dry weight of filter cake, submit for lead analysis

Data:

Leach Time, minutes		0	30	60	120	180	240
Clock Time		7:45	8:15	8:45	9:45	10:45	11:45
Solution Temperature, °C		23	24	24	24	25	25
Solution pH		2.52	1.20	1.23	1.25	1.23	1.24
Solution emf, mv		199	435	540	556	556	552
Grams HCl Added		8.00					
Grams H_2O_2 Added			1.00	1.00			
Grams HNO_3 Added		8.00					
HNO_3, milliequivalents / gram soil		0.889	0.889	0.889	0.889		
HNO_3 Consumed	100% HNO_3	153	kg/1000 kg soil added				
HCl Consumed	100% HCl	81	kg/1000 kg soil added				
H_2O_2 Consumed	100% H_2O_2	16	kg/1000 kg soil added				

Test No:	"A"-5A	Dry Wt.	Solid or Solution Analysis or Mass				
		or Volume	mg/kg, mg/L, or mg				
Sample ID	Description	g or mL	Pb				
<150 mesh tailings + middlings		37	1566				
			57	0	0	0	0
"A"-5A-1 4 hr filtrate + wash		415	95				
			39	0	0	0	0
"A"-5A-2 4 hr residue		38	588				
			23	0	0	0	0
Calculated Feed Analysis		5%	**1689**	**0**	**0**	**0**	**0**
Calculated Leach Recovery			**64%**				

AGITATION LEACH DATA SHEET

Sample:	79 grams Miramar "A," <150 mesh tails
	21 grams Miramar "A," <150 mesh mids
	100 grams Miramar "A" tailings + middlings
Instructions:	Add 100 grams of sample in 400 mL water
	Use 600 mL beaker and propeller stirrer
	Add 8 grams HNO_3
	Add 8 grams HCl
	Add 1 + 1 grams H_2O_2
	Measure pH and emf at 30, 60, and 120 minutes
	At 120 minutes, flocculate, filter, and wash residue with 100 mL water
	Combine filtrate and wash, measure volume, submit for lead analysis
	Get wet and dry weight of filter cake, submit for lead analysis

Data:

Leach Time, minutes	0	30	60	120		
Clock Time	9:00	9:30	10:00	11:00		
Solution Temperature, °C	23	23	23	22		
Solution pH	7.20	1.30	1.24	1.14		
Solution emf, mv	162	217	510	546		
Grams HCl Added	8.00					
Grams H_2O_2 Added		1.00	1.00			
Grams HNO_3 Added	8.00					
HNO_3, milliequivalents / gram soil	0.889	0.889	0.889	0.889		
HNO_3 Consumed	100% HNO_3	56	kg/1000 kg soil added			
HCl Consumed	100% HCl	30	kg/1000 kg soil added			
H_2O_2 Consumed	100% H_2O_2	6	kg/1000 kg soil added			

Test No:	"A"-6	Dry Wt.	Solid or Solution Analysis or Mass				
		or Volume	mg/kg, mg/L, or mg				
Sample ID	Description	g or mL	Pb				
<150 mesh tailings + middlings		100	5570				
			557	0	0	0	0
"A"-6-1 2 hr filtrate + wash		318	1090				
			347	0	0	0	0
"A"-6-2 2 hr residue		100	1710				
			171	0	0	0	0
Calculated Feed Analysis		0%	**5176**	**0**	**0**	**0**	**0**
Calculated Leach Recovery			**67%**				

AGITATION LEACH DATA SHEET

Sample: Residue from Test Miramar "A"-6 46.9 grams

Instructions: Add 46.9 grams of sample in 200 mL water
Use 600 mL beaker and propeller stirrer
Add 4 grams HNO_3
Add 4 grams HCl
Add 0.5 + 0.5 grams H_2O_2
Measure pH and emf at 30, 60, and 120 minutes
At 120 minutes, flocculate, filter, and wash residue with 100 mL water
Combine filtrate and wash, measure volume, submit for lead analysis
Get wet and dry weight of filter cake, submit for lead analysis

Data:

Leach Time, minutes			0	30	60	120		
Clock Time			11:15	11:45	12:15	1:15		
Solution Temperature, °C			23	23	23	23		
Solution pH			1.75	1.16	1.07	1.04		
Solution emf, mv			499	440	517	538		
Grams HCl Added			4.00					
Grams H_2O_2 Added				0.50	0.50			
Grams HNO_3 Added			4.00					
HNO_3, milliequivalents / gram soil			0.444	0.444	0.444	0.444		
HNO_3 Consumed			100% HNO_3	60	kg/1000 kg soil added			
HCl Consumed			100% HCl	32	kg/1000 kg soil added			
H_2O_2 Consumed			100% H_2O_2	6	kg/1000 kg soil added			

Test No:	"A"-6A	Dry Wt.	Solid or Solution Analysis or Mass				
		or Volume	mg/kg, mg/L, or mg				
Sample ID	Description	g or mL	Pb				
<150 mesh tailings + middlings		47	1712				
			80	0	0	0	0
"A"-6A-1 2 hr filtrate + wash		275	185				
			51	0	0	0	0
"A"-6A-2 2 hr residue		47	875				
			41	0	0	0	0
Calculated Feed Analysis		0%	1960	0	0	0	0
Calculated Leach Recovery			55%				

AGITATION LEACH DATA SHEET

Sample:

79 grams Miramar "A," <150 mesh tails
21 grams Miramar "A," <150 mesh mids
100 grams Miramar "A" tailings + middlings

Instructions:

Add 100 grams of sample in 600 ml water
Use 1000 mL beaker and propeller stirrer
Add 12 grams HNO_3
Add 12 grams HCl
Add 1 + 1 grams H_2O_2
Measure pH and emf at 30, 60, 120, 180, and 240 minutes
At 240 minutes, flocculate, filter, wash, repulp, and wash with 100 mL aliquots of water
Combine filtrate and wash, measure volume, submit for lead analysis
Get wet and dry weight of filter cake, submit for lead analysis

Data:

Leach Time, minutes	0	30	60	120	180	240
Clock Time	9:00	9:30	10:00	11:00	12:00	13:00
Solution Temperature, °C	25	27	27	27	27	27
Solution pH	6.20	1.27	1.34	1.63	1.68	1.50
Solution emf, mv	207	476	536	564	549	569
Grams HCl Added	12.00					
Grams H_2O_2 Added		1.50	1.50			
Grams HNO_3 Added	12.00					
HNO_3, milliequivalents / gram soil	1.333	1.333	1.333	1.333		
HNO_3 Consumed	100% HNO_3	84	kg/1000 kg soil added			
HCl Consumed	100% HCl	44	kg/1000 kg soil added			
H_2O_2 Consumed	100% H_2O_2	9	kg/1000 kg soil added			

Test No:	"A"-7	Dry Wt.	Solid or Solution Analysis or Mass			
		or Volume	mg/kg, mg/L, or mg			
Sample ID	Description	g or mL	Pb			
<150 mesh tailings + middlings		100	5570			
			557	0	0	0	0
"A"-7-1 4 hr filtrate + wash		740	590			
			437	0	0	0	0
"A"-7-2 4 hr residue		96	1210			
			116	0	0	0	0
Calculated Feed Analysis		-4%	5528	0	0	0	0
Calculated Leach Recovery			79%			

AGITATION LEACH DATA SHEET

Sample: Residue from Test Miramar "A"-7 47.0 grams

Instructions: Add 47.0 grams of sample in 300 ml water
Use 600 mL beaker and propeller stirrer
Add 6 grams HNO_3
Add 6 grams HCl
Add 0.5 + 0.5 grams H_2O_2
Measure pH and emf at 30, 60, 120, 180, and 240 minutes
At 240 minutes, flocculate, filter, and wash residue with 100 mL water
Combine filtrate and wash, measure volume, submit for lead analysis
Get wet and dry weight of filter cake, submit for lead analysis

Data:

Leach Time, minutes		0	30	60	120	180	240
Clock Time		13:30	14:00	14:30	15:30	16:30	17:30
Solution Temperature, °C		27	27	27	27	27	28
Solution pH		2.27	1.28	1.26	1.14	1.04	1.00
Solution emf, mv		308	510	556	557	554	552
Grams HCl Added		6.00					
Grams H_2O_2 Added			0.50	0.50			
Grams HNO_3 Added		6.00					
HNO_3, milliequivalents / gram soil		0.667	0.667	0.667	0.667		
HNO_3 Consumed	100% HNO_3	89	kg/1000 kg soil added				
HCl Consumed	100% HCl	47	kg/1000 kg soil added				
H_2O_2 Consumed	100% H_2O_2	6	kg/1000 kg soil added				

Test No:	"A"-7A	Dry Wt.	Solid or Solution Analysis or Mass				
		or Volume	mg/kg, mg/L, or mg				
Sample ID	Description	g or mL	Pb				
<150 mesh tailings + middlings		47	1210				
			57	0	0	0	0
"A"-7A-1 4 hr filtrate + wash		312	110				
			34	0	0	0	0
"A"-7A-2 4 hr residue		46	565				
			26	0	0	0	0
Calculated Feed Analysis		-1%	**1288**	**0**	**0**	**0**	**0**
Calculated Leach Recovery			57%				

AGITATION LEACH DATA SHEET

Sample: 79 grams Miramar "A," <150 mesh tails
                      21 grams Miramar "A," <150 mesh mids
                      100 grams Miramar "A" tailings + middlings

Instructions: Add 100 grams of sample in 600 ml water
                      Use 1000 mL beaker and propeller stirrer
                      Add 12 grams HNO_3
                      Add 12 grams HCl
                      Add 1 + 1 grams H_2O_2
                      Measure pH and emf at 30, 60, and 120 minutes
                      At 120 minutes, flocculate, filter, and wash residue with 100 mL water
                      Combine filtrate and wash, measure volume, submit for lead analysis
                      Get wet and dry weight of filter cake, submit for lead analysis

Data:

Leach Time, minutes		0	30	60	120		
Clock Time		9:00	9:30	10:00	11:00		
Solution Temperature, °C		25	27	27	28		
Solution pH		6.32	1.27	1.34	1.65		
Solution emf, mv		201	491	551	573		
Grams HCl Added		12.00					
Grams H_2O_2 Added			1.50	1.50			
Grams HNO_3 Added		12.00					
HNO_3, milliequivalents / gram soil		1.333	1.333	1.333	1.333		
HNO_3 Consumed	100% HNO_3	84	kg/1000 kg soil added				
HCl Consumed	100% HCl	44	kg/1000 kg soil added				
H_2O_2 Consumed	100% H_2O_2	9	kg/1000 kg soil added				

Test No:	"A"-8	Dry Wt.	Solid or Solution Analysis or Mass				
		or Volume	mg/kg, mg/L, or mg				
Sample ID	Description	g or mL	Pb				
<150 mesh tailings + middlings		100	5570				
			557	0	0	0	0
"A"-8-1 2 hr filtrate + wash		827	510				
			422	0	0	0	0
"A"-8-2 2 hr residue		98	1290				
			126	0	0	0	0
Calculated Feed Analysis		-2%	**5482**	**0**	**0**	**0**	**0**
Calculated Leach Recovery			**77%**				

AGITATION LEACH DATA SHEET

Sample: Residue from Test Miramar "A"-8 50.0 grams

Instructions: Add 50.0 grams of sample in 300 ml water
Use 600 mL beaker and propeller stirrer
Add 6 grams HNO_3
Add 6 grams HCl
Add 0.5 + 0.5 grams H_2O_2
Measure pH and emf at 30, 60, and 120 minutes
At 120 minutes, flocculate, filter, and wash residue with 100 mL water
Combine filtrate and wash, measure volume, submit for lead analysis
Get wet and dry weight of filter cake, submit for lead analysis

Data:

Leach Time, minutes	0	30	60	120		
Clock Time	12:00	12:30	13:00	14:00		
Solution Temperature, °C	27	27	27	27		
Solution pH	2.57	1.48	1.41	1.32		
Solution emf, mv	286	547	551	547		
Grams HCl Added	6.00					
Grams H_2O_2 Added		0.50	0.50			
Grams HNO_3 Added	6.00					
HNO_3, milliequivalents / gram soil	0.667	0.667	0.667	0.667		
HNO_3 Consumed	100% HNO_3	84	kg/1000 kg soil added			
HCl Consumed	100% HCl	44	kg/1000 kg soil added			
H_2O_2 Consumed	100% H_2O_2	6	kg/1000 kg soil added			

Test No:	"A"-8A	Dry Wt.	Solid or Solution Analysis or Mass			
		or Volume	mg/kg, mg/L, or mg			
Sample ID	Description	g or mL	Pb			
<150 mesh tailings + middlings		50	1290			
			65	0	0	0	0
"A"-8A-1 2 hr filtrate + wash		346	305			
			106	0	0	0	0
"A"-8A-2 2 hr residue		49	622			
			31	0	0	0	0
Calculated Feed Analysis		-1%	**2725**	**0**	**0**	**0**	**0**
Calculated Leach Recovery			**77%**			

AGITATION LEACH DATA SHEET

Sample: 79 grams Miramar "A," <150 mesh tails
21 grams Miramar "A," <150 mesh mids
100 grams Miramar "A" tailings + middlings

Instructions: Add 100 grams of sample in 400 mL water
Use 600 mL beaker and propeller stirrer
Add 15 grams HNO_3
Add 15 grams HCl
Add 1.5 + 1.5 grams H_2O_2
Measure pH and emf at 30, 60, and 120 minutes
At 120 minutes, flocculate, filter, and wash residue with 100 mL water
Combine filtrate and wash, measure volume, submit for lead analysis
Get wet and dry weight of filter cake, submit for lead analysis

Data:

Leach Time, minutes		0	30	60	120		
Clock Time		11:45	12:15	12:45	13:15		
Solution Temperature, °C		23	23	23	23		
Solution pH		1.00	0.80	0.70	0.70		
Solution emf, mv		545	630	650	650		
Grams HCl Added		15.00					
Grams H_2O_2 Added			1.50	1.50			
Grams HNO_3 Added		15.00					
HNO_3, milliequivalents / gram soil		1.666	1.666	1.666	1.666		
HNO_3 Consumed	100% HNO_3	105	kg/1000 kg soil added				
HCl Consumed	100% HCl	56	kg/1000 kg soil added				
H_2O_2 Consumed	100% H_2O_2	9	kg/1000 kg soil added				

Test No:	"A"-9	Dry Wt.	Solid or Solution Analysis or Mass				
		or Volume	mg/kg, mg/L, or mg				
Sample ID	Description	g or mL	Pb				
<150 mesh tailings + middlings		100	5570				
			557	0	0	0	0
"A"-9-1 2 hr filtrate + wash		487	940				
			458	0	0	0	0
"A"-9-2 2 hr residue		88	995				
			88	0	0	0	0
Calculated Feed Analysis		-12%	5457	0	0	0	0
Calculated Leach Recovery			84%				

AGITATION LEACH DATA SHEET

Sample: Residue from Test Miramar "A"-9 44.2 grams

Instructions: Add 44.2 grams of sample in 200 mL water
                               Use 600 mL beaker and propeller stirrer
                               Add 7.5 grams HNO_3
                               Add 7.5 grams HCl
                               Add 0.75 + 0.75 grams H_2O_2
                               Measure pH and emf at 30, 60, and 120 minutes
                               At 120 minutes, flocculate, filter, and wash residue with 100 mL water
                               Combine filtrate and wash, measure volume, submit for lead analysis
                               Get wet and dry weight of filter cake, submit for lead analysis

Data:

Leach Time, minutes			0	30	60	120		
Clock Time			14:50	15:20	15:50	16:45		
Solution Temperature, °C			23	23	23	23		
Solution pH			0.80	0.65	0.70	0.80		
Solution emf, mv			620	640	640	640		
Grams HCl Added			7.50					
Grams H_2O_2 Added				0.75	0.75			
Grams HNO_3 Added			7.50					
HNO_3, milliequivalents / gram soil			0.833	0.833	0.833	0.833		
HNO_3 Consumed			100% HNO_3	119	kg/1000 kg soil added			
HCl Consumed			100% HCl	63	kg/1000 kg soil added			
H_2O_2 Consumed			100% H_2O_2	10	kg/1000 kg soil added			

Test No:	"A"-9A	Dry Wt.	Solid or Solution Analysis or Mass				
		or Volume	mg/kg, mg/L, or mg				
Sample ID	Description	g or mL	Pb				
<150 mesh tailings + middlings		44	995				
			44	0	0	0	0
"A"-9A-1 2 hr filtrate + wash		340	63				
			21	0	0	0	0
"A"-9A-2 2 hr residue		47	580				
			27	0	0	0	0
Calculated Feed Analysis		6%	1101	0	0	0	0
Calculated Leach Recovery			44%				

AGITATION LEACH DATA SHEET

Sample:

79 grams Miramar "A," <150 mesh tails
21 grams Miramar "A," <150 mesh mids
100 grams Miramar "A" tailings + middlings

Instructions:

Add 100 grams of sample in 400 mL water
Use 600 mL beaker and propeller stirrer
Add 18 grams HNO_3
Add 18 grams HCl
Add 2 + 2 grams H_2O_2
Measure pH and emf at 30, 60, and 120 minutes
At 120 minutes, flocculate, filter, and wash residue with 100 mL water
Combine filtrate and wash, measure volume, submit for lead analysis
Get wet and dry weight of filter cake, submit for lead analysis

Data:

Leach Time, minutes		0	30	60	120		
Clock Time		11:45	12:15	12:45	13:15		
Solution Temperature, °C		23	23	23	23		
Solution pH		0.90	0.70	0.60	0.60		
Solution emf, mv		620	643	650	660		
Grams HCl Added		18.00					
Grams H_2O_2 Added			2.00	2.00			
Grams HNO_3 Added		18.00					
HNO_3, milliequivalents / gram soil		2.000	2.000	2.000	2.000		
HNO_3 Consumed	100% HNO_3	126	kg/1000 kg soil added				
HCl Consumed	100% HCl	67	kg/1000 kg soil added				
H_2O_2 Consumed	100% H_2O_2	12	kg/1000 kg soil added				

Test No:	"A"-10	Dry Wt.	Solid or Solution Analysis or Mass			
		or Volume	mg/kg, mg/L, or mg			
Sample ID	Description	g or mL	Pb			
<150 mesh tailings + middlings		100	5570			
			557	0	0	0	0
"A"-10-1 2 hr filtrate + wash		465	1020			
			474	0	0	0	0
"A"-10-2 2 hr residue		88	935			
			82	0	0	0	0
Calculated Feed Analysis		-12%	5562	0	0	0	0
Calculated Leach Recovery			85%			

AGITATION LEACH DATA SHEET

Sample: Residue from Test Miramar "A"-10 43.8 grams

Instructions: Add 43.8 grams of sample in 200 mL water
Use 600 mL beaker and propeller stirrer
Add 9 grams HNO_3
Add 9 grams HCl
Add 1 + 1 grams H_2O_2
Measure pH and emf at 30, 60, and 120 minutes
At 120 minutes, flocculate, filter, and wash residue with 100 mL water
Combine filtrate and wash, measure volume, submit for lead analysis
Get wet and dry weight of filter cake, submit for lead analysis

Data:

Leach Time, minutes		0	30	60	120		
Clock Time		14:50	15:20	15:50	16:45		
Solution Temperature, °C		23	23	23	23		
Solution pH		0.70	0.60	0.65	0.70		
Solution emf, mv		625	640	640	640		
Grams HCl Added		9.00					
Grams H_2O_2 Added			1.00	1.00			
Grams HNO_3 Added		9.00					
HNO_3, milliequivalents / gram soil		1.000	1.000	1.000	1.000		
HNO_3 Consumed		100% HNO_3	144	kg/1000 kg soil added			
HCl Consumed		100% HCl	76	kg/1000 kg soil added			
H_2O_2 Consumed		100% H_2O_2	14	kg/1000 kg soil added			

Test No: "A"-10A		Dry Wt. or Volume	Solid or Solution Analysis or Mass mg/kg, mg/L, or mg			
Sample ID	Description	g or mL	Pb			
<150 mesh tailings + middlings		44	935			
			41	0	0	0	0
"A"-10A-1 2 hr filtrate + wash		360	52			
			19	0	0	0	0
"A"-10A-2 2 hr residue		47	565			
			26	0	0	0	0
Calculated Feed Analysis		7%	1032	0	0	0	0
Calculated Leach Recovery			41%			

AGITATION LEACH DATA SHEET

Sample:	79 grams Miramar "A," <150 mesh tails
	21 grams Miramar "A," <150 mesh mids
	100 grams Miramar "A" tailings + middlings

Instructions:	Add 100 grams of sample in 400 mL water
	Use 600 mL beaker and propeller stirrer; heat to 40°C
	Add 15 grams HNO_3
	Add 15 grams HCl
	Add 1.5 + 1.5 grams H_2O_2
	Measure pH and emf at 30, 60, and 120 minutes
	At 120 minutes, flocculate, filter, and wash residue with 100 mL water
	Combine filtrate and wash, measure volume, submit for lead analysis
	Get wet and dry weight of filter cake, submit for lead analysis

Data:

Leach Time, minutes	0	30	60	120		
Clock Time	8:30	9:05	9:35	10:40		
Solution Temperature, °C	41	38	39	39		
Solution pH	7.80	1.00	1.00	1.10		
Solution emf, mv	240	680	675	685		
Grams HCl Added	15.00					
Grams H_2O_2 Added		1.50	1.50			
Grams HNO_3 Added	15.00					
HNO_3, milliequivalents / gram soil	1.666	1.666	1.666	1.666		
HNO_3 Consumed	100% HNO_3	105	kg/1000 kg soil added			
HCl Consumed	100% HCl	56	kg/1000 kg soil added			
H_2O_2 Consumed	100% H_2O_2	9	kg/1000 kg soil added			

Test No: "A"-11		**Dry Wt.**	**Solid or Solution Analysis or Mass**			
		or Volume	**mg/kg, mg/L, or mg**			
Sample ID	**Description**	**g or mL**	**Pb**			
<150 mesh tailings + middlings		100	5570			
			557	0	0	0	0
"A"-11-1 2 hr filtrate + wash		450	1110			
			500	0	0	0	0
"A"-11-2 2 hr residue		97	885			
			85	0	0	0	0
Calculated Feed Analysis		-3%	**5850**	**0**	**0**	**0**	**0**
Calculated Leach Recovery			**85%**			

AGITATION LEACH DATA SHEET

Sample: Residue from Test Miramar "A"-11 48.0 grams

Instructions: Add 48.0 grams of sample in 200 mL water
                           Use 600 mL beaker and propeller stirrer; heat to 40°C
                           Add 7.5 grams HNO_3
                           Add 7.5 grams HCl
                           Add 0.75 + 0.75 grams H_2O_2
                           Measure pH and emf at 30, 60, and 120 minutes
                           At 120 minutes, flocculate, filter, and wash residue with 100 mL water
                           Combine filtrate and wash, measure volume, submit for lead analysis
                           Get wet and dry weight of filter cake, submit for lead analysis

Data:

Leach Time, minutes		0	30	60	120		
Clock Time		11:45	12:15	12:50	14:15		
Solution Temperature, °C		38	39	38	39		
Solution pH		1.00	0.80	0.80	0.85		
Solution emf, mv		650	648	650	635		
Grams HCl Added		7.50					
Grams H_2O_2 Added			0.75	0.75			
Grams HNO_3 Added		7.50					
HNO_3, milliequivalents / gram soil		0.833	0.833	0.833	0.833		
HNO_3 Consumed	100% HNO_3	109	kg/1000 kg soil added				
HCl Consumed	100% HCl	58	kg/1000 kg soil added				
H_2O_2 Consumed	100% H_2O_2	9	kg/1000 kg soil added				

Test No:	"A"-11A	Dry Wt.	Solid or Solution Analysis or Mass				
		or Volume	mg/kg, mg/L, or mg				
Sample ID	Description	g or mL	Pb				
<150 mesh tailings + middlings		48	885				
			42	0	0	0	0
"A"-11A-1 2 hr filtrate + wash		220	89				
			20	0	0	0	0
"A"-11A-2 2 hr residue		47	473				
			22	0	0	0	0
Calculated Feed Analysis		-3%	866	0	0	0	0
Calculated Leach Recovery			47%				

AGITATION LEACH DATA SHEET

Sample:	79 grams Miramar "A," <150 mesh tails
	21 grams Miramar "A," <150 mesh mids
	100 grams Miramar "A" tailings + middlings

Instructions:	Add 100 grams of sample in 400 mL water
	Use 600 mL beaker and propeller stirrer; heat to 60°C
	Add 15 grams HNO_3
	Add 15 grams HCl
	Add 1.5 + 1.5 grams H_2O_2
	Measure pH and emf at 30, 60, and 120 minutes
	At 120 minutes, flocculate, filter, and wash residue with 100 mL water
	Combine filtrate and wash, measure volume, submit for lead analysis
	Get wet and dry weight of filter cake, submit for lead analysis

Data:

Leach Time, minutes	0	30	60	120		
Clock Time	8:30	9:05	9:35	10:40		
Solution Temperature, °C	55	57	57	55		
Solution pH	7.80	1.10	1.10	1.00		
Solution emf, mv	255	710	710	706		
Grams HCl Added	15.00					
Grams H_2O_2 Added		1.50	1.50			
Grams HNO_3 Added	15.00					
HNO_3, milliequivalents / gram soil	1.666	1.666	1.666	1.666		
HNO_3 Consumed	100% HNO_3	105	kg/1000 kg soil added			
HCl Consumed	100% HCl	56	kg/1000 kg soil added			
H_2O_2 Consumed	100% H_2O_2	9	kg/1000 kg soil added			

Test No:	"A"-12	Dry Wt.	Solid or Solution Analysis or Mass			
		or Volume	mg/kg, mg/L, or mg			
Sample ID	**Description**	**g or mL**	**Pb**			
<150 mesh tailings + middlings		100	5570			
			557	0	0	0	0
"A"-12-1 2 hr filtrate + wash		440	1116			
			491	0	0	0	0
"A"-12-2 2 hr residue		95	691			
			66	0	0	0	0
Calculated Feed Analysis		-5%	5567	0	0	0	0
Calculated Leach Recovery			88%			

AGITATION LEACH DATA SHEET

Sample: Residue from Test Miramar "A"-12 50.0 grams

Instructions: Add 50.0 grams of sample in 200 mL water
Use 600 mL beaker and propeller stirrer; heat to 60°C
Add 7.5 grams HNO_3
Add 7.5 grams HCl
Add 0.75 + 0.75 grams H_2O_2
Measure pH and emf at 30, 60, and 120 minutes
At 120 minutes, flocculate, filter, and wash residue with 100 mL water
Combine filtrate and wash, measure volume, submit for lead analysis
Get wet and dry weight of filter cake, submit for lead analysis

Data:

Leach Time, minutes		0	30	60	120		
Clock Time		11:45	12:15	12:50	14:15		
Solution Temperature, °C		57	55	57	58		
Solution pH		1.10	0.80	0.70	0.90		
Solution emf, mv		690	675	730	735		
Grams HCl Added		7.50					
Grams H_2O_2 Added			0.75	0.75			
Grams HNO_3 Added		7.50					
HNO_3, milliequivalents / gram soil		0.833	0.833	0.833	0.833		
HNO_3 Consumed	100% HNO_3	105	kg/1000 kg soil added				
HCl Consumed	100% HCl	56	kg/1000 kg soil added				
H_2O_2 Consumed	100% H_2O_2	9	kg/1000 kg soil added				

Test No:	"A"-12A	Dry Wt.	Solid or Solution Analysis or Mass			
		or Volume	mg/kg, mg/L, or mg			
Sample ID	Description	g or mL	Pb			
<150 mesh tailings + middlings		50	691			
			35	0	0	0	0
"A"-12A-1 2 hr filtrate + wash		230	78			
			18	0	0	0	0
"A"-12A-2 2 hr residue		50	327			
			16	0	0	0	0
Calculated Feed Analysis		-1%	**684**	**0**	**0**	**0**	**0**
Calculated Leach Recovery			**52%**			

AGITATION LEACH DATA SHEET

Sample: 100 grams Miramar "B," <150 mesh tails

Instructions: Add 100 grams of sample in 200 mL water
Use 400 mL beaker and propeller stirrer
Adjust pH to 2.0 target with reagent nitric acid
Measure pH and readjust as necessary at 60, 120, and 180 minutes
At 240 minutes, flocculate, filter, and wash residue with 100 mL water
Combine filtrate and wash, measure volume, submit for lead analysis
Get wet and dry weight of filter cake, submit for lead analysis

Data:

Leach Time, minutes		0	60	120	180	240	
Clock Time		9:43	10:43	11:38	12:47	13:47	
Solution Temperature, °C		24	25	25	25	25	
Solution pH		9.51	3.25	2.51	2.30	2.15	
Grams HNO_3 Added		2.79	0.40	0.21	0.14		
HNO_3, milliequivalents / gram soil		0.310	0.354	0.378	0.393		
HNO_3 Consumed		100% HNO_3		25	kg/1000 kg soil added		

Test No:	"B"-1	Dry Wt.	Solid or Solution Analysis or Mass				
		or Volume	mg/kg, mg/L, or mg				
Sample ID	Description	g or mL	Pb				
<150 mesh tailings		100	2370				
			237	0	0	0	0
"B"-1-1 4 hr filtrate + wash		284	313				
			89	0	0	0	0
"B"-1-2 4 hr residue		97	1510				
			147	0	0	0	0
Calculated Feed Analysis		-3%	2360	0	0	0	0
Calculated Leach Recovery			38%				

AGITATION LEACH DATA SHEET

Sample: 100 grams Miramar "B," <150 mesh tails

Instructions: Add 100 grams of sample in 200 mL water
                                     Use 400 mL beaker and propeller stirrer
                                     Adjust pH to 1.5 target with reagent nitric acid
                                     Measure pH and readjust as necessary at 60, 120, and 180 minutes
                                     At 240 minutes, flocculate, filter, and wash residue with 100 mL water
                                     Combine filtrate and wash, measure volume, submit for lead analysis
                                     Get wet and dry weight of filter cake, submit for lead analysis

Data:

Leach Time, minutes			0	60	120	180	240	
Clock Time			9:48	10:48	11:43	12:50	13:48	
Solution Temperature, °C			24	25	25	25	25	
Solution pH			9.50	2.13	1.60	1.45	1.53	
Grams HNO_3 Added			3.39	0.55	0.36			
HNO_3, milliequivalents / gram soil			0.377	0.438	0.478	0.478		
HNO_3 Consumed		100% HNO_3	30	kg/1000 kg soil added				

Test No:	"B"-2	Dry Wt.	Solid or Solution Analysis or Mass				
		or Volume	mg/kg, mg/L, or mg				
Sample ID	Description	g or mL	Pb				
<150 mesh tailings		100	2370				
			237	0	0	0	0
"B"-2-1 4 hr filtrate + wash		290	416				
			121	0	0	0	0
"B"-2-2 4 hr residue		98	1150				
			113	0	0	0	0
Calculated Feed Analysis		-2%	**2332**	**0**	**0**	**0**	**0**
Calculated Leach Recovery			**52%**				

AGITATION LEACH DATA SHEET

Sample: 100 grams Miramar "B," <150 mesh tails

Instructions: Add 100 grams of sample in 200 mL water
Use 400 mL beaker and propeller stirrer
Adjust pH to 1.2 target with reagent nitric acid
Measure pH and readjust as necessary at 60, 120, and 180 minutes
At 240 minutes, flocculate, filter, and wash residue with 100 mL water
Combine filtrate and wash, measure volume, submit for lead analysis
Get wet and dry weight of filter cake, submit for lead analysis

Data:

Leach Time, minutes		0	60	120	180	240	
Clock Time		9:53	10:53	11:45	12:52	13:49	
Solution Temperature, °C		24	25	25	25	25	
Solution pH		9.50	1.67	1.29	1.23	1.24	
Grams HNO$_3$ Added		3.70	0.81	0.36	0.13		
HNO$_3$, milliequivalents / gram soil		0.411	0.501	0.541	0.555		
HNO$_3$ Consumed		100% HNO$_3$	35	kg/1000 kg soil added			

Test No:	"B"-3	Dry Wt.	Solid or Solution Analysis or Mass				
		or Volume	mg/kg, mg/L, or mg				
Sample ID	Description	g or mL	Pb				
<150 mesh tailings		100	2370				
			237	0	0	0	0
"B"-3-1 4 hr filtrate + wash		275	488				
			134	0	0	0	0
"B"-3-2 4 hr residue		97	970				
			94	0	0	0	0
Calculated Feed Analysis		-3%	**2287**	**0**	**0**	**0**	**0**
Calculated Leach Recovery			**59%**				

AGITATION LEACH DATA SHEET

Sample: 100 grams Miramar "B," <150 mesh tails

Instructions: Add 100 grams of sample in 200 mL water
                               Use 400 mL beaker and propeller stirrer
                               Add 10 grams $Fe_2(SO_4)3xH_2O$
                               Adjust pH to 1.5 target with reagent nitric acid
                               Measure pH and readjust as necessary at 30, 60, 120, and 180 minutes
                               At 240 minutes, flocculate, filter, and wash residue with 100 mL water
                               Combine filtrate and wash, measure volume, submit for lead analysis
                               Get wet and dry weight of filter cake, submit for lead analysis

Data:

Leach Time, minutes			0	30	60	120	180	240
Clock Time			12:45	13:15	13:45	14:45	15:45	16:45
Solution Temperature, °C			24	25	25	25	25	
Solution pH			2.34	1.48	1.49	1.50	1.49	
Grams HNO_3 Added			3.85	0.32	0.39	0.00		
HNO_3, milliequivalents / gram soil			0.428	0.463	0.507	0.507		
HNO_3 Consumed			100% HNO_3	32	kg/1000 kg soil added			

Test No:	"B"-4	Dry Wt.	Solid or Solution Analysis or Mass				
		or Volume	mg/kg, mg/L, or mg				
Sample ID	Description	g or mL	Pb				
<150 mesh tailings		100	2370				
			237	0	0	0	0
"B"-4-1 4 hr filtrate + wash		1400	5				
			7	0	0	0	0
"B"-4-2 4 hr residue		97	2080				
			202	0	0	0	0
Calculated Feed Analysis		-3%	**2090**	**0**	**0**	**0**	**0**
Calculated Leach Recovery			**3%**				

AGITATION LEACH DATA SHEET

Sample: 100 grams Miramar "B," <150 mesh tails

Instructions: Add 100 grams of sample in 400 mL water
Use 600 mL beaker and propeller stirrer
Add HNO_3 and H_2O_2
Adjust pH to 1.5 target with reagent nitric acid
Measure pH and emf at 30, 60, 120, 180, and 240 minutes
At 240 minutes, flocculate, filter, and wash residue with 100 mL water
Combine filtrate and wash, measure volume, submit for lead analysis
Get wet and dry weight of filter cake, submit for lead analysis

Data:

Leach Time, minutes	0	30	60	90	150	210
Clock Time	13:00	13:30	14:00	14:30	15:30	16:30
Solution Temperature, °C	32	32	32	32	31	30
Solution pH	0.99	1.26	1.12	1.29	1.22	1.19
Solution emf, mv	548	558	570	556	546	
Milliliters 30% H_2O_2 Added	1.00	1.00				
Grams HNO_3 Added	7.10					
HNO_3, milliequivalents / gram soil	0.789	0.789	0.789	0.789		
HNO_3 Consumed	100% HNO_3	50	kg/1000 kg soil added			
H_2O_2 Consumed	100% H_2O_2	7	kg/1000 kg soil added (d 30% = 1.1122 g/mL)			

Test No:	"B"-5	Dry Wt.	Solid or Solution Analysis or Mass			
		or Volume	mg/kg, mg/L, or mg			
Sample ID	Description	g or mL	Pb	Sb		
<150 mesh tailings		100	2370			
			237	0	0	0	0
"B"-5-1 4 hr filtrate + wash		460	340	1		
			156	0	0	0	0
"B"-5-2 4 hr residue		95	704	41		
			67	4	0	0	0
Calculated Feed Analysis		-5%	2231	43	0	0	0
Calculated Leach Recovery			70%	11%		

AGITATION LEACH DATA SHEET

Sample: 100 grams Miramar "B," <150 mesh tails

Instructions: Add 100 grams of sample in 400 mL water
                                    Use 600 mL beaker and propeller stirrer
                                    Add HNO_3, HCL and H_2O_2
                                    Adjust pH to 1.5 target with reagent nitric acid
                                    Measure pH and emf at 30, 60, 120, 180, and 240 minutes
                                    At 240 minutes, flocculate, filter, and wash residue with 100 mL water
                                    Combine filtrate and wash, measure volume, submit for lead analysis
                                    Get wet and dry weight of filter cake, submit for lead analysis

Data:

Leach Time, minutes	0	30	60	120	180	240
Clock Time	13:00	13:30	14:00	14:30	15:30	16:30
Solution Temperature, °C	34	34	34	34	33	31
Solution pH	1.07	1.25	0.62	0.77	0.79	0.78
Solution emf, mv	560	566	569	554	541	
Milliliters 30% H_2O_2 Added	1.00	1.00				
Milliliters 37% HCl added			7.00			
Grams HNO_3 Added	7.10					
HNO_3, milliequivalents / gram soil	0.789	0.789	0.789	0.789		
HNO_3 Consumed	100% HNO_3	50	kg/1000 kg soil added			
HCL Consumed	100% HCl	31	kg/1000 kg soil added (d 37% = 1.183 g/mL)			
H_2O_2 Consumed	100% H_2O_2	7	kg/1000 kg soil added (d 30% = 1.1122 g/mL)			

Test No:	"B"-6	Dry Wt.	Solid or Solution Analysis or Mass				
		or Volume	mg/kg, mg/L, or mg				
Sample ID	Description	g or mL	Pb	Sb			
<150 mesh tailings		100	2370				
			237	0	0	0	0
"B"-6-1 4 hr filtrate + wash		450	435	1			
			196	0	0	0	0
"B"-6-2 4 hr residue		93	389	37			
			36	3	0	0	0
Calculated Feed Analysis		-7%	2320	39	0	0	0
Calculated Leach Recovery			84%	12%			

AGITATION LEACH DATA SHEET

Sample: 100 grams Miramar "B," <150 mesh tails

Instructions: Add 100 grams of sample in 400 mL water
Use 600 mL beaker and propeller stirrer
Add Concentrated Acetic Acid
Adjust pH to No target with reagent nitric acid
Measure pH and emf at 30, 60, 90, 150, and 210 minutes.
At 240 minutes, flocculate, filter, and wash residue with 100 mL water
Combine filtrate and wash, measure volume, submit for lead analysis
Get wet and dry weight of filter cake, submit for lead analysis

Data:

Leach Time, minutes	0	30	60	90	150	210
Clock Time	13:00	13:30	14:00	14:30	15:30	16:30
Solution Temperature, °C	31	32	32	32	32	31
Solution pH	5.07	5.36	5.14	3.82	3.84	3.83
Solution emf, mv	484.00	470.00	477.00	469.00	460.00	
Milliliters Acetic Acid added	2.30		15.00			
Grams HNO_3 Added						
HNO_3, milliequivalents / gram soil	0.000	0.000	0.000	0.000		
HNO_3 Consumed	100% HNO_3	0	kg/1000 kg soil added			
Acetic Acid Consumed	Acetic Acid	182	kg/1000 kg soil added (d100% = 1.0498 g/mL)			

Test No:	"B"-7	Dry Wt.	Solid or Solution Analysis or Mass			
		or Volume	mg/kg, mg/L, or mg			
Sample ID	Description	g or mL	Pb			
<150 mesh tailings		100	2370			
			237	0	0	0	0
"B"-7-1 4 hr filtrate + wash		465	235	1		
			109	0	0	0	0
"B"-7-2 4 hr residue		96	1160	44		
			111	4	0	0	0
Calculated Feed Analysis		-4%	2208	47	0	0	0
Calculated Leach Recovery			50%	10%		

AGITATION LEACH DATA SHEET

Sample: 100 grams Miramar "B," <150 mesh tails

Instructions: Add 100 grams of sample in 400 mL water
Use 600 mL beaker and propeller stirrer
Add Concentrated Citric Acid
Adjust pH to No target with reagent nitric acid
Measure pH and emf at 30, 60, 90, 150, and 210 minutes.
At 240 minutes, flocculate, filter, and wash residue with 100 mL water
Combine filtrate and wash, measure volume, submit for lead analysis
Get wet and dry weight of filter cake, submit for lead analysis

Data:

Leach Time, minutes	0	30	60	90	150	210
Clock Time	13:00	13:30	14:00	14:30	15:30	16:30
Solution Temperature, °C	32	33	34	32	34	31
Solution pH	3.24	3.40	3.17	2.70	2.66	2.67
Solution emf, mv	469	461	453	436	444	
Grams Citric Acid Added	7.70		7.70			
Grams HNO_3 Added						
HNO_3, milliequivalents / gram soil	0.000	0.000	0.000	0.000		
HNO_3 Consumed	100% HNO_3	0	kg/1000 kg soil added			
Citric Acid Consumed	Citric acid	118	kg/1000 kg soil added (d 60% = 1.2738 g/mL)			
Citric Acid Consumed	Citric acid	237	kg/1000 kg soil added (d 100% = 1.542 g/mL)			

Test No:	"B"-8	Dry Wt.	Solid or Solution Analysis or Mass				
		or Volume	mg/kg, mg/L, or mg				
Sample ID	Description	g or mL	Pb	Sb			
<150 mesh tailings		100	2370				
			237	0	0	0	0
"B"-8-1 4 hr filtrate + wash		460	270	9			
			124	4	0	0	0
"B"-8-2 4 hr residue		97	1030	28			
			99	3	0	0	0
Calculated Feed Analysis		-3%	**2237**	**68**	**0**	**0**	**0**
Calculated Leach Recovery			**56%**	**60%**			

AGITATION LEACH DATA SHEET

Sample: 100 grams Miramar "B," <150 mesh tails

Instructions: Add 100 grams of sample in 400 mL water
Use 600 mL beaker and propeller stirrer
Add 2 grams HNO_3
Add 2 grams HCl
Add 1 + 1 grams H_2O_2
Measure pH and emf at 30, 60, 120, 180, and 240 minutes
At 240 minutes, flocculate, filter, and wash residue with 100 mL water
Combine filtrate and wash, measure volume, submit for lead analysis
Get wet and dry weight of filter cake, submit for lead analysis

Data:

Leach Time, minutes		0	30	60	120	180	240
Clock Time		8:00	8:30	9:00	10:00	11:00	12:00
Solution Temperature, °C		29	29	29	28	28	28
Solution pH		8.04	1.48	2.00	2.08	2.11	2.10
Solution emf, mv		238	396	564	577	573	545
Grams HCl Added		2.00					
Grams H_2O_2 Added			1.00	1.00			
Grams HNO_3 Added		2.00					
HNO_3, milliequivalents / gram soil		0.222	0.222	0.222	0.222		
HNO_3 Consumed	100% HNO_3	14	kg/1000 kg soil added				
HCl Consumed	100% HCl	7	kg/1000 kg soil added				
H_2O_2 Consumed	100% H_2O_2	6	kg/1000 kg soil added				

Test No:	"B"-9	Dry Wt.	Solid or Solution Analysis or Mass				
		or Volume	mg/kg, mg/L, or mg				
Sample ID	Description	g or mL	Pb	Sb			
<150 mesh tailings		100	2370				
			237	0	0	0	0
"B"-9-1 4 hr filtrate + wash		413	251				
			104	0	0	0	0
"B"-9-2 4 hr residue		99	1273	45			
			125	4	0	0	0
Calculated Feed Analysis		-2%	**2291**	**44**	**0**	**0**	**0**
Calculated Leach Recovery			**45%**	**0%**			

AGITATION LEACH DATA SHEET

Sample: 100 grams Miramar "B," <150 mesh tails

Instructions: Add 100 grams of sample in 400 mL water
                                Use 600 mL beaker and propeller stirrer
                                Add 4 grams HNO_3
                                Add 4 grams HCl
                                Add 1 + 1 grams H_2O_2
                                Measure pH and emf at 30, 60, 120, 180, and 240 minutes
                                At 240 minutes, flocculate, filter, and wash residue with 100 mL water
                                Combine filtrate and wash, measure volume, submit for lead analysis
                                Get wet and dry weight of filter cake, submit for lead analysis

Data:

Leach Time, minutes		0	30	60	120	180	240
Clock Time		8:00	8:30	9:00	10:00	11:00	12:00
Solution Temperature, °C		29	29	29	28	28	28
Solution pH		8.10	1.12	1.19	1.21	1.22	1.18
Solution emf, mv		224	449	554	570	562	543
Grams HCl Added		4.00					
Grams H_2O_2 Added			1.00	1.00			
Grams HNO_3 Added		4.00					
HNO_3, milliequivalents / gram soil		0.444	0.444	0.444	0.444		
HNO_3 Consumed	100% HNO_3	28	kg/1000 kg soil added				
HCl Consumed	100% HCl	15	kg/1000 kg soil added				
H_2O_2 Consumed	100% H_2O_2	6	kg/1000 kg soil added				

Test No:	"B"-10	Dry Wt.	Solid or Solution Analysis or Mass				
		or Volume	mg/kg, mg/L, or mg				
Sample ID	Description	g or mL	Pb	Sb			
<150 mesh tailings		100	2370				
			237	0	0	0	0
"B"-10-1 4 hr filtrate + wash		410	426				
			175	0	0	0	0
"B"-10-2 4 hr residue		98	599	40			
			59	4	0	0	0
Calculated Feed Analysis		-2%	2332	39	0	0	0
Calculated Leach Recovery			75%	0%			

AGITATION LEACH DATA SHEET

Sample: 100 grams Miramar "B," <150 mesh tails

Instructions: Add 100 grams of sample in 400 mL water
Use 600 mL beaker and propeller stirrer
Add 8 grams HNO_3
Add 8 grams HCl
Add 1 + 1 grams H_2O_2
Measure pH and emf at 30, 60, 120, 180, and 240 minutes
At 240 minutes, flocculate, filter, and wash residue with 100 mL water
Combine filtrate and wash, measure volume, submit for lead analysis
Get wet and dry weight of filter cake, submit for lead analysis

Data:

Leach Time, minutes	0	30	60	120	180	240
Clock Time	8:00	8:30	9:00	10:00	11:00	12:00
Solution Temperature, °C	29	29	29	28	28	28
Solution pH	8.88	0.90	0.97	0.99	1.00	0.95
Solution emf, mv	224	499	540	555	551	539
Grams HCl Added	8.00					
Grams H_2O_2 Added		1.00	1.00			
Grams HNO_3 Added	8.00					
HNO_3, milliequivalents / gram soil	0.889	0.889	0.889	0.889		
HNO_3 Consumed	100% HNO_3	56	kg/1000 kg soil added			
HCl Consumed	100% HCl	30	kg/1000 kg soil added			
H_2O_2 Consumed	100% H_2O_2	6	kg/1000 kg soil added			

Test No:	"B"-11	Dry Wt.	Solid or Solution Analysis or Mass				
		or Volume	mg/kg, mg/L, or mg				
Sample ID	Description	g or mL	Pb	Sb			
<150 mesh tailings		100	2370				
			237	0	0	0	0
"B"-11-1 4 hr filtrate + wash		420	477				
			200	0	0	0	0
"B"-11-2 4 hr residue		97	434	36			
			42	4	0	0	0
Calculated Feed Analysis		-3%	2426	35	0	0	0
Calculated Leach Recovery			83%	0%			

AGITATION LEACH DATA SHEET

Sample: 100 grams Miramar "B," <150 mesh tails

Instructions: Add 100 grams of sample in 400 mL water
                                Use 600 mL beaker and propeller stirrer
                                Add 2 grams HNO_3
                                Add 4 grams HCl
                                Add 1 + 1 grams H_2O_2
                                Measure pH and emf at 30, 60, 120, 180, and 240 minutes
                                At 240 minutes, flocculate, filter, and wash residue with 100 mL water
                                Combine filtrate and wash, measure volume, submit for lead analysis
                                Get wet and dry weight of filter cake, submit for lead analysis

Data:

Leach Time, minutes		0	30	60	120	180	240
Clock Time		8:00	8:30	9:00	10:00	11:00	12:00
Solution Temperature, °C		29	29	29	28	28	28
Solution pH		8.73	1.24	1.37	1.43	1.45	1.44
Solution emf, mv		222	464	553	569	566	549
Grams HCl Added		4.00					
Grams H_2O_2 Added			1.00	1.00			
Grams HNO_3 Added		2.00					
HNO_3, milliequivalents / gram soil		0.222	0.222	0.222	0.222		
HNO_3 Consumed	100% HNO_3	14	kg/1000 kg soil added				
HCl Consumed	100% HCl	15	kg/1000 kg soil added				
H_2O_2 Consumed	100% H_2O_2	6	kg/1000 kg soil added				

Test No:	"B"-12	Dry Wt.	Solid or Solution Analysis or Mass				
		or Volume	mg/kg, mg/L, or mg				
Sample ID	Description	g or mL	Pb	Sb			
<150 mesh tailings		100	2370				
			237	0	0	0	0
"B"-12-1 4 hr filtrate + wash		420	379				
			159	0	0	0	0
"B"-12-2 4 hr residue		98	770	43			
			75	4	0	0	0
Calculated Feed Analysis		-2%	2346	42	0	0	0
Calculated Leach Recovery			68%	0%			

AGITATION LEACH DATA SHEET

Sample: 100 grams Miramar "B," <150 mesh tails

Instructions: Add 100 grams of sample in 400 mL water
Use 600 mL beaker and propeller stirrer
Add 2 grams HNO_3
Add 8 grams HCl
Add 1 + 1 grams H_2O_2
Measure pH and emf at 30, 60, 120, 180, and 240 minutes
At 240 minutes, flocculate, filter, and wash residue with 100 mL water
Combine filtrate and wash, measure volume, submit for lead analysis
Get wet and dry weight of filter cake, submit for lead analysis

Data:

Leach Time, minutes	0	30	60	120	180	240
Clock Time	12:00	12:30	13:00	14:00	15:00	16:00
Solution Temperature, °C	24	27	28	28	27	27
Solution pH	8.12	1.16	1.15	1.16	1.18	1.19
Solution emf, mv	332	322	522	540	538	533
Grams HCl Added	8.00					
Grams H_2O_2 Added		1.00	1.00			
Grams HNO_3 Added	2.00					
HNO_3, milliequivalents / gram soil	0.222	0.222	0.222	0.222		
HNO_3 Consumed	100% HNO_3	14	kg/1000 kg soil added			
HCl Consumed	100% HCl	30	kg/1000 kg soil added			
H_2O_2 Consumed	100% H_2O_2	6	kg/1000 kg soil added			

Test No:	"B"-13	Dry Wt.	Solid or Solution Analysis or Mass			
		or Volume	mg/kg, mg/L, or mg			
Sample ID	Description	g or mL	Pb	Sb		
<150 mesh tailings		100	2370			
			237	0	0	0	0
"B"-13-1 4 hr filtrate + wash		454	389			
			177	0	0	0	0
"B"-13-2 4 hr residue		98	596	42		
			58	4	0	0	0
Calculated Feed Analysis		-2%	2349	41	0	0	0
Calculated Leach Recovery			75%	0%		

AGITATION LEACH DATA SHEET

Sample: 100 grams Miramar "B," <150 mesh tails

Instructions: Add 100 grams of sample in 400 mL water
                               Use 600 mL beaker and propeller stirrer
                               Add 0 grams HNO_3
                               Add 8 grams HCl
                               Add 1 + 1 grams H_2O_2
                               Measure pH and emf at 30, 60, 120, 180, and 240 minutes
                               At 240 minutes, flocculate, filter, and wash residue with 100 mL water
                               Combine filtrate and wash, measure volume, submit for lead analysis
                               Get wet and dry weight of filter cake, submit for lead analysis

Data:

Leach Time, minutes			0	30	60	120	180	240
Clock Time			12:00	12:30	13:00	14:00	15:00	16:00
Solution Temperature, °C			24	27	28	28	27	27
Solution pH			8.73	1.21	1.22	1.25	1.26	1.28
Solution emf, mv			315	349	526	552	545	541
Grams HCl Added			8.00					
Grams H_2O_2 Added				1.00	1.00			
Grams HNO_3 Added								
HNO_3, milliequivalents / gram soil			0.000	0.000	0.000	0.000		
HNO_3 Consumed			100% HNO_3	0	kg/1000 kg soil added			
HCl Consumed			100% HCl	30	kg/1000 kg soil added			
H_2O_2 Consumed			100% H_2O_2	6	kg/1000 kg soil added			

Test No:	"B"-14	Dry Wt.	Solid or Solution Analysis or Mass				
		or Volume	mg/kg, mg/L, or mg				
Sample ID	Description	g or mL	Pb	Sb			
<150 mesh tailings		100	2370				
			237	0	0	0	0
"B"-14-1 4 hr filtrate + wash		460	379				
			174	0	0	0	0
"B"-14-2 4 hr residue		96	608	42			
			59	4	0	0	0
Calculated Feed Analysis		-4%	2330	40	0	0	0
Calculated Leach Recovery			75%	0%			

AGITATION LEACH DATA SHEET

Sample: 100 grams Miramar "B," <150 mesh tails

Instructions: Add 100 grams of sample in 400 mL water
Use 600 mL beaker and propeller stirrer
Add 4 grams HNO_3
Add 4 grams HCl
Add 1 + 1 grams $NaClO_3$
Measure pH and emf at 30, 60, 120, 180, and 240 minutes
At 240 minutes, flocculate, filter, and wash residue with 100 mL water
Combine filtrate and wash, measure volume, submit for lead analysis
Get wet and dry weight of filter cake, submit for lead analysis

Data:

Leach Time, minutes	0	30	60	120	180	240
Clock Time	12:00	12:30	13:00	14:00	15:00	16:00
Solution Temperature, °C	24	27	28	28	27	27
Solution pH	8.90	1.17	1.18	1.20	1.21	1.21
Solution emf, mv	305	361	460	450	456	415
Grams HCl Added	4.00					
Grams $NaClO_3$ Added		1.00	1.00			
Grams HNO_3 Added	4.00					
HNO_3, milliequivalents / gram soil	0.444	0.444	0.444	0.444		
HNO_3 Consumed	100% HNO_3	28	kg/1000 kg soil added			
HCl Consumed	100% HCl	15	kg/1000 kg soil added			
$NaClO_3$ Consumed	$NaClO_3$	20	kg/1000 kg soil added			

Test No:	"B"-15	Dry Wt. or Volume	Solid or Solution Analysis or Mass mg/kg, mg/L, or mg				
Sample ID	Description	g or mL	Pb	Sb			
<150 mesh tailings		100	2370				
			237	0	0	0	0
"B"-15-1 4 hr filtrate + wash		458	368				
			169	0	0	0	0
"B"-15-2 4 hr residue		96	664	45			
			64	4	0	0	0
Calculated Feed Analysis		-4%	2322	43	0	0	0
Calculated Leach Recovery			73%	0%			

AGITATION LEACH DATA SHEET

Sample: 100 grams Miramar "B," <150 mesh tails

Instructions: Add 100 grams of sample in 400 mL water
Use 600 mL beaker and propeller stirrer
Add 4 grams HCl
Add 4 grams Citric acid
Add 1 + 1 grams H_2O_2
Measure pH and emf at 30, 60, 120, 180, and 240 minutes
At 240 minutes, flocculate, filter, and wash residue with 100 mL water
Combine filtrate and wash, measure volume, submit for lead analysis
Get wet and dry weight of filter cake, submit for lead analysis

Data:

Leach Time, minutes		0	30	60	120	180	240
Clock Time		12:00	12:30	13:00	14:00	15:00	16:00
Solution Temperature, °C		24	27	28	28	27	27
Solution pH		8.66	1.60	1.70	1.74	1.78	1.79
Solution emf, mv		304	290	533	562	558	555
Grams Citric Acid Added		4.00					
Grams H_2O_2 Added			1.00	1.00			
Grams HCl Added		4.00					
HCl, milliequivalents / gram soil		0.406	0.406	0.406	0.406		
HCl Consumed		100% HCl	15	kg/1000 kg soil added			
Citric Acid Consumed		Citric Acid	40	kg/1000 kg soil added			
H_2O_2 Consumed		100% H_2O_2	6	kg/1000 kg soil added			

Test No:	"B"-16	Dry Wt.	Solid or Solution Analysis or Mass				
		or Volume	mg/kg, mg/L, or mg				
Sample ID	Description	g or mL	Pb	Sb			
<150 mesh tailings		100	2370				
			237	0	0	0	0
"B"-16-1 4 hr filtrate + wash		467	274				
			128	0	0	0	0
"B"-16-2 4 hr residue		98	1060	35			
			103	3	0	0	0
Calculated Feed Analysis		-3%	2313	34	0	0	0
Calculated Leach Recovery			55%	0%			

AGITATION LEACH DATA SHEET

Sample: 100 grams Miramar "B," <150 mesh tails

Instructions: Add 100 grams of sample in 400 mL water
Use 600 mL beaker and propeller stirrer
Add 8 grams HNO_3
Add 8 grams HCl
Add 1 + 1 grams H_2O_2
Measure pH and emf at 30, 60, and 120 minutes
At 120 minutes, flocculate, filter, and wash residue with 100 mL water
Combine filtrate and wash, measure volume, submit for lead analysis
Get wet and dry weight of filter cake, submit for lead analysis

Data:

Leach Time, minutes	0	30	60	120		
Clock Time	9:00	9:30	10:00	10:30		
Solution Temperature, °C	22	23	23	22		
Solution pH	7.83	1.20	1.10	1.10		
Solution emf, mv	155	379	522	550		
Grams HCl Added	8.00					
Grams H_2O_2 Added		1.00	1.00			
Grams HNO_3 Added	8.00					
HNO_3, milliequivalents / gram soil	0.889	0.889	0.889	0.889		
HNO_3 Consumed	100% HNO_3	56	kg/1000 kg soil added			
HCl Consumed	100% HCl	30	kg/1000 kg soil added			
H_2O_2 Consumed	100% H_2O_2	6	kg/1000 kg soil added			

Test No: "B"-17		Dry Wt.	Solid or Solution Analysis or Mass				
		or Volume	mg/kg, mg/L, or mg				
Sample ID	Description	g or mL	Pb	Sb			
<150 mesh tailings		100	2370				
			237	0	0	0	0
"B"-17-1 2 hr filtrate + wash		362	470				
			170	0	0	0	0
"B"-17-2 2 hr residue		100	682				
			68	0	0	0	0
Calculated Feed Analysis		0%	2383	0	0	0	0
Calculated Leach Recovery			71%				

AGITATION LEACH DATA SHEET

Sample: Residue from Test Miramar "B"-17 47.8 grams

Instructions: Add 47.8 grams of sample in 200 mL water
Use 600 mL beaker and propeller stirrer
Add 4 grams HNO_3
Add 4 grams HCl
Add 0.5 + 0.5 grams H_2O_2
Measure pH and emf at 30, 60, and 120 minutes
At 120 minutes, flocculate, filter, and wash residue with 100 mL water
Combine filtrate and wash, measure volume, submit for lead analysis
Get wet and dry weight of filter cake, submit for lead analysis

Data:

Leach Time, minutes	0	30	60	120		
Clock Time	11:15	11:45	12:15	13:15		
Solution Temperature, °C	23	23	23	23		
Solution pH	1.73	1.05	0.98	0.99		
Solution emf, mv	502	466	526	546		
Grams HCl Added	4.00					
Grams H_2O_2 Added		0.50	0.50			
Grams HNO_3 Added	4.00					
HNO_3, milliequivalents / gram soil	0.444	0.444	0.444	0.444		
HNO_3 Consumed	100% HNO_3	59	kg/1000 kg soil added			
HCl Consumed	100% HCl	31	kg/1000 kg soil added			
H_2O_2 Consumed	100% H_2O_2	6	kg/1000 kg soil added			

Test No: "B"-17A		Dry Wt.	Solid or Solution Analysis or Mass				
		or Volume	mg/kg, mg/L, or mg				
Sample ID	Description	g or mL	Pb	Sb			
<150 mesh tailings		48	682				
			33	0	0	0	0
"B"-17A-1 2 hr filtrate + wash		265	76				
			20	0	0	0	0
"B"-17A-2 2 hr residue		48	313				
			15	0	0	0	0
Calculated Feed Analysis		0%	734	0	0	0	0
Calculated Leach Recovery			57%				

AGITATION LEACH DATA SHEET

Sample: 55 grams Twentynine Palms LR-23, <150 mesh tails
                     45 grams Twentynine Palms LR-23, <150 mesh mids
                     100 grams Twentynine Palms LR-23, tailings + middlings

Instructions: Add 100 grams of sample in 200 mL water
                     Use 400 mL beaker and propeller stirrer
                     Adjust pH to 2.0 target with reagent nitric acid
                     Measure pH and readjust as necessary at 30, 60, 120, and 180 minutes
                     At 240 minutes, flocculate, filter, and wash residue with 100 mL water
                     Combine filtrate and wash, measure volume, submit for lead analysis
                     Get wet and dry weight of filter cake, submit for lead analysis

Data:

Leach Time, minutes		0	30	60	120	180	240
Clock Time		8:30	9:06	9:30	10:30	11:30	12:30
Solution Temperature, °C		24		25	25	25	25
Solution pH		9.40	3.25	2.76	2.88	2.77	2.47
Grams HNO_3 Added		6.56	0.75	0.34	0.48	0.37	
HNO_3, milliequivalents / gram soil		0.729	0.812	0.850	0.903	0.944	
HNO_3 Consumed		100% HNO_3	60	kg/1000 kg soil added			

Test No:	"LR23"-1	Dry Wt.	Solid or Solution Analysis or Mass				
		or Volume	mg/kg, mg/L, or mg				
Sample ID	Description	g or mL	Pb				
<150 mesh tailings + middlings		100	15,100				
			1510	0	0	0	0
"LR23"-1-1 4 hr filtrate + wash		320	3536				
			1132	0	0	0	0
"LR23"-1-2 4 hr residue		94	3838				
			360	0	0	0	0
Calculated Feed Analysis		-6%	**14,919**	**0**	**0**	**0**	**0**
Calculated Leach Recovery			**76%**				

AGITATION LEACH DATA SHEET

Sample: 55 grams Twentynine Palms LR-23, <150 mesh tails
                        45 grams Twentynine Palms LR-23, <150 mesh mids
                        100 grams Twentynine Palms LR-23, tailings + middlings

Instructions: Add 100 grams of sample in 200 mL water
                        Use 400 mL beaker and propeller stirrer
                        Adjust pH to 2.0 target with reagent nitric acid
                        Measure pH and readjust as necessary at 30, 60, 120, and 180 minutes
                        At 240 minutes, flocculate, filter, and wash residue with 100 mL water
                        Combine filtrate and wash, measure volume, submit for lead analysis
                        Get wet and dry weight of filter cake, submit for lead analysis

Data:

Leach Time, minutes	0	30	60	120	180	240
Clock Time	8:30	9:06	9:30	10:30	11:30	12:30
Solution Temperature, °C	24		25	25	25	25
Solution pH	9.40	3.25	2.76	2.88	2.77	2.47
Grams HNO_3 Added	6.56	0.75	0.34	0.48	0.37	
HNO_3, milliequivalents / gram soil	0.729	0.812	0.850	0.903	0.944	
HNO_3 Consumed	100% HNO_3	60	kg/1000 kg soil added			

Test No:	"LR23"-1(2)	Dry Wt.	Solid or Solution Analysis or Mass			
		or Volume	mg/kg, mg/L, or mg			
Sample ID	Description	g or mL	Pb			
<150 mesh tailings + middlings		100	15,100			
			1510	0	0	0	0
"LR23"-1(2)-1 4 hr filtrate + wash		320	3536			
			1132	0	0	0	0
"LR23"-1(2)-2 4 hr residue		94	3838			
			360	0	0	0	0
Calculated Feed Analysis		-6%	**14,919**	**0**	**0**	**0**	**0**
Calculated Leach Recovery			**76%**			

AGITATION LEACH DATA SHEET

Sample: 55 grams Twentynine Palms LR-23, <150 mesh tails
45 grams Twentynine Palms LR-23, <150 mesh mids
100 grams Twentynine Palms LR-23, tailings + middlings

Instructions: Add 100 grams of sample in 200 mL water
Use 400 mL beaker and propeller stirrer
Adjust pH to 1.5 target with reagent nitric acid
Measure pH and readjust as necessary at 30, 60, 120, and 180 minutes
At 240 minutes, flocculate, filter, and wash residue with 100 mL water
Combine filtrate and wash, measure volume, submit for lead analysis
Get wet and dry weight of filter cake, submit for lead analysis

Data:

Leach Time, minutes			0	30	60	120	180	240
Clock Time			8:35	9:08	9:31	10:33	11:37	12:37
Solution Temperature, °C			24	24	25	25	25	25
Solution pH			9.40	2.26	2.31	2.13	1.65	1.57
Grams HNO_3 Added			7.77	0.58	0.37	0.34	0.21	
HNO_3, milliequivalents / gram soil			0.863	0.928	0.969	1.007	1.030	
HNO_3 Consumed			100% HNO_3	65	kg/1000 kg soil added			

Test No:	"LR23"-2	Dry Wt.	Solid or Solution Analysis or Mass				
		or Volume	mg/kg, mg/L, or mg				
Sample ID	Description	g or mL	Pb				
<150 mesh tailings + middlings		100	15,100				
			1510	0	0	0	0
"LR23"-2-1 4 hr filtrate + wash		305	4247				
			1295	0	0	0	0
"LR23"-2-2 4 hr residue		93	2651				
			248	0	0	0	0
Calculated Feed Analysis		-7%	**15,431**	**0**	**0**	**0**	**0**
Calculated Leach Recovery			**84%**				

AGITATION LEACH DATA SHEET

Sample: 55 grams Twentynine Palms LR-23, <150 mesh tails
                             45 grams Twentynine Palms LR-23, <150 mesh mids
                             100 grams Twentynine Palms LR-23, tailings + middlings

Instructions: Add 100 grams of sample in 200 mL water
                            Use 400 mL beaker and propeller stirrer
                            Adjust pH to 1.2 target with reagent nitric acid
                            Measure pH and readjust as necessary at 30, 60, 120, and 180 minutes
                            At 240 minutes, flocculate, filter, and wash residue with 100 mL water
                            Combine filtrate and wash, measure volume, submit for lead analysis
                            Get wet and dry weight of filter cake, submit for lead analysis

Data:

Leach Time, minutes	0	30	60	120	180	240
Clock Time	8:37	9:09	9:32	10:34	11:38	12:38
Solution Temperature, °C	24	24	25	25	25	25
Solution pH	9.40	1.78	1.42	1.16	1.23	1.25
Grams HNO_3 Added	8.75	0.49	0.63	0.07	0.07	
HNO_3, milliequivalents / gram soil	0.972	1.027	1.096	1.104	1.112	
HNO_3 Consumed	100% HNO_3	70	kg/1000 kg soil added			

Test No:	"LR23"-3	Dry Wt. or Volume	Solid or Solution Analysis or Mass mg/kg, mg/L, or mg			
Sample ID	Description	g or mL	Pb			
<150 mesh tailings + middlings		100	15,100			
			1510	0	0	0	0
"LR23"-3-1 4 hr filtrate + wash		323	4273			
			1380	0	0	0	0
"LR23"-3-2 4 hr residue		93	2092			
			195	0	0	0	0
Calculated Feed Analysis		-7%	**15,749**	**0**	**0**	**0**	**0**
Calculated Leach Recovery			**88%**			

AGITATION LEACH DATA SHEET

Sample: 55 grams Twentynine Palms LR-23, <150 mesh tails
45 grams Twentynine Palms LR-23, <150 mesh mids
100 grams Twentynine Palms LR-23, tailings + middlings

Instructions: Add 100 grams of sample in 200 mL water
Use 400 mL beaker and propeller stirrer
Add 10 grams $Fe_2(SO_4)3H_2O$
Adjust pH to 1.2 target with reagent nitric acid
Measure pH and readjust as necessary at 30, 60, 120, and 180 minutes
At 240 minutes, flocculate, filter, and wash residue with 100 mL water
Combine filtrate and wash, measure volume, submit for lead analysis
Get wet and dry weight of filter cake, submit for lead analysis

Data:

Leach Time, minutes	0	30	60	120	180	240
Clock Time	12:45	13:15	13:45	14:45	15:45	16:45
Solution Temperature, °C	24	24	25	25	25	25
Solution pH	2.57	1.92	1.49	1.47	1.49	
Grams HNO_3 Added	3.48	1.52	1.02	1.03	0.40	
HNO_3, milliequivalents / gram soil	0.387	0.555	0.669	0.783	0.828	
HNO_3 Consumed	100% HNO_3	52	kg/1000 kg soil added			

Test No:	"LR23"-4	Dry Wt.	Solid or Solution Analysis or Mass			
		or Volume	mg/kg, mg/L, or mg			
Sample ID	Description	g or mL	Pb			
<150 mesh tailings + middlings		100	15,100			
			1510	0	0	0	0
"LR23"-4-1 4 hr filtrate + wash		970	10			
			10	0	0	0	0
"LR23"-4-2 4 hr residue		98	12,700			
			1248	0	0	0	0
Calculated Feed Analysis		-2%	12,581	0	0	0	0
Calculated Leach Recovery			1%			

AGITATION LEACH DATA SHEET

Sample: 55 grams Twentynine Palms LR-23, <150 mesh tails
                     45 grams Twentynine Palms LR-23, <150 mesh mids
                     100 grams Twentynine Palms LR-23, tailings + middlings

Instructions: Add 100 grams of sample in 400 mL water
                     Use 600 mL beaker and propeller stirrer
                     Add 8 grams HNO_3
                     Add 8 grams HCl
                     Add 1 + 1 grams H_2O_2
                     Measure pH and emf at 30, 60, 120, 180, and 240 minutes
                     At 240 minutes, flocculate, filter, and wash residue with 100 mL water
                     Combine filtrate and wash, measure volume, submit for lead analysis
                     Get wet and dry weight of filter cake, submit for lead analysis

Data:

Leach Time, minutes		0	30	60	120	180	240
Clock Time		10:15	10:45	11:15	12:15	13:15	14:15
Solution Temperature, °C		25	26	26	25	25	25
Solution pH		7.98	1.21	1.24	1.32	1.23	1.33
Solution emf, mv		195	213	534	548	537	531
Grams HCl Added		8.00					
Grams H_2O_2 Added			1.00	1.00			
Grams HNO_3 Added		8.00					
HNO_3, milliequivalents / gram soil		0.889	0.889	0.889	0.889		
HNO_3 Consumed		100% HNO_3	56	kg/1000 kg soil added			
HCl Consumed		100% HCl	30	kg/1000 kg soil added			
H_2O_2 Consumed		100% H_2O_2	6	kg/1000 kg soil added			

Test No:	"LR23"-5	Dry Wt.	Solid or Solution Analysis or Mass				
		or Volume	mg/kg, mg/L, or mg				
Sample ID	Description	g or mL	Pb				
<150 mesh tailings + middlings		100	15,100				
			1510	0	0	0	0
"LR23"-5-1 4 hr filtrate + wash		501	2845				
			1425	0	0	0	0
"LR23"-5-2 4 hr residue		97	1117				
			108	0	0	0	0
Calculated Feed Analysis		-3%	15,334	0	0	0	0
Calculated Leach Recovery			93%				

AGITATION LEACH DATA SHEET

Sample: Residue from Test Twentynine Palms "LR23"-5 50.5 grams

Instructions: Add 50.5 grams of sample in 400 mL water
Use 600 mL beaker and propeller stirrer
Add 8 grams HNO_3
Add 8 grams HCl
Add 1 + 1 grams H_2O_2
Measure pH and emf at 30, 60, 120, 180, and 240 minutes
At 240 minutes, flocculate, filter, and wash residue with 100 mL water
Combine filtrate and wash, measure volume, submit for lead analysis
Get wet and dry weight of filter cake, submit for lead analysis

Data:

Leach Time, minutes	0	30	60	120	180	240
Clock Time	7:45	8:15	8:45	9:45	10:45	11:45
Solution Temperature, °C	23	23	24	24	25	25
Solution pH	2.70	1.24	1.25	1.27	1.25	1.25
Solution emf, mv	204	419	536	553	553	548
Grams HCl Added	8.00					
Grams H_2O_2 Added		1.00	1.00			
Grams HNO_3 Added	8.00					
HNO_3, milliequivalents / gram soil	0.889	0.889	0.889	0.889		
HNO_3 Consumed	100% HNO_3	111	kg/1000 kg soil added			
HCl Consumed	100% HCl	59	kg/1000 kg soil added			
H_2O_2 Consumed	100% H_2O_2	12	kg/1000 kg soil added			

Test No: "LR23"-5A		Dry Wt.	Solid or Solution Analysis or Mass			
		or Volume	mg/kg, mg/L, or mg			
Sample ID	Description	g or mL	Pb			
<150 mesh tailings + middlings		51	1117			
			56	0	0	0	0
"LR23"-5A-1 4 hr filtrate + wash		387	332			
			128	0	0	0	0
"LR23"-5A-2 4 hr residue		45	167			
			8	0	0	0	0
Calculated Feed Analysis		-11%	2693	0	0	0	0
Calculated Leach Recovery			94%			

AGITATION LEACH DATA SHEET

Sample: 55 grams Twentynine Palms LR-23, <150 mesh tails
                         45 grams Twentynine Palms LR-23, <150 mesh mids
                         100 grams Twentynine Palms LR-23, tailings + middlings

Instructions: Add 100 grams of sample in 400 mL water
                         Use 600 mL beaker and propeller stirrer
                         Add 8 grams HNO_3
                         Add 8 grams HCl
                         Add 1 + 1 grams H_2O_2
                         Measure pH and emf at 30, 60, and 120 minutes
                         At 120 minutes, flocculate, filter, and wash residue with 100 mL water
                         Combine filtrate and wash, measure volume, submit for lead analysis
                         Get wet and dry weight of filter cake, submit for lead analysis

Data:

Leach Time, minutes		0	30	60	120		
Clock Time		9:00	9:30	10:00	10:30		
Solution Temperature, °C		23	23	23	23		
Solution pH		7.99	1.15	1.16	1.17		
Solution emf, mv		150	178	513	555		
Grams HCl Added		8.00					
Grams H_2O_2 Added			1.00	1.00			
Grams HNO_3 Added		8.00					
HNO_3, milliequivalents / gram soil		0.889	0.889	0.889	0.889		
HNO_3 Consumed	100% HNO_3	56	kg/1000 kg soil added				
HCl Consumed	100% HCl	30	kg/1000 kg soil added				
H_2O_2 Consumed	100% H_2O_2	6	kg/1000 kg soil added				

Test No:	"LR23"-6	Dry Wt.	Solid or Solution Analysis or Mass				
		or Volume	mg/kg, mg/L, or mg				
Sample ID	Description	g or mL	Pb				
<150 mesh tailings + middlings		100	15,100				
			1510	0	0	0	0
"LR23"-6-1 2 hr filtrate + wash		382	2185				
			835	0	0	0	0
"LR23"-6-2 2 hr residue		94	3200				
			301	0	0	0	0
Calculated Feed Analysis		-6%	**11,355**	**0**	**0**	**0**	**0**
Calculated Leach Recovery			**74%**				

AGITATION LEACH DATA SHEET

Sample: Residue from Test Twentynine Palms "LR23"-6 47.0 grams

Instructions: Add 47.0 grams of sample in 400 mL water
Use 600 mL beaker and propeller stirrer
Add 4 grams HNO_3
Add 4 grams HCl
Add 0.5 + 0.5 grams H_2O_2
Measure pH and emf at 30, 60, and 120 minutes
At 120 minutes, flocculate, filter, and wash residue with 100 mL water
Combine filtrate and wash, measure volume, submit for lead analysis
Get wet and dry weight of filter cake, submit for lead analysis

Data:

Leach Time, minutes	0	30	60	120		
Clock Time	11:15	11:45	12:15	13:15		
Solution Temperature, °C	24	23	23	23		
Solution pH	1.92	1.03	0.97	0.98		
Solution emf, mv	490	464	529	546		
Grams HCl Added	4.00					
Grams H_2O_2 Added		0.50	0.50			
Grams HNO_3 Added	4.00					
HNO_3, milliequivalents / gram soil	0.444	0.444	0.444	0.444		
HNO_3 Consumed	100% HNO_3	60	kg/1000 kg soil added			
HCl Consumed	100% HCl	31	kg/1000 kg soil added			
H_2O_2 Consumed	100% H_2O_2	6	kg/1000 kg soil added			

Test No:	"LR23"-6A	Dry Wt.	Solid or Solution Analysis or Mass				
		or Volume	mg/kg, mg/L, or mg				
Sample ID	Description	g or mL	Pb				
<150 mesh tailings + middlings		47	3200				
			150	0	0	0	0
"LR23"-6A-1 2 hr filtrate + wash		280	361				
			101	0	0	0	0
"LR23"-6A-2 2 hr residue		47	360				
			17	0	0	0	0
Calculated Feed Analysis		-1%	2507	0	0	0	0
Calculated Leach Recovery			86%				

AGITATION LEACH DATA SHEET

Sample: 100 grams Twentynine Palms LR-30, <150 mesh tails

Instructions: Add 100 grams of sample in 200 mL water
                                  Use 400 mL beaker and propeller stirrer
                                  Adjust pH to 2.0 target with reagent nitric acid
                                  Measure pH and readjust as necessary at 30, 60, 120, and 180 minutes
                                  At 240 minutes, flocculate, filter, and wash residue with 100 mL water
                                  Combine filtrate and wash, measure volume, submit for lead analysis
                                  Get wet and dry weight of filter cake, submit for lead analysis

Data:

Leach Time, minutes			0	30	60	120	180	240
Clock Time			8:40	9:11	9:34	10:35	11:39	12:40
Solution Temperature, °C			24	24	25	25	25	25
Solution pH			12.00	2.87	2.43	2.51	1.98	2.05
Grams HNO_3 Added			7.40	0.42	0.26	0.32	0.07	
HNO_3, milliequivalents / gram soil			0.822	0.869	0.898	0.933	0.941	
HNO_3 Consumed			100% HNO_3	59	kg/1000 kg soil added			

Test No:	"LR30"-1	Dry Wt.	Solid or Solution Analysis or Mass				
		or Volume	mg/kg, mg/L, or mg				
Sample ID	Description	g or mL	Pb				
<150 mesh tailings		100	24,300				
			2430	0	0	0	0
"LR30"-1-1 4 hr filtrate + wash		324	6435				
			2085	0	0	0	0
"LR30"-1-2 4 hr residue		94	3314				
			310	0	0	0	0
Calculated Feed Analysis		-7%	23,948	0	0	0	0
Calculated Leach Recovery			87%				

AGITATION LEACH DATA SHEET

Sample: 100 grams Twentynine Palms LR-30, <150 mesh tails

Instructions: Add 100 grams of sample in 200 mL water
Use 400 mL beaker and propeller stirrer
Adjust pH to 1.5 target with reagent nitric acid
Measure pH and readjust as necessary at 30, 60, 120, and 180 minutes
At 240 minutes, flocculate, filter, and wash residue with 100 mL water
Combine filtrate and wash, measure volume, submit for lead analysis
Get wet and dry weight of filter cake, submit for lead analysis

Data:

Leach Time, minutes		0	30	60	120	180	240
Clock Time		8:44	9:12	9:35	10:36	11:40	12:41
Solution Temperature, °C		24	24	25	25	25	25
Solution pH		12.00	1.11	1.21	1.36	1.49	1.51
Grams HNO_3 Added		9.01				0.07	
HNO_3, milliequivalents / gram soil		1.001	1.001	1.001	1.001	1.009	
HNO_3 Consumed		100% HNO_3	64	kg/1000 kg soil added			

Test No: "LR30"-2		Dry Wt.	Solid or Solution Analysis or Mass			
		or Volume	mg/kg, mg/L, or mg			
Sample ID	Description	g or mL	Pb			
<150 mesh tailings		100	24,300			
			2430	0	0	0	0
"LR30"-2-1 4 hr filtrate + wash		320	6838			
			2188	0	0	0	0
"LR30"-2-2 4 hr residue		94	2601			
			243	0	0	0	0
Calculated Feed Analysis		-7%	24,314	0	0	0	0
Calculated Leach Recovery			90%			

AGITATION LEACH DATA SHEET

Sample: 100 grams Twentynine Palms LR-30, <150 mesh tails

Instructions: Add 100 grams of sample in 200 mL water
Use 400 mL beaker and propeller stirrer
Adjust pH to 1.2 target with reagent nitric acid
Measure pH and readjust as necessary at 30, 60, 120, and 180 minutes
At 240 minutes, flocculate, filter, and wash residue with 100 mL water
Combine filtrate and wash, measure volume, submit for lead analysis
Get wet and dry weight of filter cake, submit for lead analysis

Data:

Leach Time, minutes		0	30	60	120	180	240
Clock Time		8:46	9:12	9:35	10:37	11:41	12:43
Solution Temperature, °C		24	24	25	25	25	25
Solution pH		12.00	0.64	0.70	1.00	1.05	1.12
Grams HNO$_3$ Added		10.54					
HNO$_3$, milliequivalents / gram soil		1.171	1.171	1.171	1.171	1.171	
HNO$_3$ Consumed		100% HNO$_3$	67	kg/1000 kg soil added			

Test No:	"LR30"-3	Dry Wt.	Solid or Solution Analysis or Mass				
		or Volume	mg/kg, mg/L, or mg				
Sample ID	Description	g or mL	Pb				
<150 mesh tailings		110	24,300				
			2673	0	0	0	0
"LR30"-3-1 4 hr filtrate + wash		327	7473				
			2444	0	0	0	0
"LR30"-3-2 4 hr residue		102	2204				
			225	0	0	0	0
Calculated Feed Analysis		-7%	24,264	0	0	0	0
Calculated Leach Recovery			92%				

AGITATION LEACH DATA SHEET

Sample: 100 grams Twentynine Palms LR-30, <150 mesh tails

Instructions: Add 100 grams of sample in 200 mL water
Use 400 mL beaker and propeller stirrer
Add 10 grams $Fe_2(SO_4)3H_2O$
Adjust pH to 1.2 target with reagent nitric acid
Measure pH and readjust as necessary at 30, 60, 120, and 180 minutes
At 240 minutes, flocculate, filter, and wash residue with 100 mL water
Combine filtrate and wash, measure volume, submit for lead analysis
Get wet and dry weight of filter cake, submit for lead analysis

Data:

Leach Time, minutes		0	30	60	120	180	240
Clock Time		12:45	13:15	13:45	14:45	15:45	16:45
Solution Temperature, °C		24	24	25	25	25	25
Solution pH		2.82	1.78	1.48	1.49	1.52	
Grams HNO_3 Added		5.78	1.25	0.69	0.37		
HNO_3, milliequivalents / gram soil		0.642	0.781	0.858	0.899	0.899	
HNO_3 Consumed		100% HNO_3	51	kg/1000 kg soil added			

Test No:	"LR30"-4	Dry Wt.	Solid or Solution Analysis or Mass				
		or Volume	mg/kg, mg/L, or mg				
Sample ID	**Description**	**g or mL**	**Pb**				
<150 mesh tailings		110	24,300				
			2673	0	0	0	0
"LR30"-4-1 4 hr filtrate + wash		900	10				
			9	0	0	0	0
"LR30"-4-2 4 hr residue		98	19,400				
			1897	0	0	0	0
Calculated Feed Analysis		-11%	**17,330**	**0**	**0**	**0**	**0**
Calculated Leach Recovery			**0%**				

AGITATION LEACH DATA SHEET

Sample: 100 grams Twentynine Palms LR-30, <150 mesh tails

Instructions: Add 100 grams of sample in 400 mL water
Use 600 mL beaker and propeller stirrer
Add 8 grams HNO_3
Add 8 grams HCl
Add 1 + 1 grams H_2O_2
Measure pH and emf at 30, 60, 120, 180, and 240 minutes
At 240 minutes, flocculate, filter, and wash residue with 100 mL water
Combine filtrate and wash, measure volume, submit for lead analysis
Get wet and dry weight of filter cake, submit for lead analysis

Data:

Leach Time, minutes		0	30	60	120	180	240
Clock Time		10:15	10:45	11:15	12:15	13:15	14:15
Solution Temperature, °C		25	26	26	25	25	25
Solution pH		8.63	1.22	1.25	1.33	1.21	1.33
Solution emf, mv		184	200	526	540	525	522
Grams HCl Added		8.00					
Grams H_2O_2 Added			1.00	1.00			
Grams HNO_3 Added		8.00					
HNO_3, milliequivalents / gram soil		0.889	0.889	0.889	0.889		
HNO_3 Consumed	100% HNO_3	56	kg/1000 kg soil added				
HCl Consumed	100% HCl	30	kg/1000 kg soil added				
H_2O_2 Consumed	100% H_2O_2	6	kg/1000 kg soil added				

Test No:	"LR30"-5	Dry Wt.	Solid or Solution Analysis or Mass				
		or Volume	mg/kg, mg/L, or mg				
Sample ID	Description	g or mL	Pb				
<150 mesh tailings		100	24,300				
			2430	0	0	0	0
"LR30"-5-1 4 hr filtrate + wash		536	3935				
			2109	0	0	0	0
"LR30"-5-2 4 hr residue		90	3087				
			277	0	0	0	0
Calculated Feed Analysis		-10%	23,864	0	0	0	0
Calculated Leach Recovery			88%				

AGITATION LEACH DATA SHEET

Sample: Residue from Test Twentynine Palms "LR30"-5 44.1 grams

Instructions:

Add 44.1 grams of sample in 400 mL water
Use 600 mL beaker and propeller stirrer
Add 8 grams HNO_3
Add 8 grams HCl
Add 1 + 1 grams H_2O_2
Measure pH and emf at 30, 60, 120, 180, and 240 minutes
At 240 minutes, flocculate, filter, and wash residue with 100 mL water
Combine filtrate and wash, measure volume, submit for lead analysis
Get wet and dry weight of filter cake, submit for lead analysis

Data:

Leach Time, minutes	0	30	60	120	180	240
Clock Time	7:45	8:15	8:45	9:45	10:45	11:45
Solution Temperature, °C	23	23	24	24	25	25
Solution pH	2.71	1.35	1.30	1.31	1.31	1.30
Solution emf, mv	204	359	529	546	541	540
Grams HCl Added	8.00					
Grams H_2O_2 Added		1.00	1.00			
Grams HNO_3 Added	8.00					
HNO_3, milliequivalents / gram soil	0.90	0.90	0.90	0.90		
HNO_3 Consumed	100% HNO_3	127	kg/1000 kg soil added			
HCl Consumed	100% HCl	67	kg/1000 kg soil added			
H_2O_2 Consumed	100% H_2O_2	14	kg/1000 kg soil added			

Test No: "LR30"-5A		Dry Wt. or Volume	Solid or Solution Analysis or Mass mg/kg, mg/L, or mg				
Sample ID	Description	g or mL	Pb				
<150 mesh tailings		44	3087				
			136	0	0	0	0
"LR30"-5A-1 4 hr filtrate + wash		413	332				
			137	0	0	0	0
"LR30"-5A-2 4 hr residue		44	252				
			11	0	0	0	0
Calculated Feed Analysis		-1%	3359	0	0	0	0
Calculated Leach Recovery			93%				

AGITATION LEACH DATA SHEET

Sample: 100 grams Twentynine Palms LR-30, <150 mesh tails

Instructions: Add 100 grams of sample in 400 mL water
Use 600 mL beaker and propeller stirrer
Add 8 grams HNO_3
Add 8 grams HCl
Add 1 + 1 grams H_2O_2
Measure pH and emf at 30, 60, and 120 minutes
At 120 minutes, flocculate, filter, and wash residue with 100 mL water
Combine filtrate and wash, measure volume, submit for lead analysis
Get wet and dry weight of filter cake, submit for lead analysis

Data:

Leach Time, minutes			0	30	60	120		
Clock Time			9:00	9:30	10:00	10:30		
Solution Temperature, °C			23	23	23	23		
Solution pH			8.55	1.11	1.13	1.13		
Solution emf, mv			140	160	504	550		
Grams HCl Added			8.00					
Grams H_2O_2 Added				1.00	1.00			
Grams HNO_3 Added			8.00					
HNO_3, milliequivalents / gram soil			0.889	0.889	0.889	0.889		
HNO_3 Consumed	100% HNO_3		56	kg/1000 kg soil added				
HCl Consumed	100% HCl		30	kg/1000 kg soil added				
H_2O_2 Consumed	100% H_2O_2		6	kg/1000 kg soil added				

Test No:	"LR30"-6	Dry Wt.	Solid or Solution Analysis or Mass				
		or Volume	mg/kg, mg/L, or mg				
Sample ID	Description	g or mL	Pb				
<150 mesh tailings		100	24,300				
			2430	0	0	0	0
"LR30"-6-1 4 hr filtrate + wash		395	4195				
			1657	0	0	0	0
"LR30"-6-2 4 hr residue		93	6043				
			562	0	0	0	0
Calculated Feed Analysis		-7%	22,190	0	0	0	0
Calculated Leach Recovery			75%				

AGITATION LEACH DATA SHEET

Sample: Residue from Test Twentynine Palms "LR30"-5 46.0 grams

Instructions: Add 46.0 grams of sample in 200 mL water
Use 600 mL beaker and propeller stirrer
Add 4 grams HNO_3
Add 4 grams HCl
Add 0.5 + 0.5 grams H_2O_2
Measure pH and emf at 30, 60, and 120 minutes
At 120 minutes, flocculate, filter, and wash residue with 100 mL water
Combine filtrate and wash, measure volume, submit for lead analysis
Get wet and dry weight of filter cake, submit for lead analysis

Data:

Leach Time, minutes	0	30	60	120		
Clock Time	11:15	11:45	12:15	13:15		
Solution Temperature, °C	24	23	23	23		
Solution pH	1.97	1.01	0.95	0.96		
Solution emf, mv	468	454	528	543		
Grams HCl Added	4.00					
Grams H_2O_2 Added		0.50	0.50			
Grams HNO_3 Added	4.00					
HNO_3, milliequivalents / gram soil	0.45	0.45	0.45	0.45		
HNO_3 Consumed	100% HNO_3	61	kg/1000 kg soil added			
HCl Consumed	100% HCl	32	kg/1000 kg soil added			
H_2O_2 Consumed	100% H_2O_2	7	kg/1000 kg soil added			

Test No: "LR30"-6A		Dry Wt.	Solid or Solution Analysis or Mass				
		or Volume	mg/kg, mg/L, or mg				
Sample ID	Description	g or mL	Pb				
<150 mesh tailings		46	6044				
			278	0	0	0	0
"LR30"-6A-1 4 hr filtrate + wash		259	1070				
			277	0	0	0	0
"LR30"-6A-2 4 hr residue		45	640				
			29	0	0	0	0
Calculated Feed Analysis		-2%	6655	0	0	0	0
Calculated Leach Recovery			91%				

AGITATION LEACH DATA SHEET

Sample: 100 grams Twentynine Palms LR-30, <150 mesh tails

Instructions: Add 100 grams of sample in 600 ml water
Use 1000 mL beaker and propeller stirrer
Add 12 grams HNO_3
Add 12 grams HCl
Add 1 + 1 grams H_2O_2
Measure pH and emf at 30, 60, and 120 minutes
At 120 minutes, flocculate, filter, and wash residue with 100 mL water
Combine filtrate and wash, measure volume, submit for lead analysis
Get wet and dry weight of filter cake, submit for lead analysis

Data:

Leach Time, minutes	0	30	60	120		
Clock Time	9:00	9:30	10:00	11:00		
Solution Temperature, °C	25	27	27	27		
Solution pH	7.57	1.31	1.37	1.69		
Solution emf, mv	208	220	568	580		
Grams HCl Added	12.00					
Grams H_2O_2 Added		1.50	1.50			
Grams HNO_3 Added	12.00					
HNO_3, milliequivalents / gram soil	1.333	1.333	1.333	1.333		
HNO_3 Consumed	100% HNO_3	84	kg/1000 kg soil added			
HCl Consumed	100% HCl	44	kg/1000 kg soil added			
H_2O_2 Consumed	100% H_2O_2	9	kg/1000 kg soil added			

Test No:	"LR30"-7	Dry Wt.	Solid or Solution Analysis or Mass			
		or Volume	mg/kg, mg/L, or mg			
Sample ID	Description	g or mL	Pb			
<150 mesh tailings		100	24,300			
			2430	0	0	0	0
"LR30"-7-1 2 hr filtrate + wash		795	2860			
			2274	0	0	0	0
"LR30"-7-2 2 hr residue		43	1250			
			54	0	0	0	0
Calculated Feed Analysis		-57%	**23,275**	**0**	**0**	**0**	**0**
Calculated Leach Recovery			**98%**			

AGITATION LEACH DATA SHEET

Sample: Residue from Test Twentynine Palms "LR30"-7 42.0 grams

Instructions:
Add 42.0 grams of sample in 300 ml water
Use 600 mL beaker and propeller stirrer
Add 4 grams HNO_3
Add 4 grams HCl
Add 0.5 + 0.5 grams H_2O_2
Measure pH and emf at 30, 60, and 120 minutes
At 120 minutes, flocculate, filter, and wash residue with 100 mL water
Combine filtrate and wash, measure volume, submit for lead analysis
Get wet and dry weight of filter cake, submit for lead analysis

Data:

Leach Time, minutes	0	30	60	120		
Clock Time	12:00	12:30	13:00	14:00		
Solution Temperature, °C	27	27	27	27		
Solution pH	2.64	1.46	1.38	1.30		
Solution emf, mv	285	541	552	554		
Grams HCl Added	6.00					
Grams H_2O_2 Added		0.50	0.50			
Grams HNO_3 Added	6.00					
HNO_3, milliequivalents / gram soil	0.667	0.667	0.667	0.667		
HNO_3 Consumed	100% HNO_3	100	kg/1000 kg soil added			
HCl Consumed	100% HCl	53	kg/1000 kg soil added			
H_2O_2 Consumed	100% H_2O_2	7	kg/1000 kg soil added			

Test No: "LR30"-7A		Dry Wt.	Solid or Solution Analysis or Mass				
		or Volume	mg/kg, mg/L, or mg				
Sample ID	Description	g or mL	Pb				
<150 mesh tailings		42	1250				
			53	0	0	0	0
"LR30"-7A-1 2 hr filtrate + wash		363	135				
			49	0	0	0	0
"LR30"-7A-2 2 hr residue		42	176				
			7	0	0	0	0
Calculated Feed Analysis		0%	1342	0	0	0	0
Calculated Leach Recovery		87%					

AGITATION LEACH DATA SHEET

Sample: 100 grams Twentynine Palms LR-30, <150 mesh tails

Instructions: Add 100 grams of sample in 400 mL water
                               Use 600 mL beaker and propeller stirrer
                               Add 8 grams HNO_3
                               Add 8 grams HCl
                               Add 1 + 1 grams H_2O_2
                               Measure pH and emf at 30, 60, and 120 minutes
                               At 120 minutes, flocculate, filter, and wash residue with 100 mL water
                               Combine filtrate and wash, measure volume, submit for lead analysis
                               Get wet and dry weight of filter cake, submit for lead analysis

Data:

Leach Time, minutes		0	30	60	120		
Clock Time		9:00	9:30	10:00	11:00		
Solution Temperature, °C		25	27	27	28		
Solution pH		7.73	1.35	1.42	1.73		
Solution emf, mv		195	196	581	590		
Grams HCl Added		8.00					
Grams H_2O_2 Added			1.00	1.00			
Grams HNO_3 Added		8.00					
HNO_3, milliequivalents / gram soil		0.889	0.889	0.889	0.889		
HNO_3 Consumed		100% HNO_3	56	kg/1000 kg soil added			
HCl Consumed		100% HCl	30	kg/1000 kg soil added			
H_2O_2 Consumed		100% H_2O_2	6	kg/1000 kg soil added			

Test No:	"LR30"-8	Dry Wt.	Solid or Solution Analysis or Mass				
		or Volume	mg/kg, mg/L, or mg				
Sample ID	Description	g or mL	Pb				
<150 mesh tailings		100	24,300				
			2430	0	0	0	0
"LR30"-8-1 2 hr filtrate + wash		528	4120				
			2175	0	0	0	0
"LR30"-8-2 2 hr residue		93	2010				
			187	0	0	0	0
Calculated Feed Analysis		-7%	**23,623**	**0**	**0**	**0**	**0**
Calculated Leach Recovery			**92%**				

AGITATION LEACH DATA SHEET

Sample: Residue from Test Twentynine Palms "LR30"-8 46.0 grams

Instructions: Add 46.0 grams of sample in 200 mL water
Use 600 mL beaker and propeller stirrer
Add 4 grams HNO_3
Add 4 grams HCl
Add 0.5 + 0.5 grams H_2O_2
Measure pH and emf at 30, 60, and 120 minutes
At 120 minutes, flocculate, filter, and wash residue with 100 mL water
Combine filtrate and wash, measure volume, submit for lead analysis
Get wet and dry weight of filter cake, submit for lead analysis

Data:

Leach Time, minutes	0	30	60	120		
Clock Time	12:00	12:30	13:00	14:00		
Solution Temperature, °C	27	27	27	27		
Solution pH	2.65	1.44	1.36	1.27		
Solution emf, mv	280	562	565	565		
Grams HCl Added	4.00					
Grams H_2O_2 Added		0.50	0.50			
Grams HNO_3 Added	4.00					
HNO_3, milliequivalents / gram soil	0.444	0.444	0.444	0.444		
HNO_3 Consumed	100% HNO_3	61	kg/1000 kg soil added			
HCl Consumed	100% HCl	32	kg/1000 kg soil added			
H_2O_2 Consumed	100% H_2O_2	7	kg/1000 kg soil added			

Test No: "LR30"-8A		Dry Wt. or Volume	Solid or Solution Analysis or Mass mg/kg, mg/L, or mg				
Sample ID	**Description**	**g or mL**	**Pb**				
<150 mesh tailings		46	2010				
			92	0	0	0	0
"LR30"-8A-1 2 hr filtrate + wash		278	110				
			31	0	0	0	0
"LR30"-8A-2 2 hr residue		46	242				
			11	0	0	0	0
Calculated Feed Analysis		0%	**906**	**0**	**0**	**0**	**0**
Calculated Leach Recovery			**73%**				

AGITATION LEACH DATA SHEET

Sample: 100 grams Twentynine Palms LR-30, <150 mesh tails

Instructions: Add 100 grams of sample in 400 mL water
Use 600 mL beaker and propeller stirrer
Add 8 grams HNO_3
Add 8 grams HCl
Add 1 + 1 grams H_2O_2
Measure pH and emf at 30, 60, and 120 minutes
At 120 minutes, flocculate, filter, and wash residue with 100 mL water
Combine filtrate and wash, measure volume, submit for lead analysis
Get wet and dry weight of filter cake, submit for lead analysis

Data:

Leach Time, minutes		0	30	60	120		
Clock Time							
Solution Temperature, °C							
Solution pH							
Solution emf, mv							
Grams HCl Added		8.00					
Grams H_2O_2 Added			1.00	1.00			
Grams HNO_3 Added		8.00					
HNO_3, milliequivalents / gram soil		0.889	0.889	0.889	0.889		
HNO_3 Consumed	100% HNO_3	56	kg/1000 kg soil added				
HCl Consumed	100% HCl	30	kg/1000 kg soil added				
H_2O_2 Consumed	100% H_2O_2	6	kg/1000 kg soil added				

Test No:	"LR30"-7(3)	Dry Wt.	Solid or Solution Analysis or Mass				
		or Volume	mg/kg, mg/L, or mg				
Sample ID	Description	g or mL	Pb				
<150 mesh tailings		100	24,300				
			2430	0	0	0	0
"LR30"-7(3)-1 2 hr filtrate + wash							
			0	0	0	0	0
"LR30"-7(3)-2 2 hr residue							
			0	0	0	0	0
Calculated Feed Analysis		-100%	0%	0%	0%	0%	0%
Calculated Leach Recovery							

APPENDIX B

Demonstration Process Flowsheet

CONTENTS

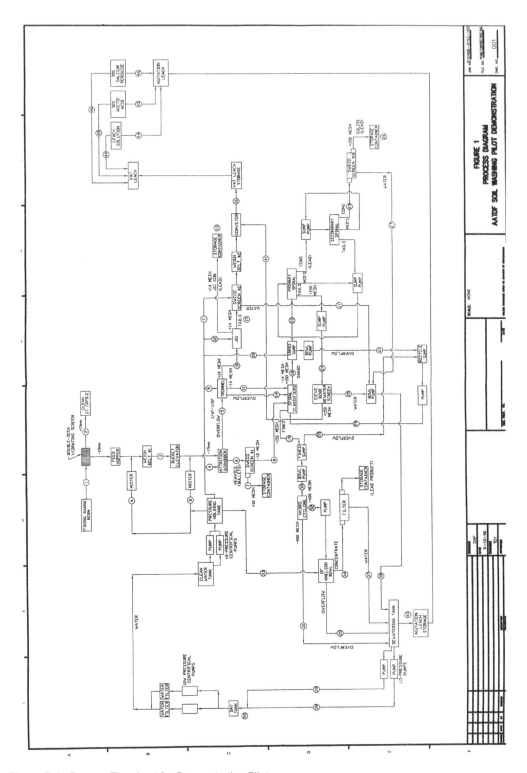

Figure B.1 Process Flowsheet for Demonstration Pilot.

Operating Rates and Mass Balance for NAS Miramar Pilot Plant

CONTENTS

C.1 SPECIFICATIONS FOR MASS BALANCE

This section describes key process operating conditions that were used to prepare the mass balance in Table C.1. All mass split, screen fraction, and gravity treatment values were from CMRI's treatability study for NAS Miramar soil (Table A.1). The values were averages of the results for Miramar Samples A and B.

Treatment Plant Feed

Initially, a vibrating grizzly screen and a double-deck-vibrating screen removed the large stones and debris and the +19.0 mm (+3/4 in.) particles from the firing range soil. The +19.0 mm fraction was assumed to be 25% of the total soil mass. The soil moisture content was about 10%. The feed to the attrition scrubber was the −19.0 mm soil fraction. For the base case, it is assumed that the scrubber feed rate would be 907 kilograms of dry soil per hour (kg/hr) or 2000 pounds per hour (lbs/hr).

Attrition Scrubber

Description	Soil Particle Size Range	Mass Split %(dwb)
Auger Product (Bullets & Stones)	−19.0 mm, +4.76 mm (−3/4 in., +4 mesh)	7.70
Scrubber Overflow	−4.76 mm (−4 mesh)	92.30

Mass Split: Based on the dry weight of the attrition scrubber feed.

The pulp density of the slurry in the attrition scrubber was 60% solids. The actual pulp density of the scrubber slurry during the Miramar demonstration was about 28% solids. It was assumed that the attrition scrubber could suspend only −4.76 mm soil particles. Thus, the +4.76 mm (+4 mesh) soil settled to the bottom of the scrubber. The auger removed the coarse soil and lead particles from the bottom. It was also assumed that the pulp density of the auger product was 80% solids.

Table C.1 Mass Balance for the Design of the NAS Miramar Demonstration Plant

Unit Operation	Flow	Description	Mass Flows		Distributions		Lead Content in Solid			Comments
			Solid kg/hour	Liquid kg/hour	Solid wt %	Liquid wt%	kg/hour	mg/kg	Distribution wt%	
Double-Deck Screen (25.4 mm Openings)	1	Feed: Soil from FR Berm	1209.58	134.40	90.00	10.00	10.8605	8979	101.413	Mass Split, +/-19.0 mm: 25% & 75%
	2	+19.0 mm Soil	302.39	33.60	90.00	10.00	0.1512	500	1.413	**To Clean Soil Stockpile**
	3	-19.0 mm Soil	907.18	100.80	90.00	10.00	10.7093	11,805	100.000	Feed to Attrition Scrubber
Attrition Scrubber	3	Feed: -19.0 mm Soil	907.18	100.80	90.00	10.00	10.7093	11,805	100.000	Mass Split, +/- 4.76 mm: 7.70% & 92.30%
	4	Feed: Dilution Water	0.00	474.89	0.00	100.00	0.0000	0	0.000	Water From Day Tank
	5	Overflow: -4.76 mm (-4 mesh) Soil	837.35	558.23	60.00	40.00	5.5582	6638	51.900	Feed to 16 mesh Trommel Screen
	6	Auger Product: Bullets/Stones	69.84	17.46	80.00	20.00	5.1511	73,757	48.100	Feed to Vibrating Screen #1
Vibrating Screen #1 (Remove Water from Auger Product)	6	Feed: Auger Bullets/Stones	69.84	17.46	80.00	20.00	5.1511	73,757	48.100	Screen Efficiency at 100%
	7	Oversize: +1.19mm (+16 mesh) Product	69.84	3.68	95.00	5.00	5.1511	73,757	48.100	**Lead Product (to storage barrel)**
	8	Undersize: -1.19 mm (-16 mesh) Slurry	0.00	13.78	0.00	100.00	0.0000	0	0.000	Feed to Spiral Classifier
Trommel Scrubber (16 mesh Screen)	5	Feed: -4.76 mm (-4 mesh Soil)	837.35	558.23	60.00	40.00	5.5582	6638	51.900	Screen Efficiency at 100%
	9	Feed: Screen Wash Water	0.00	235.67	0.00	100.00	0.0000	0	0.000	Water From Day Tank (S/L=4/1)
	10	Oversize: -4.76 mm, +1.19 mm (-4 mesh, +16 mesh) Soil	58.92	14.73	80.00	20.00	2.3298	39,543	21.755	Feed to Jig: +1.19 mm (+16 mesh)
	11	Undersize: -1.19 mm (-16 mesh) Soil	758.97	759.69	49.98	50.02	3.1477	4147	29.392	Feed to Spiral Classifier

Equipment	No.	Stream								Description
	14	Trommel Overflow −1.19 mm (−16 mesh)	19.46	19.48	49.98	50.02	0.0807	4147	0.754	Overflow at 2.50% of −1.19 mm (−16 mesh) Slurry
Mineral Jig	10	Feed: +1.19 mm (+16 mesh) Soil	58.92	14.73	80.00	20.00	2.3298	39,543	21.755	Mass Split, con/tail: 34.61% & 65.39%
	12	Feed: Dilution Water	0.00	95.55	0.00	100.00	0.0000	0	0.000	Water From Day Tank
	13	Jig Concentrate	20.39	20.39	50.00	50.00	2.2125	108,487	20.660	**Lead Product (to storage barrel)**
	15	Jig Tailings	38.52	89.89	30.00	70.00	0.1173	3045	1.095	Feed to Vibrating Screen #2
Vibrating Screen #2 (Remove Water from Jig Tailings)	15	Feed: Jig +1.19 mm (+16 mesh) Tails	38.52	89.89	30.00	70.00	0.1173	3045	1.095	Screen Efficiency at 100%
	C	Feed: Screen Wash Water	0.00	0.00	0.00	100.00	0.0000	0	0.000	Water From Day Tank
	16	Oversize: Jig Tailings	38.52	16.51	70.00	30.00	0.1173	3045	1.095	**To Vat Leach Feed Stockpile**
	17	Undersize: Water	0.00	73.38	0.00	100.00	0.0000	0	0.000	To Classifier Recycle Sump
Spiral Classifier	8	Feed: Water From Screen #1	0.00	13.78	0.00	100.00	0.0000	0	0.000	Classifier Efficiency at 100%
	11	Feed: −1.19 mm (−16 mesh) Soil	758.97	759.69	49.98	50.02	3.1477	4147	29.392	
	14	Feed: Trommel Overflow	19.46	19.48	49.98	50.02	0.0807	4147	0.754	
	29	Feed: Slurry from Recycle Sump	19.96	176.55	10.16	89.84	0.0884	4430	0.826	
	18	Sands: −1.19 mm, +105 μm (−16 mesh, +150 mesh)	413.72	177.31	70.00	30.00	1.462419	3535	13.656	Feed to Gravity Separator:+105 μm (+150 mesh)
	19	Fines: −105 μm (−150 mesh Soil)	384.67	792.20	32.69	67.31	1.8544	4821	17.316	Feed to Hydrocyclone −105 μm (−150 mesh)

continued

Table C.1 (continued) Mass Balance for the Design of the NAS Miramar Demonstration Plant

Unit Operation	Flow	Description	Mass Flows Solid kg/hour	Mass Flows Liquid kg/hour	Distributions Solid wt%	Distributions Liquid wt%	Lead Content in Solid kg/hour	Lead Content in Solid mg/kg	Lead Content in Solid Distribution wt%	Comments
Sands Sump	18	Feed: –1.19 mm, +105 μm (–16 mesh, +150 mesh) Sands	413.72	177.31	70.00	30.00	1.4624	3535	13.656	Add Water to Separator Feed
	20	Feed: Gravity Separator Water Sump Overflow	0.00	3157.42	0.00	100.00	0.0000	0	0.000	Water Flow to Sands Sump at +52 Lpm
	21	Sump Overflow	10.34	83.37	11.04	88.96	0.036561	3535	0.341	To Recycle Sump at 2.50% Total Flow
	22	Feed to Gravity Separator	403.37	3251.36	11.04	88.96	1.4259	3535	13.314	Feed to Reichert Spiral at ~55 Lpm
Recycle Sump	17	Feed: From Screen #2	0.00	73.38	0.00	100.00	0.0000	0	0.000	
	21	Feed: From Sands Sump Overflow	10.34	83.37	11.04	88.96	0.0366	3535	0.341	
	28	Feed: Fines Sump Overflow	9.62	19.80	32.69	81.07	0.0519	5392	0.484	
	29	Recycle to Spiral Classifier	19.96	176.55	10.16	89.84	0.0884	4430	0.826	Recycle to Spiral Classifier
Gravity Separators (Two Reichert Spiral Concentrators)	22	Feed: –1.19 mm, +105 μm (–16 mesh, +150 mesh)	403.37	3251.36	11.04	88.96	1.4259	3535	13.314	Feed: Spiral Classifier Sand
	23	Separator Concentrate	47.59	269.66	15.00	85.00	1.2843	26,988	11.992	Feed to Screen #3
	24	Separator Tailings	355.79	2981.70	10.66	89.34	0.1416	398	1.322	Feed to Sieve Bend Screen
Sieve Bend Screen (Remove Water from Separator Tailings)	24	Feed: Separator Tailings	355.79	2981.70	10.66	89.34	0.1416	398	1.322	Screen Efficiency at 100%
	25	Undersize: Water	0.00	2918.91	0.00	100.00	0.0000	0	0.000	Feed to Bowl Pump
	41	Oversize: +105 μm (+150 mesh) Soil	355.79	62.79	85.00	15.00	0.1416	398	1.322	**To Vat Leach Feed Stockpile**
Bowl Pump (Sieve-Bend Screen)	25	Feed: Sieve Bend Water	0.00	2918.91	0.00	15.00	0.0000	0	0.000	

Equipment	No.	Stream Description								Notes
	27	Feed: Screen #3 Wash Water	0.00	257.76	0.00	100.00	0.0000	0	0.000	Feed to Dewatering Tank
	38	Flow into Dewatering Tank	0.00	3176.68	0.00	100.00	0.0000	0	0.000	Feed to Dewatering Tank
Vibrating Screen #3 (Remove Water from Separator Concentrate)	23	Feed: Separator Concentrate	47.59	269.66	15.00	85.00	1.2843	26,988	11.992	Screen Efficiency at 100%
	26	Oversize: +105 μm (+150 mesh) Soil	47.59	11.90	**80.00**	20.00	1.2843	26,988	11.992	**Lead Product (to storage barrel)**
	27	Undersize: Water	0.00	257.76	0.00	100.00	0.0000	0	0.000	Feed to Dewatering Tank
Fines Sump	19	Feed: −105 μm (−150 mesh) Soil (Fines)	384.67	792.20	32.69	67.31	1.8544	4821	17.316	Feed: Spiral Classifier Overflow
	28	Fines Sump Overflow	**9.62**	19.80	**32.69**	67.31	0.0519	**5392**	0.484	To Recycle Sump at 2.50% of Total Flow
	30	Hydrocyclone Feed	375.05	772.39	32.69	67.31	1.8025	5392	16.832	Feed to Hydrocyclone
Hydrocyclone #1 (Separate at 400 mesh)	30	Feed: −105 μm (−150 mesh) Soil	375.05	772.39	32.69	67.31	1.8025	5392	16.832	Mass Split, +/−37 μm (+/−400 mesh): 64.2% & 35.8%
	31	Overflow: −37 μm (−400) mesh Soil	134.33	575.44	18.93	81.07	0.8590	6395	8.021	Feed to Dewatering Tank
	32	Underflow: +37 μm (+400) mesh Soil	**240.72**	196.95	**55.00**	45.00	**0.9435**	3920	8.811	Feed to Knelson Bowl
Knelson Concentrator	32	Feed: Cyclone #1 Underflow	240.72	196.95	55.00	45.00	0.9435	3920	8.811	Feed Solid: −105 μm, +37 μm (−150 mesh, +400 mesh)
	33	Feed: Fluidizing & Wash Water	0.00	6813.00	0.00	100.00	0.0000	0	0.000	Water Flow from Day Tank at ~115 Lpm
	34	KC Concentrate	**8.58**	3.68	**70.00**	30.00	**0.2592**	**30,209**	2.420	Feed to Filter Press
	35	KC Tailings	232.14	7006.28	3.21	96.79	0.6843	2948	6.390	Feed to Dewatering Tank

continued

Table C.1 (continued) Mass Balance for the Design of the NAS Miramar Demonstration Plant

Unit Operation	Flow	Description	Mass Flows		Distributions		Lead Content in Solid			Comments
			Solid kg/hour	Liquid kg/hour	Solid wt %	Liquid wt%	kg/hour	mg/kg	Distribution wt%	
Filter Press (Remove Water from Bowl Concentrate)	34	Feed: KC Concentrate	8.58	3.68	70.00	30.00	0.2592	30,209	2.420	From Knelson Bowl
	36	Filter Cake	8.58	0.95	90.00	10.00	0.2592	30,209	2.420	**Lead Product (to storage barrel)**
	37	Filtrate Water	0.00	2.72	0.00	100.00	0.0000	0	0.000	Feed to Dewatering Tank
Dewatering Tank	31	Feed: Hydrocyclone Overflow	134.33	575.44	18.93	81.07	0.8590	6395	8.021	Hydrocyclone Overflow
	35	Feed: KC Tailings	232.14	7006.28	3.21	96.79	0.6843	2948	6.390	Knelson Bowl Tailings
	37	Feed: Filter Press Filtrate	0.00	2.72	0.00	100.00	0.0000	0	0.000	Filtrate from Knelson Bowl Concentrate
	38	Feed: Sieve-Bend Water	0.00	3176.68	0.00	100.00	0.0000	0	0.000	Water from Spiral Separators
	39	Overflow (<1 μm)	0.00	10,516.8	0.00	100.00	0.0000	0	0.000	Water to Day Tank
	40	Underflow	366.47	244.32	60.00	40.00	1.5433	4211	14.411	**Feed to Agitation Leach**
Day Tank	39	Feed: Dewater Tank Overflow	0.00	10,516.8	0.00	100.00	0.0000	0	0.000	From Dewatering Tank
	50	Feed: Make-up Water	0.00	259.73	0.00	100.00	0.0000	0	0.000	
	#Flows	Recycle Water to Gravity Plant (#Flows = 4, 9, 12, C, 20, 33)	0.00	10,776.5	0.00	100.00	0.0000	0	0.000	

Solid, Water, and Lead Balance

								Notes
Inflow Contaminated Soil	1209.58	134.40	90.00	10.00	10.8605	8979	101.413	
Day Tank Inflow Water	0.00	10,516.8	0.00	100.00	0.0000	0	0.000	
Day Tank Make-up Water	0.00	259.73	0.00	100.00	0.0000	0	0.000	
Total Inflow	*1209.58*	*10,910.9*			*10.8605*	*8979*	*101.413*	
Outflow Coarse Washed Soil	696.71	112.90	86.06	13.94	0.4101	589	3.830	To Vat Leach Feed Stockpile
Outflow Lead Products	146.40	36.92	79.86	20.14	8.9071	60,841	83.171	To Lead Product Storage Barrels
Dewatering Tank Overflow	0.00	10,516.8	0.00	100.00	0.0000	0	0.000	
Dewatering Tank Underflow	366.47	244.32	60.00	40.00	1.5433	4211	14.411	To Agitation Leach Tanks
Total Outflow	*1209.58*	*10,910.9*			*10.8605*	*8979*	*101.413*	

Agitation Leach

								Notes
40 Feed: Thickener #1 Underflow	366.47	244.32	60.00	40.00	1.5433	4211	14.411	Pb Extraction = 88.1%
44 Feed: Leach Solution	0.00	550.35	0.00	100.00	0.0000	0	0.000	Leach Solution Pb Content = 0.00 g/L
45 30% Acetic Acid (Solution)	0.00	52.55	0.00	100.00	0.0000	0	0.000	Acid = 13.04 kg/1000 kg Feed Soil (Flow #1)
46 30% Calcium Peroxide (Solution)	0.00	7.89	0.00	100.00	0.0000	0	0.000	Peroxide = 1.95 kg/1000 kg Feed
42 Agitation Leach Solution	0.00	610.79	0.00	100.00	1.3601	2227	12.700	Leach Solution Pb Content = 2.23 g/L
43 Agitation Leach Tailings	366.47	244.32	60.00	40.00	0.1832	500	1.711	Pb in Solid (Wet Tailing) = 500 mg/kg
Total Feed to Leach	366.47	855.10	30.00	70.00	1.5433	4211	14.411	
Total Products from Leach	366.47	855.10	30.00	70.00	1.5433	4211	14.411	

continued

Table C.1 (continued) Mass Balance for the Design of the NAS Miramar Demonstration Plant

Unit Operation	Flow	Description	Mass Flows		Distributions		Lead Content in Solid			Comments
			Solid kg/hour	Liquid kg/hour	Solid wt %	Liquid wt%	kg/hour	mg/kg	Distribution wt%	
Vat Leach	2	Feed: +19 mm Soil	302.39	33.60	90.00	10.00	0.1512	500	1.413	Pb Extraction = 15.0%
	16	Feed: Jig Tailings	38.52	16.51	70.00	30.00	0.1173	3045	1.095	
	41	Feed: Reichert Separator Tailings	355.79	62.79	85.00	15.00	0.1416	398	1.322	
	44	Feed: Leach Solution	0.00	305.13	0.00	100.00	0.0000	0	0.000	Leach Solution Pb Content = 0.00 g/L
	45	30% Acetic Acid (Solution)	0.00	2.39	0.00	100.00	0.0000	0	0.000	Acetic Acid = 0.593 kg/1000 kg Feed (Flow #1)
	46	30% Calcium Peroxide (Solution)	0.00	0.36	0.00	100.00	0.0000	0	0.000	Ca Peroxide = 0.089 kg/1000 kg
	47	Vat Leach Solution	0.00	245.73	0.00	100.00	0.0617	251	0.576	Leach Solution Pb Content = 0.251 g/L
	48	Vat Leach Tailings	696.71	175.04	79.92	20.08	0.3484	500	3.253	Pb in Solid (Wet Tailing) = 500 mg/kg
		Total Feed to Leach	696.71	420.77	62.35	37.65	0.4101	589	3.830	
		Total Products from Leach	696.71	420.77	62.35	37.65	0.4101	589	3.829	
Lead Recovery from Leach Solutions	42	Feed: Agitation Leach Solution	0.0000	610.79	0.00	100.00	1.3601	2227	12.700	Leach Solution Pb Content = 2.27 g/L
	47	Feed: Vat Leach Solution	0.0000	245.73	0.00	100.00	0.0617	251	0.576	Leach Solution Pb Content = 0.281 g/L
	Fe	Feed: Scrap Iron	0.4258	0.00	100.00	0.00	0.0000	0	0.000	Iron (Fe) Utilization = 90%
	49	Precipitate: Lead Metal	1.4644	0.00	100.00	0.00	1.4218	970,924	13.277	To Lead Product Storage Barrels

							Recycled to Agitation and Vat Leach	Bleed Solution Adjusted to 0.0 kg/hr
44	Treated Solution	0.0000	855.48	0.00	100.00	0.0000	0	0.000
	Bleed Solution	0.0000	0.00	0.00	100.00	0.0000	0	0.000

$Pb + CaO_2 + 4CH_3COOH = Pb(CH_3COOH)_2 + 2H_2O$

$CaO_2/Pb =$ 0.347860

$CH_3COOH/Pb =$ 1.159210

$Pb(CH_3COO)_2/Pb =$ 1.569905

$Ca(CH_3COO)_2 =$ 0.763332

$H_2O/Pb =$ 0.086941

$Pb^{+2} + Fe^0 = Pb^0 + Fe^{+2}$

$Fe/Pb =$ 0.26952

Trommel Screen

Description	Soil Particle Size Range	Mass Split %(dwb)
Trommel Oversize Flow	−4.76 mm, +1.19 mm (−4 mesh, +16 mesh)	6.49
Trommel Undersize Flow	−1.19 mm (−16 mesh)	85.81

Mass Split: Based on the dry weight of the attrition scrubber feed.

The screen oversize fraction, −4.76 mm, +1.19 mm, contained the humate material. The design feed to the trommel, −4.76 mm scrubber fraction, was about 840 kg/hr. At 100% screen efficiency, the mass of the −1.19 mm soil passing though the screen was about 760 kg/hr. The balance assumed a wash water flow of 4.0 kg/kg of soil passing and a pulp density for the +1.19 mm (+16 mesh) soil at 80% solids.

Mineral Jig

Description	Soil Particle Size Range	Mass Split %(dwb)
Tailings (Overflow Product)	−4.76 mm, +1.19 mm (−4 mesh, +16 mesh)	4.25
Concentrate (Hutch Product)	−4.76 mm, +1.19 mm (−4 mesh, +16 mesh)	2.25

Mass Split: Based on the dry weight of the attrition scrubber feed.

The feed to the mineral jig was about 59 kg/hr of the −4.76 mm, +1.19 mm fraction. The humate fraction remained with the mineral jig tailings. The jig produced a concentrate that contained 10.8% lead (Pb). The mass balance assumed a pulp density for the tailings at 30% solids and for the concentrate at 50% solids.

Spiral Classifiers

Description	Soil Particle Size Range	Mass Split %(dwb)
Spiral Overflow (Slimes Fraction)	−105 µm (−150 mesh)	41.34
Spiral Underflow (Sands Fraction)	−1.19 mm, +105 µm (−16 mesh, +150 mesh)	44.46

Mass Split: Based on the dry weight of the attrition scrubber feed.

The pulp density for the sands fraction was assumed to be 70% solids. The total feed to the classifiers was about 760 kg/hr (−1.19 mm trommel fraction). The total overflow rate of the −105 µm fraction was 385 kg/hr at a pulp density of 30% solids. To produce a −105 µm product, the design calculations required that the overflow rate from the classifiers not exceed 430 kg/hr at a pulp density of 33% solids.

Reichert Gravity Separators (Sands Sump)

Description	Soil Particle Size Range	Mass Split %(dwb)
Concentrate (Heavy Fraction)	−1.19 mm, +105 µm (−16 mesh, +150 mesh)	5.25
Tailings (Light Fraction)	−1.19 mm, +105 µm (−16 mesh, +150 mesh)	39.22

Mass Split: Based on the dry weight of the attrition scrubber feed.

The feed to the Reichert separator contained the sand fraction for the spiral classifiers (−1.19 mm, +105 µm trommel fraction). The mass flow of dry soil was about 404 kg/hr. About 53 liters per minute (Lpm) of water was added to the slurry in the sands sump. The water addition adjusted the slurry feed rate into the separators to the desirable 57 Lpm (about 15 gpm). At this flow rate, the pulp density was 11% solids. The Reichert separator produced a concentrate that contained 2.7% Pb.

Hydrocyclone

Description	Soil Particle Size Range	Mass Split %(dwb)
Overflow (Slimes Fraction)	−105 µm , +37 µm (−150 mesh, +400 mesh)	64.18
Underflow (Sands Fraction)	−37 µm (−400 mesh)	35.82
Mass Split: Based on the dry weight of the hydrocyclone feed.		

The mass split was based on CMRI's hydrocyclone treatability tests in Table A.4 for Miramar A and B samples. The treatability results in this table were from the tests that had an inlet gauge pressure of 69 kPa (10 psig). The hydrocyclone feed was the overflow slurry from the spiral classifier containing the −105 µm soil. The total mass of the feed slurry was about 1150 kg/hr (about 16 Lpm). The pulp density of the feed slurry was about 33% solids. It is assumed that the underflow slurry contained 55% solids.

Knelson Bowl Concentrator

Description	Soil Particle Size Range	Mass Split %(dwb)
Concentrate (Heavy Fraction)	−105 µm , +37 µm (−150 mesh, +400 mesh)	1.47
Tailings (Light Fraction)	−105 µm , +37 µm (−150 mesh, +400 mesh)	39.87
Mass Split: Based on the dry weight of the attrition scrubber feed.		

The underflow slurry from the hydrocyclone was fed to the Knelson bowl concentrator at a rate of 440 kg/hr. This slurry contained 55% solids. It was assumed that the pulp density of the concentrate leaving the concentrator was about 70% solids. The Knelson bowl produced a lead concentrate that contained 2.4% Pb.

Dewatering Tank

The dewatering tank received the hydrocyclone overflow, Knelson concentrator tailings, filter press filtrate, and sieve-bend screen underflow. The tank was selected to handle a total slurry flow of 10,900 kg/hr (about 180 Lpm). Proper selection of the tanks produced a settled slurry containing 60% solids and an overflow containing a small amount of suspended solids. The settled slurry was fed to the agitation leach. The overflow was recycled to the day tank.

Agitation Leach

The operating conditions listed below are for the agitation leach:

- Feed: Dewatering Tank Underflow Slurry
- Mass Feed Slurry: 611 kg/hr at 60% solids (or 365 kg solids/hr)
- Agitation Leach Slurry: 30% solids
- Acetic Acid Addition: 13.0 kg/1000 kg dry soil (10 times stoichiometric requirement)
- Calcium Peroxide Addition: 2.0 kg/1000 kg dry soil (5 times stoichiometric requirement)
- Lead content of Leach Tailings: 0.050% Pb
- Settled Leach Tailings: 60% solids (calculated)

Vat Leach

The operating conditions below were for the vat leach:

- Feed: +19 mm soil at 90% solids, mineral jig tailings at 70% solids, and Reichert separator tailings at 85% solids
- Total Wet Solid Feed: 809 kg/hr at 86% solids (or 697 kg solids/hr)
- Liquid to Solid Ratio: 0.60 kg total liquid per kg dry soil
- Acetic Acid: 0.60 kg/1000 kg dry soil (10 times stoichiometric requirement)
- Calcium Peroxide: 0.090 kg/1000 kg dry feed soil (5 times stoichiometric requirement)
- Lead content of Leach Tailings: 0.050% Pb
- Leach Tailings: 80% solids (calculated)

Lead Recovery

The operating conditions below were for the lead recovery process:

- Powdered Iron (Fe) Treatment of Agitation and Vat Leach Solutions
- Efficiency of Powdered Iron Replacement Reaction: 90%

Supporting Calculations for Equipment Capacities

CONTENTS

D.1 ATTRITION SCRUBBER

Design Data

Mass Balance Data		kg/hr		% Distribution	
Feed	Feed Rate	Solid	Water	Solid	Water
Feed: –19.0 mm Solids	1007.98 kg/hr	907.18	100.80	90.0	10.0
Feed: Water	474.89 L/hr	0.00	474.89	0.0	100.0
Auger Product (–19.0 mm, +4.76 mm particles)		69.84	17.46	80.0	20.0
Scrubber Overflow (–4.76 mm particles)		837.35	558.23	60.0	40.0
Totals Mass Input and Output		907.18	575.69		

Demonstration Data		kg/hr		% Distribution	
Feed	Feed Rate	Solid	Water	Solid	Water
Feed: –19.0 mm Solids	1007.7 kg/hr	907.0[1]	100.7	90.0	10.0
Feed: Water	2271.2 L/hr	0.0	2271.2	0.0	100.0
Totals Mass Input and Output		907.0	2371.9	27.7	72.3

[1] Feed Rate at 2000 pounds dry solid per hour

Demonstration Attrition Scrubber Specifications

- Number of Scrubber Cells = 2
- T = Tank Length = Tank Width = 0.61 m (2.0 ft)
- Z = Slurry Depth = Depth (at overflow weir) = 0.762 m (2.50 ft)
- Z/T = 0.762 m/0.610 m = 1.25
- Type of Impeller – pitched blade turbine with four blades at 45° pitch

Table D.1 Number of Impellers for Solid Suspension

Number of Impellers	Impeller Clearance		Maximum Ratio, Z/T
	Bottom	Upper	
1	Z/4	—	1.2
2	Z/4	(2/3) Z	1.8

Z = Liquid Depth; T = Tank Diameter

- From Table D.1: Since Z/T was greater than 1.20, two impellers were required with opposed blade pitch. Therefore, n = Number of Impellers per Tank = 2.
- Distance Lower Impeller Off Bottom = Z/4 = 0.762 m/4 = 0.191 m (0.625 ft)
- Distance Upper Impeller Off Bottom = (2 × Z)/3 = (2 × (0.762 m))/3 = 0.508 m (1.67 ft)
- Impeller Spacing = (0.508m – 0.191 m) = 0.317 m (1.04 ft)
- D = Impeller Diameter = 30.5 cm (1.0 ft)
- H = Impeller Height = 10.1 cm (0.33 ft)
- N = Shaft Speed = 425 rpm
- Drive Motor Power = 14.9 kW (20 Hp)

Attrition Scrubber Design Calculations

References: Svedala, 1996, pp. 5:3–5:20; Chemineer, 1985 and Gates et al., 1976.

Attrition Scrubber Slurry Volume (V)

V = (2 cells)(0.61m)(0.61m)(0.76m)(1000 L/m^3) = 566 liters (20 ft^3)

Svedala Basic Product Handbook

The attrition scrubber's optimum retention time (RT) must be 10 to 15 minutes with a 60 to 70% slurry pulp density.

Mass Balance Soil Feed Rate

RT = Retention Time = 15 min
Slurry Solid (Pulp) Density = 60.0% = 0.600 kg solid per kg slurry
Slurry Water = (1.000 kg – 0.600 kg) = 0.400 kg (L)/kg slurry

$$\text{Slurry Volume} = \frac{\dfrac{(0.600 \text{ kg solid/kg slurry})}{(2.60 \text{ kg/L})} + (0.400 \text{ L/kg slurry})}{(0.600 \text{ kg solid/kg slurry})} = 1.051 \text{ L/kg solid}$$

$$\text{Slurry Flow Rate} = \frac{V}{RT} = \frac{566 \text{ L slurry}}{15 \text{ min}} = 37.7 \text{ L slurry/min} (1.33 \text{ ft}^3\text{/min})$$

ORF = Optimum Feed Rate

$$\text{ORF} = \frac{\text{Slurry Flow Rate}}{\text{Slurry Volume}} = \frac{(37.7 \text{ L/min})(60 \text{ min/hr})}{(1.051 \text{ L/kg solid})} = 2152 \text{ kg dry solid/hr} (2.37 \text{ t/hr})$$

$$\text{ORF} = \frac{(2152 \text{ kg dry solid/hr})}{(90 \text{ kg dry solid/100 kg wet solid})} = 2391 \text{ kg wet solids/hr} (2.64 \text{ t/hr})$$

Mass Balance Data: Mixer Design to Suspend Soil Particles in Water at 60% Solids

Number of Tanks = 2
V = Slurry Volume = 566 L = 0.566 m^3

$$V/Tank = \frac{0.566 \text{ m}^3}{2} = 0.283 \text{ m}^3$$

$$D/T = \frac{0.305 \text{ m}}{0.610 \text{ m}} = 0.50$$

% Solids Slurry = 60.0%

Mass Balance Data: Mixer Design to Suspend Soil Particles in Water at 60% Solids

$(S_g)_s$ = Specific Gravity Solid = 2.60 kg/L
$(S_g)_l$ = Specific Gravity Water = 1.00 kg/L
$(S_g)_s - (S_g)_l = (2.60 - 1.00) = 1.60$ kg/L

$$(S_g)_{slurry} = \frac{1.0 \text{ kg slurry}}{\left[\frac{(0.600 \text{ kg solid/kg slurry})}{(2.60 \text{ kg solid/L solid})} + 0.400 \text{ L water/kg slurry}\right]} = 1.58 \text{ kg/L}$$

U_t = Terminal Settling Velocity
U_d = Design Settling Velocity
$U_d = (U_t)(F_w)$
From Figure D.1, F_w at 60.0% solids was 2.20
The overflow of attrition scrubber contained –4760 μm (–4 mesh) soil particles
From Figure D.2, U_t was about 26.5 m/min (about 87 ft/min) for 4760 μm (4 mesh) particles
  with an $(S_g)_s - (S_g)_l$ 1.60 kg/L
$U_d = (26.5 \text{ m/min})(2.20) = 58.3$ m/min (191.3 ft/min)
From Table D.2, scale of agitation was set at 9:

1. Near uniform solids suspension
2. Slurry uniformity of solids to 98% of slurry height
3. Overflow slurry draw-off

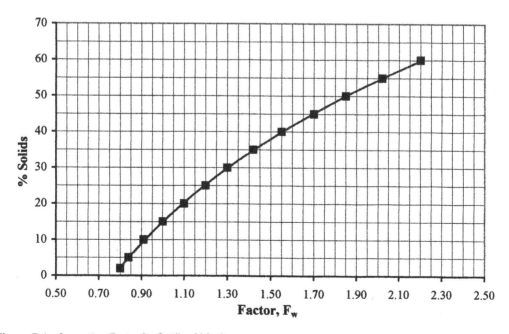

Figure D.1 Correction Factor for Settling Velocity.

Table D.2 Set Degree of Agitation for Solid Suspensions

Suspension Scale	Description
3	1. Suspend all of the solids of the design-settling velocity completely off the tank bottom. 2. Provide slurry uniformity to at least one-third of the slurry height. 3. Be suitable for slurry draw-off at low exit elevations.
6	1. Provide concentration uniformity of solids to 95% of the slurry height. 2. Be suitable for slurry draw-off up to 80% of slurry height.
9	1. Provide slurry uniformity of solids to 98% of the slurry height. 2. Be suitable for slurry draw-off by means of overflow.

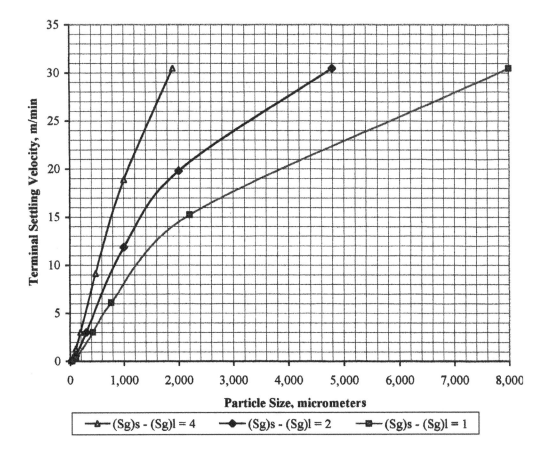

Figure D.2 Terminal Settling Velocity as a Function of Particle Size.

From Figure D.3, ϕ equals 3.2×10^{12} when D/T = 0.50 and the scale of agitation is 9.

1. Calculate N (rpm) using the formula

$$\phi = \frac{(N^{3.75})(D^{2.81})}{U_d}, \text{ where D in centimeters and } U_d \text{ in m/min}$$

2. $\phi = 3.2 \times 10^{12} = \dfrac{(N^{3.75})(30.48 \text{ cm})^{2.81}}{58.3 \text{ m/min}}$

3. N = 495 rpm (operating N at 425 rpm)

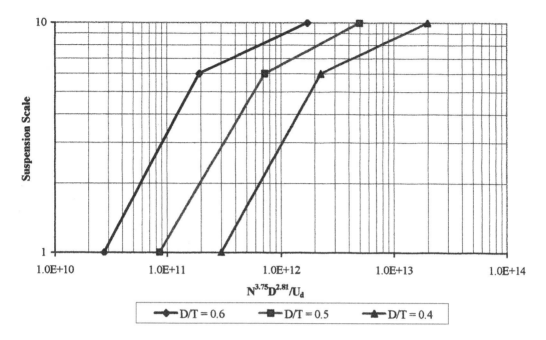

Figure D.3 Plot of Suspension Scale for a Single Pitched Blade Tubine.

The required motor power to drive two agitators with four impellers (n) was calculated:

1. $\text{kilowatts} = \dfrac{(n)(N^3)(D^5)(S_g)_{\text{slurry}}}{(1061)^5} = \dfrac{(4)(495 \text{ rpm})^3(30.5 \text{ cm})^5(1.58 \text{ kg/L})}{(1061)^5}$

2. kilowatts = 15.05 kW (20.16 Hp)

Demonstration Slurry Volume Per Unit Mass (V/kg)

Specific Gravity of Dry Solid in Soil = 2.6 kg/L
Slurry Solid (Pulp) Density = 27.7% = 0.277 kg solid/kg slurry
Slurry Water = 1.000 kg – 0.277 kg = 0.723 kg (L)/kg slurry

$$\text{Slurry Volume} = \dfrac{\dfrac{(0.277 \text{ kg solid/kg slurry})}{(2.60 \text{ kg/L})} + (0.723 \text{ L/kg slurry})}{(0.277 \text{ kg solid/kg slurry})} = 3.00 \text{ L/kg solid}$$

Demonstration Scrubber Retention Time (RT)

Minimum Soil Feed, Demonstration Data = 907.0 kg solid/hr
Volumetric Feed Rate = F = (907.0 kg solid/hr)(3.00 L slurry/kg solid) = 2720 L slurry/hr

$$\text{RT} = \dfrac{V}{F} = \dfrac{(566 \text{ L})}{(2720 \text{ L/hr})(60 \text{ min/hr})} = 12.5 \text{ min}$$

Demonstration Data: Mixer Design to Suspend Soil Particles in Water at 27% Solids

Number of Tanks = 2
V = 566 L = 0.566 m³

$$V/Tank = \frac{0.566 \text{ m}^3}{2} = 0.283 \text{ m}^3$$

Shaft Speed = N = 425 rpm
Drive Motor Power = 14.9 kW (20 Hp)
D/T = 0.50
% Solids Slurry = 27.7%
$(S_g)_l$ = Specific Gravity Water = 1.00
$(S_g)_s$ = Specific Gravity Solid = 2.60
$(S_g)_s - (S_g)_l = (2.60 - 1.00) = 1.60$

$$(S_g)_{slurry} = \frac{1.0}{\left[\dfrac{(0.277 \text{ kg solid/kg slurry})}{(2.60 \text{ kg solid/L solid})} + 0.723 \text{ L water/kg slurry}\right]} = 1.20 \text{ kg/L}$$

U_t = Terminal Settling Velocity
U_d = Design Settling Velocity
$U_d = (U_t)(F_w)$
From Figure D.1, F_w at 27.7% solids was 1.25
The overflow of attrition scrubber contained −4760 μm (−4 mesh) soil particles
From Figure D.2, U_t was about 26.5 m/min (about 87 ft/min) for 4760 μm (4 mesh) particles
with an $(S_g)_s - (S_g)_l = 1.60$ kg/L
$U_d = (26.5 \text{ m/min})(1.25) = 33.1 \text{ m/min} (108.6 \text{ ft/min})$

Demonstration Data: Mixer Design to Suspend Soil Particles in Water at 27% Solids

From Table D.2, scale of agitation was set at 9:

1. Near uniform solids suspension
2. Slurry uniformity of solids to 98% of slurry height
3. Overflow slurry draw-off

From Figure D.3, ϕ equals 3.2×10^{12} when D/T = 0.50 and the scale of agitation is 9.

1. Calculate N (rpm) using the formula

   $$\phi = \frac{(N^{3.75})(D^{2.81})}{U_d}, \text{ where D in centimeters and } U_d \text{ in m/min}$$

2. $\phi = 3.2 \times 10^{12} = \dfrac{(N^{3.75})(30.48 \text{ cm})^{2.81}}{33.1 \text{ m/min}}$

3. N = 425 rpm (operating N at 425 rpm)

The required motor power to drive two agitators with four impellers (n) was calculated:

1. kilowatts $= \dfrac{(n)(N^3)(D^5)(S_g)_{slurry}}{(1061)^5} = \dfrac{(4)(425 \text{ rpm})^3(30.5 \text{ cm})^5(1.20 \text{ kg/L})}{(1061)^5}$

2. kilowatts = 7.22 kW (9.68 Hp)

Summary of Design Results

	Slurry %Solids	Solid Feed kg/hr	Maximum Particle Size μm	Shaft Speed rpm	Required Motor Power kW (Hp)
Mass Balance Data	60.0	2152	4760	495	15.1 (20.2)
Demonstration Data	27.7	+907	4760	425	7.2 (9.7)

Mass Balance Data Calculated at 60.0% Solids

- At 495 rpm, the agitators in the two scrubber tanks should suspend 4760 μm (4 mesh) soil particles at 60.0% solids.
- However, the specified drive motor at 14.9 kW might not have the capacity to mix the 60.0% solid slurry and suspend the 4762 μm (4 mesh) soil particles.

Field Data Calculated at 27.7% Solids

- The retention time of 12.5 min for the two scrubber cells was appropriate, but the solid pulp density for the slurry was well below optimum. Therefore, scrubbing effectiveness would be decreased.
- At 425 rpm, the agitators in the two scrubber tanks should suspend 4760 μm (4 mesh) soil particles.
- The specified drive motor at 14.9 kW had the capacity to mix the 27.7% solid slurry and to suspend the 4760 μm (4 mesh) soil particles.

D.2 TROMMEL

Design Data

Mass Balance Data		kg/hr		% Distribution	
Feed	Feed Rate	Solid	Water	Solid	Water
Feed: −4.76 mm Slurry	1395.58 kg/hr	837.35	558.23	60.0	40.0
Feed: Wash Water	235.67 L/hr	0.00	235.67	0.0	100.0
−4.76 mm, +1.19 mm Oversize Product		58.92	14.73	80.0	20.0
−1.19 mm Undersize Product		758.97	759.69	50.0	50.0
−1.19 mm Trommel Overflow		19.46	19.48	50.0	50.0
Totals Mass Input/Output		837.35	793.90		

Demonstration Trommel Specifications

- Screen Opening   = 1.19 mm (16 mesh, U.S. Standard Series)
- Diameter   = 0.457 m (1.50 ft)
- Screen Length   = 0.762 m (2.50 ft)
- Area of Screen Surface = 1.094 m² (11.8 ft²)

Trommel Design Calculations

References: Denver, 1954; Portec, 1992.

Mass Balance Trommel Capacity and Required Screen Area

Screen Opening Area = (1.19 mm)² = 1.42 mm²
Trommel Screen Capacity of −1.19 mm (16 mesh) Soil, Table D.3 or Figure D.4:
Capacity = 3760 kg/hr-m²
Minus 1.19 mm Soil in Feed to Trommel = 759 kg/hr (1701 lb/hr)

$$\text{Required Effective Area of Trommel Screen} = \frac{759 \text{ kg/hr}}{3760 \text{ kg/hr} \cdot \text{m}^2}$$

Required Effective Area = 0.202 m² (2.17 ft²)

Table D.3 CMRI Data for Trommel Screen

Screen Size inch/mesh	Screen Opening mm	Screen Opening (mm)²	Solid Passing Screen Opening[1] Handbook kg/hr	Solid Passing Screen Opening[1] Calculated kg/hr-m²	Solid Passing Screen Opening[1] Calculated kg/hr	Area of Data Screen	
1 in.	25.40	645.16	10,090	25,878	10,017	D, m	0.762
3/4 in.	19.05	362.90	8427	21,589	8357	L, m	1.524
1/2 in.	12.70	161.29	6500	16,722	6473	Area, m²	0.387
3/8 in.	9.52	90.73		13,950	5400	**Area of BOR Screen**	
1/4 in.	6.30	40.32	4044	10,805	4182	D, m	0.457
8 mesh	2.38	5.66		5823	2254	L, m	0.762
16 mesh	1.19	1.42		3763	1465	Area, m²	0.116

1. Solid rates, kg/hr-m², were calculated using formula from Figure D.4. Solid rates, kg/hr, were calculated by multiplying the value in kg/hr-m² by 0.387 m².

Effective Area of Data Trommel Screen

$$\text{Effective Screen Width} = \frac{\text{Screen Diameter, m}}{3} = \frac{0.762 \text{ m}}{3} = 0.254 \text{ m}$$

Effective Screen Area, m² = (Effective Screen Width, m)(Screen Length, m)
Effective Screen Area = (0.254 m)(1.524 m) = 0.387 m²

Mass Balance Screen Size

Effective Screen Area, m² = (Effective Screen Width, m)(Screen Length, m)

$$\text{Effective Screen Width, m} = \frac{(\text{Effective Screen Area, m}^2)}{(\text{Screen Length, m})} = \frac{0.202 \text{ m}^2}{0.559 \text{ m}} = 0.361 \text{ m } (\sim 1.18 \text{ ft})$$

Screen Diameter, m = (3)(Effective Screen Width, m) = (3)(0.361) = 1.083 m (3.55 ft)
Selected Trommel Screen Diameter = 1.219 m (4.00 ft or 48 in.)
Selected Screen Length, m = 0.559 m (1.83 ft or 22.0 in.)

$$\text{Effective Screen Area, m}^2 = \frac{(1.219 \text{ m})}{3} \times (0.559 \text{ m}) = 0.227 \text{ m}^2 \ (2.44 \text{ ft}^2)$$

Demonstration Trommel Capacity

Screen Opening Area = (1.19 mm)² = 1.42 mm²
Trommel Screen Capacity of −1.19 mm (16 mesh) Soil, Table D.3 or Figure D.4:
Capacity = 3760 kg/hr-m²

Demonstration Screen Size:

Diameter = 0.457 m (1.50 ft)
Screen Length = 0.762 m (2.50 ft)

$$\text{Effective Screen Width} = \frac{\text{Screen Diameter, m}}{3} = \frac{0.457 \text{ m}}{3} = 0.152 \text{ m}$$

Effective Screen Area, m² = (Effective Screen Width, m)(Screen Length, m)
Effective Screen Area = (0.152 m)(0.762 m) = 0.116 m² (1.25 ft²)
Maximum Trommel Capacity of −1.19 mm Soil = (3760 kg/hr-m²)(0.116 m²)
Maximum Capacity = 435 kg/hr (965 lb/hr)

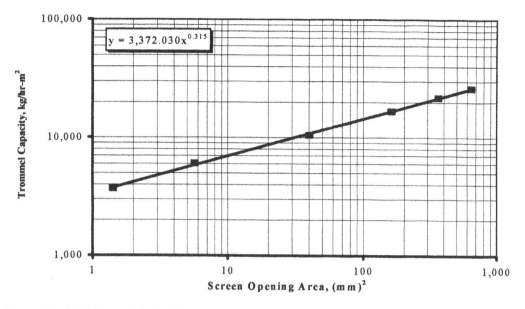

Figure D.4 CMRI Trommel Design Data.

Summary of Design Results

	Screen Diameter m (ft)	Screen Length m (ft)	Effective Screen Area m² (ft²)	Screen Capacity kg/hr
Mass Balance Data	1.22 (4.0)	0.56 (1.8)	0.227 (2.45)	855
Demonstration Data	0.47 (1.5)	0.76 (2.5)	0.116 (1.25)	435

- The demonstration trommel with an effective screen area of 0.116 m² (1.25 ft²) was too small to handle a soil feed rate of 840 kg/hr (or 760 kg/hr of −1.19 mm soil)
- The design capacity of the trommel screen was 855 kg/hr of −1.19 mm soil. The recommended trommel screen easily handled the mass balance rate of 760 kg/hr.
- A single-deck vibrating screen with high-pressure water wash should replace the trommel screen.

D.3 MINERAL JIG

Design Data

Mass Balance Data		kg/hr		% Distribution	
Feed	Feed Rate	Solid	Water	Solid	Water
Feed: +1.19 mm Slurry	73.65 kg/hr	58.92	14.73	80.0	20.0
Feed: Dilution Water	95.55 L/hr	0.00	95.55	0.0	100.0
+1.19 mm Overflow Product (light fraction)		38.52	89.89	30.0	70.0
+1.19 mm Hutch Product (heavy fraction)		20.39	20.39	50.0	50.0
Totals Mass Input/Output		58.92	110.28		

Demonstration Mineral Jig Specifications

- Type Mineral Jig: Simplex 10 in. × 10 in. (25.4 cm × 25.4 cm)
- Mineral Jig Bed Area = 0.0645 m² (0.694 ft²)

Mineral Jig Design Calculations

Reference: Svedala, 1996, pp. 4:11 to 4:13

Mass Balance Mineral Jig Capacity

Mass Balance Mineral Jig Feed = 58.92 kg/hr (130 lb/hr)
Capacity from Figure D.5
Specific Gravity of Lead Metal = 11.4 kg/L
Minimum Mineral Jig Capacity = 7500 kg/hr-m^2 (1540 lb/hr-ft^2)

$$\text{Minimum Jig Area, m}^2 = \frac{\text{Mass Balance Feed Rate}}{\text{Minimum Jig Capacity}} = \frac{58.92 \text{ kg/hr}}{7500 \text{ kg/hr/m}^2} = 0.00786 \text{ m}^2 \text{ (0.0846 ft}^2)$$

Minimum Jig Size from Table D.4:

Simplex 4 in. × 6 in. (10.2 cm × 15.2 cm), minimum size available
Bed Area = (10.2 cm)(15.2 cm) = 155 cm^2 = 0.016 m^2 (0.17 ft^2)
Expected Capacity = (0.016 m^2)(7500 kg/hr-m^2) = 120 kg/hr (265 lb/hr)
Expected Media Flow = 0.1 to 0.3 m^3/hr (0.4 to 1.3 gpm)
Design Feed Water = 0.20 m^3/hr (0.85 gpm)

Demonstration Mineral Jig Capacity

Capacity from Figure D.5
Specific Gravity of Lead Metal = 11.4 kg/L
Minimum Mineral Jig Capacity = 7500 kg/hr-m^2 (1540 lb/hr-ft^2)
Jig Bed Area = 0.0645 m^2 (0.694 ft^2)
Comparable Jig Size from Table D.4:

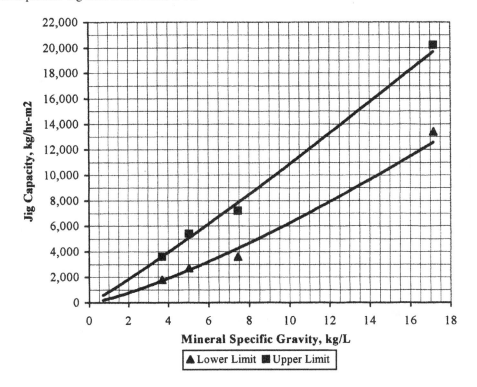

Figure D.5 Jig Capacity Curves.

Table D.4 Gravity Separation Mineral Jig, Sizing

Jig Type[1]	Bed Area		Media Flow (water)	
	m²	ft²	m³/hr	gpm
Simplex 4 × 6	0.016	0.17	0.1 to 0.3	0.4 to 1.3
Simplex 8 × 12	0.062	0.67	0.5 to 0.8	2.2 to 3.5
Simplex 12 × 18	0.139	1.50	1.3 to 2.0	5.7 to 8.7
Simplex 16 × 24	0.248	2.67	1.5 to 3.0	6.6 to 13.2
Simplex 24 × 36	0.557	6.00	5.0 to 7.0	22 to 31

1. Type number refers to width and length of particle bed area (inches).

Simplex 8 in. × 12 in. (20.3 cm × 30.4 cm)
Bed Area = 0.062 m² (0.67 ft²)
Expected Media Flow = 0.5 to 0.8 m³/hr (2.2 to 3.5 gpm)

Demonstration Jig Capacity = (7500 kg/hr/m²)(0.0645 m²) = 485 kg/hr (1065 lb/hr)
Mass Balance Media Flow = (0.0956 + 0.0147 m³/hr) = 0.110 m³/hr (0.485 gpm)

Summary of Design Results

	Jig Type	Jig Bed Area m² (ft²)	Media Flow m³/hr (gpm)	Jig Capacity kg/hr
Mass Balance Data	Simplex 4 × 6	0.016 (0.17)	0.11 (0.49)	120
Demonstration Data	Simplex 10 × 10	0.065 (0.69)	0.20 (0.85)	485

• Therefore, the mineral jig that was used in the demonstration pilot-plant demonstration had considerable excess capacity.

Gravity Separation - Limitations

Particle sizes — 200 mm (8 in.), using dense media, to 50 µm (270 mesh)) or in certain cases down to 5 µm.
Particle density (specific gravity) — 1.2 g/cm³ to 20 g/cm³
Particle density differences for selective separation — > 1.0 g/cm³

Gravity Separation

Illustration of Classic Mineral Jig operation

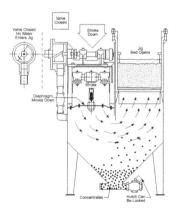 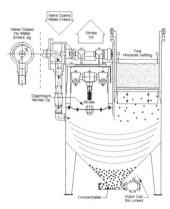

1. On the down stroke of the diaphragm mechanism, the fluidized bed redistributes the particles.
   • Lighter particles move up in the bed
   • Heavier particles move down in the bed

2. On the upward stroke of the diaphragm mechanism, the bed rests and the heavier particles flow down through the bed into the concentrate zone.

Mineral Jig Design

- Mainly used for –6mm feed particles, a "through the bed" jig
- Right or left-hand layout
- Simplex (one) mineral jig and duplex (two) mineral jigs as illustrated below
- Heavy-duty diaphragm
- Adjustable diaphragm stroke
- Water valve synchronized with diaphragm mechanism

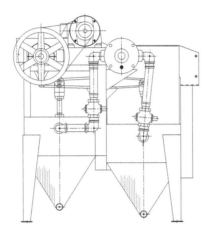

D.4 SPIRAL CLASSIFIER

Design Data

Mass Balance Data		kg/hr		% Distribution	
Feed	Feed Rate	Solid	Water	Solid	Water
Feed: –1.19 mm Slurry	1767.89 kg/hr	798.39	969.51	45.2	54.8
Feed: Wash Water	0.00 L/hr	0.00	0.00	0.0	0.0
–1.19 mm, +105 μm Sand Product		413.72	177.31	70.0	30.0
–105 μm Slimes (Overflow) Product		384.67	792.20	32.7	67.3
Totals Mass Input/Output		798.39	969.51		

Mass Balance Spiral Classifier Specifications

- Particle Size of Separation = 105 μm (150 mesh)
- Solid Feed Rate = 798 kg solids/hr (0.880 ton/hr)
- O/F C = Overflow Capacity = 385 kg solids/hr (0.421 ton/hr)
- Sand C = Sand Capacity = 414 kg/hr (0.425 t/hr)
- O/F%S = Overflow Pulp Density = 32.7% solids
- Specific Gravity of Solids $(S_g)_S$ = 2.60 kg/L (162 lb/ft^3)

Calculation of Overflow (O/F) Pool Area

Reference: Svedala, 1996, pp. 3:11–3:13

O/F Volume Rate (O/F V)

$$O/F\ V = \frac{(O/FC)}{(S_g)_s} + \frac{(O/FC)(100\% - O/F\%S)}{(O/F\%S)} = \frac{(385\ kg/hr)}{2.60\ kg/L} + \frac{(385\ kg/hr)(100\% - 32.7\%)}{(32.7\%)}$$

O/F V = 940 L/hr (0.940 m³/hr)

O/F Volume % Solids (O/F V%S)

$$O/F\ V\%S = \frac{100}{\left[1 + (S_g)_s \times \frac{(100 - O/F\%Solids)}{(O/F\%Solids)}\right]} = \frac{100}{\left[1 + 2.60 \times \frac{(100 - 32.7)}{32.7}\right]} = 15.7\%$$

From the figure on page 241 in this appendix section (Svedala, 1996), the settling rate of 105 μm (150 mesh) particles at 16% O/FV%S was about 15 m/hr (0.82 ft/min)

O/F Pool Area (O/F PA)

Specific Gravity Correction:

$$\text{Corrected Settling Rate} = (15\ m/hr) \times \left[\frac{((S_g)_s - 1)}{1.65}\right]^{0.5} = (15\ m/hr) \times \left[\frac{(2.60 - 1)}{1.65}\right]^{0.5}$$

Corrected Settling Rate = 14.8 m/hr. (0.81 ft/min)

O/F Pool Area (O/F PA)

$$O/F\ PA = \frac{(O/F\ V)}{(0.7)(\text{Corrected Settling Rate})} = \frac{(0.940\ m^3/hr)}{(0.7)(14.8\ m/hr)} = 0.0907\ m^2\ (0.976\ ft^2)$$

The factor, 0.7, is for the disturbance in the settling pool caused by the rotating spiral.

Calculation of Sand Raking Capacity and Sand Compression Pool Area

Reference: Svedala, 1996, pp. 3:11–3:13

Sand Raking Capacity (Sand V)

Sand Raking Volume = Sand V = m³/hr
Sand Capacity, kg/hr = Sand C = 414 kg/hr
Sand % Solids = Sand %S = 70%

$$\text{Sand V} = \left[\frac{(\text{Sand C})}{(S_g)_s} + \frac{(\text{Sand C})(100\% - \text{Sand }\%S)}{(\text{Sand }\%S)}\right]/(1000\ L/m^3)$$

$$\text{Sand V} = \left[\frac{(414\ kg/hr)}{(2.60\ kg/L)} + \frac{(414\ kg/hr)(100\% - 70\%)}{(70\%)}\right]/(1000\ L/m^3)$$

Sand V = 0.337 m³/hr(11.9 ft³/hr)

Sand Compression Pool Area (Sand CPA)

From the figure on page 241 in this appendix section, the settling rate at 40%V for 105 μm (150 mesh) particles was about 1.8 m/hr.
Specific Gravity Correction:

$$\text{Corrected Settling Rate} = (1.8 \text{ m/hr}) \times \left[\frac{((S_g)_s - 1)}{1.65}\right]^{0.5} = (1.8 \text{ m/hr}) \times \left[\frac{(2.60 - 1)}{1.65}\right]^{0.5}$$

Corrected Settling Rate = 1.77 m/h (0.096 ft/hr)

Compression Zone Volume = CZV

$$\text{CZV} = \left[\frac{(\text{Sand C, t/hr})}{\dfrac{(S_g)_s \times (0.7)(0.8)(40\%)}{(100\%)}}\right] = \left[\frac{(414 \text{ kg/hr})/(1000 \text{ kg/t})}{\dfrac{((2.60)(0.7)(0.8)(40\%))}{(100\%)}}\right] = 0.711 \text{ m}^3/\text{hr}$$

The factor, 0.7, is for the disturbance in the settling pool caused by the rotating spiral. The factor, 0.8, refers to the smaller pool area available at the compression level.

$$\text{Sand CPA} = \frac{(\text{CZV})}{(\text{Corrected Settling Rate})} = \frac{(0.711 \text{ m}^3/\text{hr})}{(1.77 \text{ m/hr})} = 0.40 \text{ m}^2(4.3 \text{ ft}^2)$$

Calculation of Spiral Classifier Capacity

Reference: Weiss, 1985, pp. 3D-46 to 3D-52.

Mass Balance Classifier Pool Area

Mass Balance Overflow Capacity = 385 kg/hr (842 lb/hr)
Using Figure D.6:
Classifier Overflow Particle Size = 105 μm (150 mesh)
Overflow Capacity per unit Pool Area = 26,700 kg/24 hr/m^2

$$\text{Overflow Capacity per unit Pool Area} = \left[\frac{(26,700 \text{ kg/24 hr/m}^2)}{(24)}\right] = 1112 \text{ kg/hr/m}^2$$

$$\text{Pool Area} = \frac{(\text{Mass Balance Rate, kg/hr})}{(1112 \text{ kg/hr/m}^2)} = \frac{(385 \text{ kg/hr})}{(1112 \text{ kg/hr/m}^2)} = 0.346 \text{ m}^2(3.72 \text{ ft}^2)$$

Demonstration Classifier Overflow Capacity

Two 150 mm (6 in.) spiral classifiers with each settling pool measuring 0.305 m length by 0.366 m width (1 ft by 1.2 ft).
Total Pool Area = 0.111 m^2 (1.20 ft^2)

$$\text{Overflow Capacity per unit Pool Area} = \left[\frac{(26,700 \text{ kg/24 hrs/m}^2)}{(24)}\right] = 1112 \text{ kg/hr/m}^2$$

Overflow Capacity of Spiral Classifier = (1112 kg/hr/m^2)(0.111 m^2)
Overflow Capacity = 125 kg/hr (275 lb/hr)

Therefore, for the two classifiers:

Total Overflow Capacity = (2)(125 kg/hr) = 250 kg/hr (550 lb/hr)

Mass Balance Overflow Solid (105 µm (150 mesh) fraction) = 385 kg/hr (842 lb/hr)

$$\text{Classifier Capacity} = (\text{Mass Balance Classifier Feed}) \times \frac{\text{Demonstration Overflow}}{\text{Mass Balance Overflow}}$$

$$\text{Classifier Capacity} = (798 \text{ kg/hr}) \times \frac{250 \text{ kg/hr}}{385 \text{ kg/hr}} = 520 \text{ kg/hr}$$

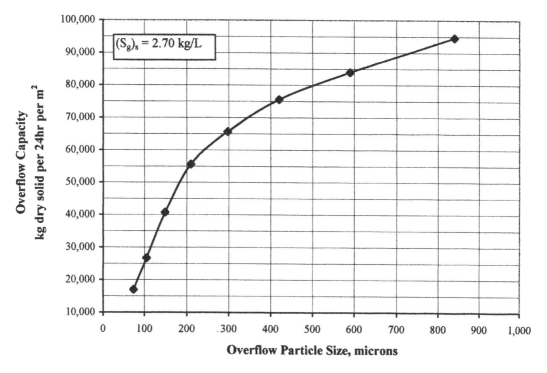

Figure D.6 Overflow Capacity of Spiral Classifiers.

Summary of Design Results

	Units	Diameter Spiral mm (in.)	Classifier Design/Model	Total Max. Pool Area m² (ft²)	Classifier Capacity kg/hr (lb/hr)
Mass Balance Data	2	300 (12)	SP/Simplex/150	0.640 (6.89)	1275 (2815)
Demonstration Data	2	150 (6)	SP/Simplex/150	0.222 (2.40)	520 (1150)

- The classifier pool area of 0.35 m² (3.7 ft²) is similar to the sand compression pool area of 0.40 m² (4.3 ft²). However, the sand compression pool area governs the classifier size. From Table D.5, one 400 mm (16 in.) or two 300 mm (12 in.) classifiers is recommended. The average sand raking capacity for each 300 mm classifier was 1135 kg/hr.
- The spiral classifiers in the demonstration pilot plant were sized as 150 mm (6 in.) units, each with a pool area of about 0.111 m² (~1.20 ft²). The total pool area was 0.222 m² (2.35 ft²). Based on the compression pool area requirement of 0.40 m² (4.3 ft²), the two 150 mm spiral classifiers were not large enough to treat the 748 kg/hr feed (1760 lb/hr).
- The mass balance overflow capacity of 385 kg/hr (842 lb/hr) exceeds the calculated overflow capacity of 250 kg/hr (550 lb/hr) for the two 150 mm (6 in.) classifiers.

- The total estimated capacity for the two 150 mm (6 in.) classifiers was about 520 kg/hr (1150 lb/hr).
- The average raking capacity for each 150 mm (6 in.) spiral classifier was 870 kg/hr (1920 lb/hr). This capacity was sufficient to handle the mass balance sand capacity of 414 kg/hr (913 lb/hr).

Table D.5 Spiral Classifier – Pool Area And Raking Capacity

Diameter Spiral		Spiral rpm	Classifier	Raking Capacity kg/hr (m³/hr)		Tank	Available Pool Area m²	
in.	mm	min/max	Design	min	max	Model	min	max
6	150	35/65	SP/Simplex	640 (0.40)	1100 (0.69)	150	0.070	0.110
9	230	23/46	SP/Simplex	720 (0.45)	1450 (0.91)	150	0.100	0.200
12	300	15/20	SP/Simplex	1000 (0.63)	1270 (0.80)	150	0.200	0.320
16	400	10/15	SP/Simplex	1090 (0.68)	1630 (1.02)	150	0.320	0.500
24	600	6/16	SP/Simplex	5700 (3.6)	13,600 (9.6)	100	1.07	1.55
		6/16	DP/Simplex	11,300 (6.6)	27,300 (17.6)	125	1.54	2.08
						150	2.05	2.64
30	750	5/14	SP/Simplex	9100 (5.0)	17,300 (14.0)	100	1.76	2.37
		5/14	DP/Simplex	18,150 (9.5)	33,550 (26.6)	125	2.52	3.21
36	900	4/11	SP/Simplex	15,400 (8.0)	30,850 (22.0)	100	2.41	3.36
		4/11	DP/Simplex	30,850 (16.0)	61,700 (44.0)	125	3.49	4.53
						150	4.66	5.81

All classifiers have "Straight Tanks" with a maximum slope of 29.1 cm per m (3.5 in. per ft)

Model 100 – 100% Spiral submergence in hindered settling zone

Model 125 – 125% Spiral submergence in hindered settling zone

Model 150 – 150% Spiral submergence in hindered settling zone

Simplex – Single spiral in tank

SP – Single pitch spiral

DP – Double pitch spiral

Spiral Classifier Models

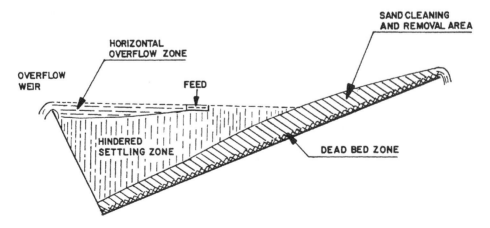

Treatment zones within a spiral classifier.

Reference: Svedala, 1996, pp. 3:11–3:13

Model	Spiral Submergence	Classifier Pool Areas	Particle Size Range	Tank Design
100 Simplex	100%	small to medium	0.8 to 0.2 mm (20 to 65 mesh)	straight of modified flare
125 Simplex	125%	medium to large	0.4 to 0.1 mm (35 to 150 mesh)	modified or full flare
150 Simplex	150%	large to very large	0.2 to 0.075 mm 65 to 200 mesh	modified or full flare

Model 100, 125, or 150 Duplex classifier is used for large areas that are required by hydraulic loading capacity and overflow particle size demands.

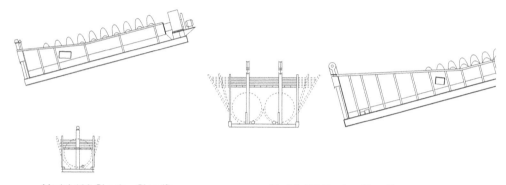

Model 100 Simplex Classifier   Model 150 Duplex Classifier

Particle Settling Rate for Spiral Classifiers

Reference: Svedala, 1996, pp. 3:11–3:13
Settling Rate (m/hr) versus particle size (micrometers) at 0 to 40%V

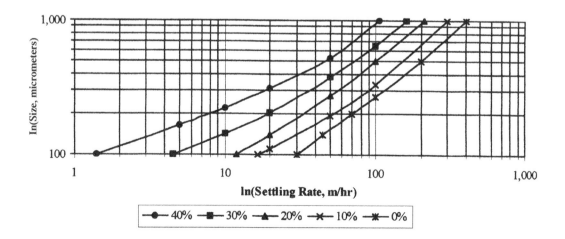

Applications for Spiral Classifiers

Reference: Svedala, 1996, pp. 3:11–3:13

- Small to medium mineral processing circuits
- Two stage circuits consisting of primary classifier and secondary hydrocyclones
- Desliming to upgrade products like barite and salt crystal
- Recovery of sand product in sewage treatment plants
- Dewatering of process materials like hydrocyclone underflow (sand) product

Sizing of Spiral Classifiers

Reference: Svedala, 1996, pp. 3:11–3:13

Calculation of overflow pool area

1. Select the particle size in micrometers (mesh) that is required for separation in the classifier
2. Select or calculate the weight percent solids in the overflow – O/F %Solids
3. Calculate the volume percent solids – $\%V = \dfrac{100}{1 + \dfrac{(S_g)_s \times (100 - O/F\%Solids)}{O/F\%Solids}}$
4. Calculate the volumetric flow rate for the overflow in m^3/hr – O/F V
5. For the particle size of separation, select the hindered settling rate in m/hr from the figure on page 241. For cases other than the $(S_g)_S$ at 2.65 multiply the settling rate by the Stokes law factor $[(((S_g)_S - 1)/1.65)^{0.5}]$.
6. Calculate required overflow pool area by dividing the overflow flow rate with (0.7 × corrected settling rate). The 0.7 value is the spiral disturbance factor.
7. Select the classifier size from Table D.5 that has a pool area equal to or greater than the calculated area.

Calculation of spiral sand transport capacity

1. Calculate the sand transport volume in m^3/hr that will be removed from the pool by the spiral conveyor.
2. Compare the sand transport volume to the rake capacity of the selected classifier in Table D-5.
3. If the sand transport volume for the selected classifier is too low, use a double start spiral or a larger classifier.
4. If the sand transport volume can not be calculated, use total feed solid feed rate as the "worse case."
5. Depending on particle size and solid specific gravity, the wet sand product will have a solid content of 80 to 86 weight percent.

Calculate sand fraction compression pool area

As solid particles settle in the classifier pool, crowding increases, decreasing the settling rate of the particles (hindered settling theory). In an undersized classifier, particle build-up usually occurs in the tank. The particles are too small to be removed by the spiral conveyor, but too large to overflow the classifier weir. As the build-up continues, the particles will eventually "surge" into the overflow stream. Hence, the classifier operates in an unstable condition, causing poor particle separation.

The minimum tank area needed to avoid the "surge" condition is defined as the "sand compression pool area." It can be estimated as follows:

1. From the figure on page 241, the 40%V settling rate is established for the maximum particle size in the overflow. If necessary, the rate is adjusted for other solid specific gravity than 2.65.
2. The sand compression volume is calculated by dividing the dry solid rate with $[0.7 \times 0.8 \times 0.4 \times (S_g)_S]$. The 0.7 factor is defined above. The 0.8 factor accounts for the smaller pool area at the compression zone level.
3. The compression volume is divided by the settling rate which gives the minimum pool area for the compression zone.
4. The smallest classifier tank with this or bigger pool area is selected.
5. The largest classifier is selected according to the three sizing criteria.

Example

Dry solid feed rate – 15 t/hr or 15,000 kg/hr
Overflow (O/F) slurry capacity – 10 t/hr or 10,000 kg/hr
Overflow weight % solids – 25%
Sand weight % solid – 80%
Specific gravity of solid particles $(S_g)_S$ – 3.0 g/cm³
Liquid phase – water at room temperature
Maximum particle size in overflow –105 μm (150 mesh)
Calculation of overflow pool area

$$\text{O/F volume \% solids} = \%V\ S = \frac{100}{\left(1 + (S_g)_s \times \frac{(100 - 25)}{25}\right)} = \frac{100}{1 + 3.0 \times \frac{(100 - 25)}{25}} = 10.0\%V$$

$$\text{O/F pulp volume} = \text{O/F } V = \frac{10,000}{3.0} + \frac{(10,000)(100 - 25)}{25} = 33,330 \text{ L/hr} \equiv 33.3 \text{ m}^3/\text{hr}$$

Particle settling rate from figure on page 241 at 105 μm and 10.0 %V = 15.0 m/hr

$$\text{Specific gravity correction} = 15.0 \times \left(\frac{3.0 - 1}{1.65}\right)^{0.5} = 16.5 \text{ m/hr}$$

$$\text{O/F minimum pool area} = \text{O/F } PA = \frac{33.3}{(16.5)(0.7)} = 2.88 \text{ m}^2$$

Minimum classifier size from Table D5 – 750 mm (30 in.) DP/Simplex 125, Straight Tank (2.52 to 3.21 m²)
Calculation of spiral sand transport volume

$$\text{Sand slurry volume} = \text{Sand } V = \frac{(15,000 - 10,000)}{3.0} + \frac{(15,000 - 10,000)(100 - 80)}{80}$$

$$= 2920 \text{ L/hr} = 2.92 \text{ m}^3/\text{hr}$$

Acceptable classifier size from Table D5 – 750 mm (30 in.) DP/Simplex 125, Straight Tank (9.5 m³/hr)
Calculation of sand fraction compression pool area
The settling rate at 40%V for 105 μm particles from figure on page 241 is about 1.5 m/hr.

$$\text{Specific gravity correction} = 1.5 \times \left(\frac{3.0 - 1}{1.65}\right)^{0.5} = 1.65 \text{ m/hr}$$

$$\text{Compression zone volume rate} = CZV = \frac{(15,000 - 10,000)}{\frac{(3.0)(0.7)(0.8)(40\%)}{100\%}} = 7440 \text{ L/hr} \equiv 7.44 \text{ m}^3/\text{hr}$$

$$\text{Compression zone pool area} = CPA = \frac{7.44}{(1.65)} = 4.50 \text{ m}^2$$

Minimum classifier size from Table D5 – 900 mm (36 in.) DP/Simplex 125, Straight Tank (3.49 to 4.53 m²)

In this case, the compression zone pool area will determine the size of the classifier.

D.5 HYDROCYCLONE

Design Data

Mass Balance Data		kg/hr		% Distribution	
Feed	Feed Rate	Solid	Water	Solid	Water
Feed: −105 µm Slurry	1147.45 kg/hr	375.05	772.39	32.7	67.3
Feed: Make-up Water	0.00 L/hr	0.00	0.00	0.0	100.0
−37µm Overflow Product		134.33	575.44	18.93	81.1
−105 µm, +37µm Underflow Product		240.72	196.95	55.0	45.0
Totals Mass Input/Output		375.05	772.39		

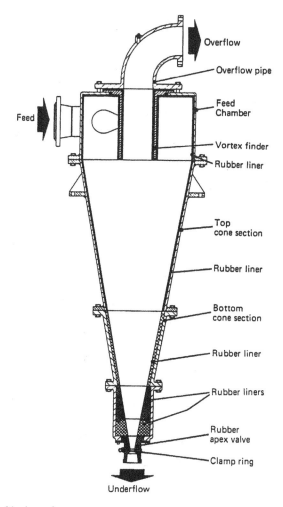

Cross-sectional view of hydrocyclone.

Demonstration Hydrocyclone Specifications

- Hydrocyclone Size: 7.6 cm (3 in.)
- Vortex Finder = 25 mm (0.98 in.) or 18 mm (0.71 in.)
- Spigot Opening = 15 mm (0.59 in.)
- Minimum Operating Gauge Pressure = 69 kPa (10 psig)

Hydrocyclone Design Calculations

References: Carpco, 1992a and Carpco, 1992b; Weiss, 1985, pp. 3D-46 to 3D-52.

Mass Balance Hydrocyclone Capacity

%Solids Feed = %Solids-F = 32.7%

$$\text{Feed \%Solids (by Volume)} = F - V\%S = \frac{100}{\left[1 + (S_g)_s \times \frac{(100 - (\%\text{Solids} - F))}{(\%\text{Solids} - F)} \right]}$$

$$\text{Feed \%Solids (by Volume)} = F - V\%S = \frac{100}{\left[1 + 2.60 \times \frac{(100 - 32.7)}{(32.7)} \right]} = 15.7\%$$

Hydrocyclone Solid Feed Rate = Slime-F = 375 kg/hr (827 lb/hr)

$$\text{Feed Slurry Volumetric Rate} = \left[\frac{(\text{Slime} - F)}{(S_g)_s} + \frac{(\text{Slime} - F)(100 - (\%\text{Solids} - F))}{(\%\text{Solids} - F)} \right]$$

$$\text{Feed Slurry Voumetric Rate} = \left[\frac{(375 \text{ kg/hr})}{(2.60 \text{ kg/L})} + \frac{(375 \text{ kg/hr})(100 - 32.7)}{(32.7)} \right]$$

Feed Slurry Volumetric Rate = 916 L/hr or 0.916 m³/hr (4.03 gpm)
%Solids Underflow = %Solids-U = 55.0%

$$\text{Underflow \%Solids (by Volume)} = U - V\%S = \frac{100}{\left[1 + (S_g)_s \times \frac{(100 - (\%\text{Solids} - U))}{(\%\text{Solids} - U)} \right]}$$

Table D.6 % Volume Split of Liquid Feed for a 50 mm (2 in.) Hydrocyclone

Spigot, mm		9.4	6.4	4.5	3.2
Diameter	14	17%	8%	3%	2%
Vortex	11	30%	16%	7%	3%
Finder, mm	8	62%	42%	23%	9%

Percentage by volume of water feed recovered in underflow
Operating Gauge Pressure = 345 kPa (50 psig)
Volume split to underflow increases at lower operating pressures.

$$\text{Underflow \%Solids (by Volume)} = U - V\%S = \frac{100}{\left[1 + 2.60 \times \frac{(100 - 55.0)}{(55.0)} \right]} = 32.0\%$$

Hydrocyclone Underflow Rate = U'Flow = 241 kg/hr (531 lb/hr)

$$\text{Underflow Slurry Volumetric Rate} = \left[\frac{(\text{U'Flow})}{(S_g)_s} + (\text{U'Flow})\frac{(100 - (\%\text{Solids} - U))}{(\%\text{Solids} - U)} \right]$$

Underflow Slurry Volumetric Rate $= \left[\dfrac{(241\ \text{kg/hr})}{(2.60\ \text{kg/L})} + \dfrac{(241\ \text{kg/hr})(100 - 55.0)}{(55.0)} \right]$

Underflow Slurry Volumetric Rate = 290 L/hr or 0.290 m³/hr (1.28 gpm)

Percentage by Volume of Feed in Underflow = %U'Flow

$\%\text{U'Flow} = \dfrac{290\ \text{L/hr}}{916\ \text{L/hr}} \times 100 - 31.7\ \text{V}\%$

Table D.6, 50 mm (2 in.) Hydrocyclone Volume Split:

Operating Gauge Pressure = 345 kPa (50 psig)

The volume split to underflow increases at lower operating pressures.

Vortex Finder = 8 mm (0.31 in.)

Spigot = 3.2 mm (0.13 in.); %U'Flow = 9%V

Spigot = 4.5 mm (0.18 in.); %U'Flow = 23%V

Vortex Finder = 11 mm (0.44 in.)

Spigot = 4.5 mm (0.18 in.); %U'Flow = 7%V

Spigot = 6.4 mm (0.25 in.); %U'Flow = 16%V

Figure D.7, 50 mm (2 in.) Hydrocyclone Capacity:

Capacity Correction, Figure D.8:

V − %S = 15.4%; Correction Factor = ~1.15

Vortex Finder = 8 mm (0.31 in.)

Operating Gauge Pressure = 69 kPa (10 psig)

Water Rate = (1.15)(0.85 m³/hr) = 1.0 m³/hr (4.4 gpm)

Operating Gauge Pressure = 207 kPa (30 psig)

Water Rate = (1.15)(1.50 m³/hr) = 1.7 m³/hr (7.6 gpm)

Vortex Finder = 11 mm (0.44 in.)

Operating Gauge Pressure = 69 kPa (10 psig)

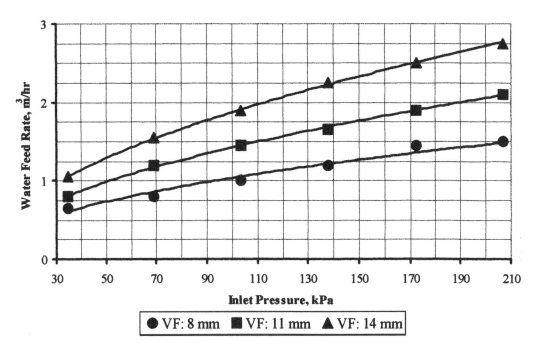

Figure D.7   Capacity Curves for 50 mm (2 in.) Hydrocyclone.

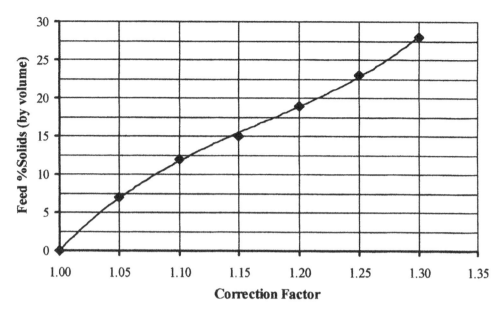

Figure D.8 Correction of capacity for % solids.

Water Rate = (1.15)(1.20 m³/hr) = 1.4 m³/hr (6.1 gpm)
Operating Gauge Pressure = 207 kPa (30 psig)
Water Rate = (1.15)(2.10 m³/hr) = 2.4 m³/hr (10.6 gpm)

Summary of Design Results

	Size mm (in.)	Gauge Pressure kPa (psig)	Vortex Finder mm (in.)	Spigot mm (in.)	Hydrocyclone Capacity m³/hr (gpm)
Mass Balance Data	50 (2)	69 (10)	8 (0.31)	3.2 (0.13)	1.0 (4.4)
		207 (30)	8 (0.31)	3.2 (0.13)	1.7 (7.6)
Demonstration Data	80 (3)	69 (10)	18 (0.71)	10.0 (0.39)	4.1 (18)
		207 (30)	18 (0.71)	10.0 (0.39)	7.4 (33)

- For the 50 mm (2 in.) hydrocyclone operating at 69 kPa (10 psig), the mass balance feed rate of 0.92 m³/hr was similar to the required rate of 1.0 m³/hr. At a mid-range capacity of 1.35 m³/hr, the hydrocyclone can treat 550 kg/hr of the −105 μm (150 mesh) soil:

$$\text{Solid Capacity} = \frac{1.0 \text{ m}^3/\text{hr} + \dfrac{(1.7 \text{ m}^3/\text{hr} - 1.0 \text{ m}^3/\text{hr})}{2}}{0.92 \text{ m}^3/\text{hr}} \times 375 \text{ kg/hr} = 550 \text{ kg/hr}$$

- For the 50 mm hydrocyclone, the volume split to the underflow was about 9%V at an operating pressure of 345 kPa (50 psig). The volume split to underflow would increase toward the required 32%V at lower operating pressures.
- The spigot opening and vortex finder can be adjusted to bring the correct percentage of feed slurry into underflow toward the 32%V value. However, the adjustment of the vortex finder will also affect the volumetric throughput.

- At 69 kPa (10 psig) operating pressure, the optimal operating rate for the 80 mm (3 in.) hydrocyclone was about 4.1 m³/hr (28 gpm). This rate was calculated using the above method and Table D.7 and Figures D.8 and D.9.
- For proper operation of the 80 mm hydrocyclone during the demonstration, recycle water had to be added to bring the flow rate of the feed to at least 4.1 m³/hr.

Table D.7 % Volume Split of Liquid Feed for a 80 mm (3 in.) Hydrocyclone

Spigot, mm		25	20	15	10
Diameter	25	56%	33%	11%	2%
Vortex	18	91%	70%	39%	12%
Finder, mm	13	100%	100%	74%	37%

Percentage by volume of water feed recovered in underflow

Operating Gauge Pressure = 207 kPa (30 psig)

Volume split to underflow increases at lower operating pressures.

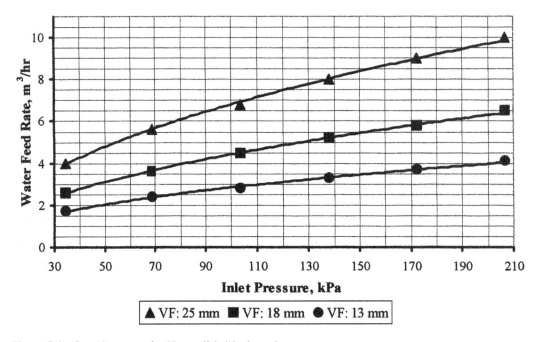

Figure D.9   Capacity curves for 80 mm (3 in.) hydrocyclone.

D.6 REICHERT SPIRAL CONCENTRATOR

Design Data

Mass Balance Data		kg/hr		% Distribution	
Feed	Feed Rate	Solid	Water	Solid	Water
Feed Sump : –1.19 mm Sand	591.03 kg/hr	413.72	177.31	70.0	30.0
Feed Sump: Dilution Water	3157.42 L/hr	0.00	3157.42	0.0	100.0
Feed Sump Overflow (recycle)		10.34	83.37	11.0	89.0
Spiral –1.19 mm Tailings (light fraction)		355.79	2981.70	10.7	89.3
Spiral –1.19 mm Concentrate (heavy fraction)		47.59	269.66	15.0	85.0
Totals Mass Input/Output		413.72	3334.73		

Demonstration Reichert Concentrator Specifications

References: Weiss, 1985, pp. 3D-46 to 3D-52; MD Mineral Technologies Brochures and Data Sheets for Spiral Concentrators (enclosed).

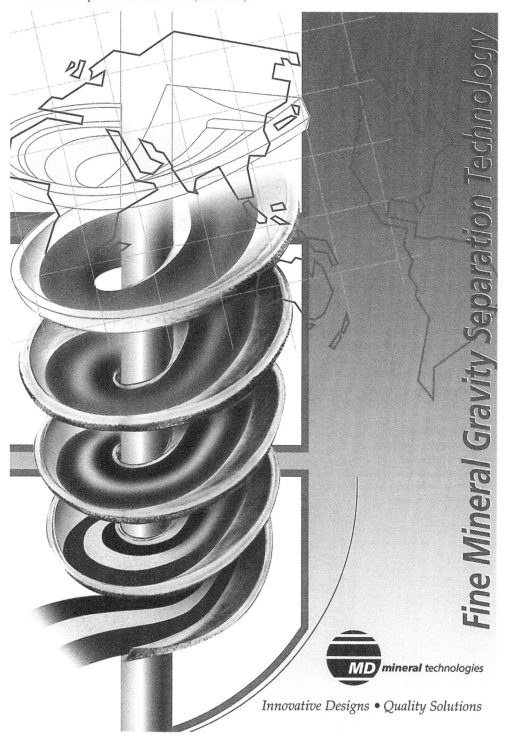

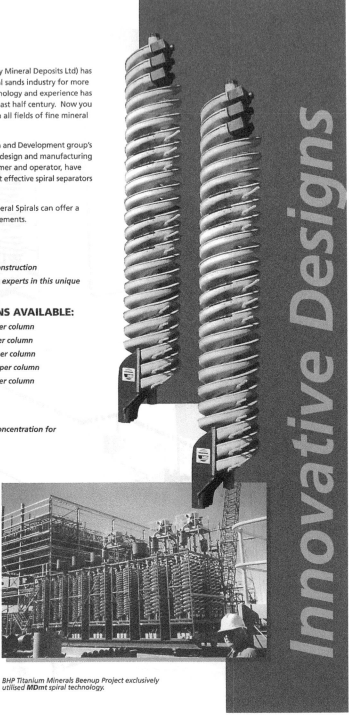

MD mineral technologies (formerly Mineral Deposits Ltd) has been synonymous with the mineral sands industry for more than 50 years. Our expertise, technology and experience has developed continuously over the past half century. Now you can benefit from our know-how in all fields of fine mineral separation and concentration.

MD mineral technologies Research and Development group's continuous efforts in new product design and manufacturing techniques ensures you, the customer and operator, have access to the most efficient and cost effective spiral separators available.

Our unequalled range of Fine Mineral Spirals can offer a solution to your separation requirements.

BENEFITS:

- *HIGH Reliability*
- *LONG Wearing Materials of Construction*
- *ACCESS to the World's leading experts in this unique process technology*

SPIRAL CONFIGURATIONS AVAILABLE:

- *Single Start - 1 spiral trough per column*
- *Twin Start - 2 spiral troughs per column*
- *Triple Start - 3 spiral troughs per column*
- *Quad Start* - 4 spiral troughs per column*
- *Five Start* - 5 spiral troughs per column*
  - ** - selected models only*

APPLICATIONS:

- *RUTILE, ILMENITE & ZIRCON Concentration for mineral sands*
- *IRON ORE Beneficiation*
- *CHROMITE Beneficiation*
- *TIN Concentration*
- *GOLD Recovery*
- *SILICA Sands processing*
- *PUMICE sands separation*

*BHP Titanium Minerals Beenup Project exclusively utilised **MD**mt spiral technology.*

Innovative Designs

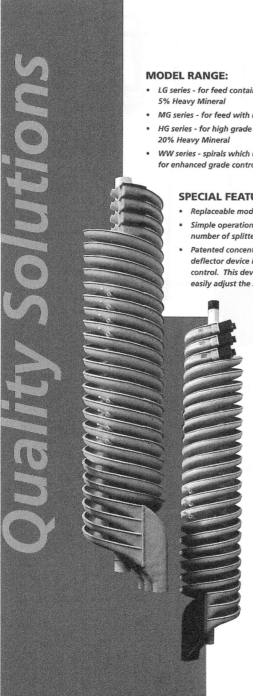

MDmt's Walkabout Spiral for in-situ exploration sampling and training demonstration.

MODEL RANGE:

- *LG series - for feed containing less than 5% Heavy Mineral*
- *MG series - for feed with up to 20% Heavy Mineral*
- *HG series - for high grade feed generally greater than 20% Heavy Mineral*
- *WW series - spirals which utilise wash water additions for enhanced grade control in specific applications*

SPECIAL FEATURES:

- *Replaceable modular cast polyurethane feed boxes*
- *Simple operation of splitters regardless of size of spiral and number of splitters required*
- *Patented concentrate divertor. Preceding each slide splitter a deflector device is installed to assist with concentrate grade control. This device diverts the flow and allows the operator to easily adjust the splitters to obtain the best concentrate product*
- *Repulping device located on the spiral trough after each auxiliary splitter to assist recovery and concentration*
- *Strong lightweight fibreglass construction with polyurethane covering and castings for corrosion and abrasion resistance*
- *Auxiliary splitters*
- *Patented ganged splitter arrangement for ease of operation*
- *Cast modular feed box offering high wear resistance and positive feedline connection*
- *Polyurethane product collector box ensuring long life and compact collection*
- *Integrated engaged launder system*
- *Efficient feed distribution system*

Your operation can benefit from the superior designs and manufacturing quality of the spiral separators that MDmt offer. Couple this with our knowledge, experience and technical support network, and you have all the backup you need. We pride ourselves on this being an integral part of our customer service.

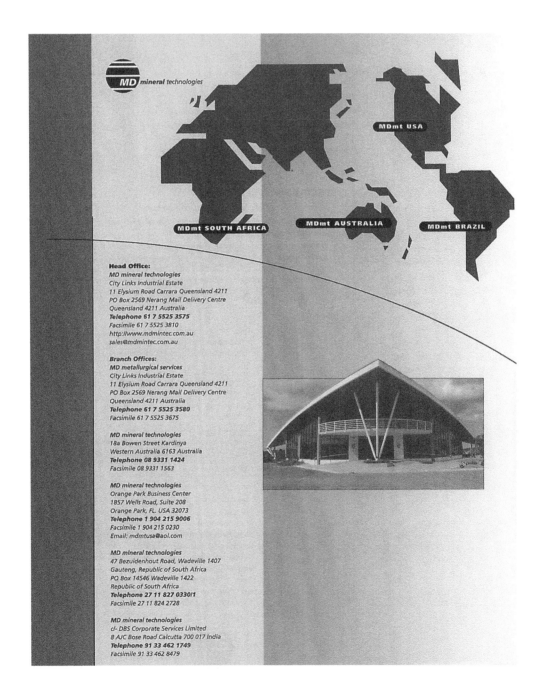

MD *mineral* technologies

MDmt USA

MDmt SOUTH AFRICA   MDmt AUSTRALIA   MDmt BRAZIL

Head Office:
MD mineral technologies
City Links Industrial Estate
11 Elysium Road Carrara Queensland 4211
PO Box 2569 Nerang Mail Delivery Centre
Queensland 4211 Australia
Telephone 61 7 5525 3575
Facsimile 61 7 5525 3810
http://www.mdmintec.com.au
sales@mdmintec.com.au

Branch Offices:
MD metallurgical services
City Links Industrial Estate
11 Elysium Road Carrara Queensland 4211
PO Box 2569 Nerang Mail Delivery Centre
Queensland 4211 Australia
Telephone 61 7 5525 3580
Facsimile 61 7 5525 3675

MD mineral technologies
18a Bowen Street Kardinya
Western Australia 6163 Australia
Telephone 08 9331 1424
Facsimile 08 9331 1563

MD mineral technologies
Orange Park Business Center
1857 Wells Road, Suite 208
Orange Park, FL. USA 32073
Telephone 1 904 215 9006
Facsimile 1 904 215 0230
Email: mdmtusa@aol.com

MD mineral technologies
47 Bezuidenhout Road, Wadeville 1407
Gauteng, Republic of South Africa
PO Box 14546 Wadeville 1422
Republic of South Africa
Telephone 27 11 827 0330/1
Facsimile 27 11 824 2728

MD mineral technologies
c/- DBS Corporate Services Limited
8 AJC Bose Road Calcutta 700 017 India
Telephone 91 33 462 1749
Facsimile 91 33 462 8479

mineral technologies

MG4C SPIRAL SEPARATOR

patented technology

MECHANICAL FEATURES

➢ High wear resistant polyurethane trough surface
➢ Two repulpers (with splash lids) for enhanced performance
➢ Number of Turns – 7
➢ Number of Starts – single, twin & triple
➢ No wash water required
➢ Auxiliary Slide Splitters
➢ Optional water/slimes off-take splitter
➢ Ganged Product Outlet Splitters
➢ Product Box and Feed Box components cast in high wear-resistant polyurethane

DESIGN DATA

Head Feed (per start)
➢ Up to 3.2 t/h solids
➢ Up to 8m³/h slurry volume
➢ Up to 55% solids pulp density
➢ Particle size range 0.03 – 2.0mm

APPLICATIONS

The principal area of application is in a roughing or scavenging duty where the feed contains up to 15-20% heavy mineral.

Specific Applications include:

Mineral Sand	Gold – Alluvial and Hard rock
Silica Sand	Tin Garnet
Tungsten	Chromite Sillimanite

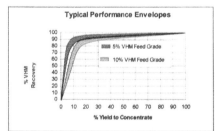

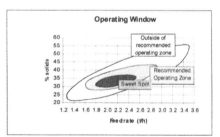

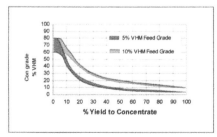

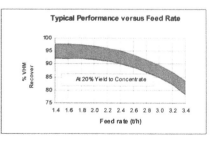

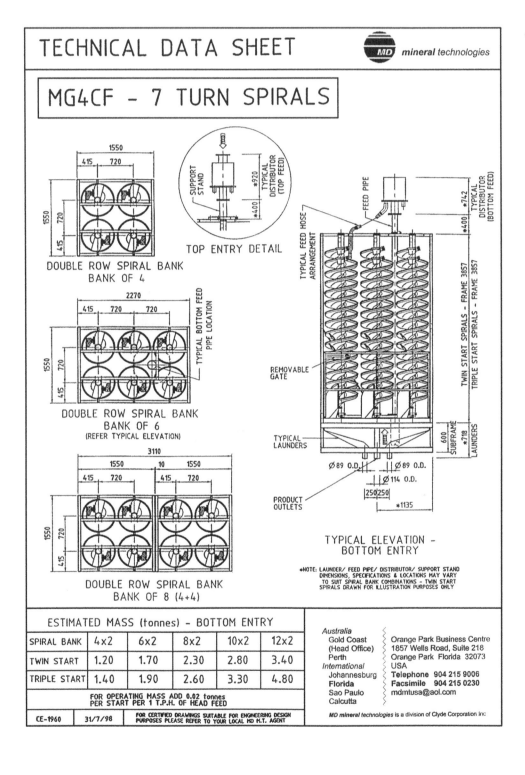

TECHNICAL DATA SHEET

MD *mineral* technologies

MG4CF – 7 TURN SPIRALS

DOUBLE ROW SPIRAL BANK
BANK OF 4

TOP ENTRY DETAIL

DOUBLE ROW SPIRAL BANK
BANK OF 6
(REFER TYPICAL ELEVATION)

DOUBLE ROW SPIRAL BANK
BANK OF 8 (4+4)

TYPICAL FEED HOSE ARRANGEMENT

FEED PIPE

TWIN START SPIRALS – FRAME 3857
TRIPLE START SPIRALS – FRAME 3857

REMOVABLE GATE

TYPICAL LAUNDERS

PRODUCT OUTLETS

TYPICAL ELEVATION –
BOTTOM ENTRY

Ø89 O.D.   Ø89 O.D.
Ø114 O.D.
250 250
*1135

*NOTE: LAUNDER/ FEED PIPE/ DISTRIBUTOR/ SUPPORT STAND DIMENSIONS, SPECIFICATIONS & LOCATIONS MAY VARY TO SUIT SPIRAL BANK COMBINATIONS – TWIN START SPIRALS DRAWN FOR ILLUSTRATION PURPOSES ONLY

ESTIMATED MASS (tonnes) – BOTTOM ENTRY

SPIRAL BANK	4×2	6×2	8×2	10×2	12×2
TWIN START	1.20	1.70	2.30	2.80	3.40
TRIPLE START	1.40	1.90	2.60	3.30	4.80

FOR OPERATING MASS ADD 0.02 tonnes
PER START PER 1 T.P.H. OF HEAD FEED

CE-1960	31/7/98	FOR CERTIFIED DRAWINGS SUITABLE FOR ENGINEERING DESIGN PURPOSES PLEASE REFER TO YOUR LOCAL MD M.T. AGENT

Australia
Gold Coast   Orange Park Business Centre
(Head Office)   1857 Wells Road, Suite 218
Perth   Orange Park Florida 32073
International   USA
Johannesburg   **Telephone** 904 215 9006
Florida   **Facsimile** 904 215 0230
Sao Paulo   mdmtusa@aol.com
Calcutta

MD mineral technologies is a division of Clyde Corporation Inc

Optimum Operational Requirements

- Feed Particle Size Range: -1.19 mm by $+74$ μm (-16 mesh by $+200$ mesh)
- Feed Slurry %Solids: 20% and 30% (good operation between 15 and 40%)
- Solid Feed Rate per Unit: 450 kg/hr (~1000 lb/hr) for fine material to 2250 kg/hr (~5000 lb/hr) for coarse material
- Feed Slurry Rate per Unit = 1.15 m³/hr to 7.62 m³/hr (6.7 gpm to 34 gpm)
- Wash Water Rate per Unit = 0.68 m³/hr to 3.40 m³/hr (3 gpm to 15 gpm)
- Feed and wash water flows must be balanced to prevent sandbar formation at low flows. At high flows, the heavy minerals are swept wide of the upper exit ports.

Reichert Concentrator Design Calculations

Average Design Requirements

$$\text{Average Feed Slurry \%Solids} = \text{\%Solids F} = \frac{(20\% + 30\%)}{2} = 25.0\%$$

$$\text{Feed \%Solids (by Volume)} = (\text{F V\%S}) = \frac{100}{\left[1 + (S_g)_s \times \left(\frac{100 - (\text{\%Solids F})}{(\text{\%Solids F})}\right)\right]}$$

$$\text{Feed \%Solids (by Volume)} = (\text{F V\%S}) = \frac{100}{\left[1 + 2.60 \times \frac{(100 - 25.0)}{(25.0)}\right]} = 11.4 \text{ V\%}$$

$$\text{Average Slurry Volumetric Rate} = \frac{(7.62 \text{ m}^3/\text{hr} - 1.15 \text{ m}^3/\text{hr})}{2}$$

Average Slurry Volumetric Rate = 4.385 m³/hr = 4385 L/hr or (19.3 gpm)
Solid Feed Rate per Unit = Sand F, kg/hr

$$\text{Feed Slurry Volumetric Rate} = \left[\frac{(\text{Sand F})}{(S_g)_s} + \frac{(\text{Sand F})(100 - (\text{\%Solids F}))}{(\text{\%Solids F})}\right]$$

$$\text{Feed Slurry Volumetric Rate} = \left[\frac{\text{Sand, F kg/hr}}{(2.60 \text{ kg/L})} + \frac{(\text{Sand F, kg/hr})(100 - 25.0)}{(25.0)}\right] = 4385 \text{ L/hr}$$

Solve the above equation for Sand F:
Sand F = 1295 kg/hr (2856 lb/hr)
Calculate Feed Slurry Volumetric Rate at 20%Solids and 30%Solids:

$$\text{Feed Slurry Volumetric Rate} = \left[\frac{1295 \text{ kg/hr}}{(2.60 \text{ kg/L})} + \frac{(1295 \text{ kg/hr})(100 - 20)}{(20)}\right] = 5678 \text{ L/hr}$$

$$\text{Feed Slurry Volumetric Rate} = \left[\frac{1295 \text{ kg/hr}}{(2.60 \text{ kg/L})} + \frac{(1295 \text{ kg/hr})(100 - 30)}{(30)}\right] = 3519 \text{ L/hr}$$

Minimum Wash Water Flow = 680 L/hr or 0.68 m³/hr (3.0 gpm)

Mass Balance Operational Data

%Solids Feed = %Solids F = 11.04%
Sand Feed Rate = Sand F = (413.7 kg/hr − 10.3 kg/hr) = 403.4 kg/hr (890 lb/hr)

$$\text{Feed \%Solids (by Volume)} = (F\ V\%S) = \cfrac{100}{\left[1 + (S_g)_s \times \cfrac{(100 - (\%\text{Solids F}))}{(\%\text{Solids F})}\right]}$$

$$\text{Feed \%Solids (by Volume)} = (F\ V\%S) = \cfrac{100}{\left[1 + 2.60 \times \cfrac{(100 - 11.04)}{(11.04)}\right]} = 4.56\ V\%$$

$$\text{Feed Slurry Volumetric Rate} = \left[\frac{(\text{Sand F})}{(S_g)_s} + \frac{(\text{Sand F})(100 - (\%\text{Solids F}))}{(\%\text{Solids F})}\right]$$

$$\text{Feed Slurry Volumetric Rate} = \left[\frac{(403.4\ \text{kg/hr})}{(2.60\ \text{kg/L})} + \frac{(403.4\ \text{kg/hr})(100 - 11.04)}{(11.04)}\right]$$

Slurry Volumetric Rate = 3406 L/hr or 3.40 m³/hr (15 gpm)

Summary of Design Results

	%Solids Feed	Solids Feed Rate kg/hr	Slurry Rate m³/hr (gpm)	Minimum Wash Water m³/hr (gpm)
Mass Balance Data	11.0	404	3.40 (15.0)	0.00 (0.0)
Average Design Data	25.0	1295	4.39 (19.3)	0.68 (3.0)

- The design operating conditions are:
  - Solid Feed Rate = 1295 kg/hr, fine material
  - Range for Slurry %Solids = 20% and 30%
  - Range for Feed Slurry = 3.52 m³/hr to 5.68 m³/hr (15.5 gpm to 25.0 gpm)
  - Range for Wash Water = 0.68 m³/hr to 3.40 m³/hr (3.0 gpm to 15.0 gpm)
- Demonstration wash water flow was set at 0 m³/hr (0 gpm).
- The Reichert concentrators in the pilot plant were not optimally utilized to concentrate the lead.

D.7 SIEVE BEND SCREEN

Design Data

Mass Balance Data Feed		kg/hr Solid	kg/hr Water	% Distribution Solid	% Distribution Water
Feed: Spiral Separator Tailings	437.67 kg/hr	355.79	2981.70	10.66	89.34
Undersize: Water		0.00	2918.91	0.00	100.00
Oversize: Tailings; +105 μm (+105 mesh)		355.79	62.79	85.00	15.00
Totals Mass Input/Output		355.79	2981.70		

Demonstration Sieve Bend Specifications

- Type: Sieve Bend Screen
- Mass Balance Sieve Bend Screen Feed Rate = 356 kg/hr (785 lb/hr)
- Mass Balance Solids Content of Feed = 10.66%
- Low Pressure Wash Water Requirement = 0.00 L/min per m³ (0.00 gpm per yd³)
- Required Bar Slot (Screen) Opening =105 μm (150 mesh)
- Required Feed Velocity = 3.05 m/s (10 ft/s)
- Required Feed Fall Height = 51 cm (20 in.)

Sieve Bend Screen Design Calculations

Reference: Weiss, 1985, pp. 3E-20 to 3E-25

Sieve Bend Screen Capacity

Screen Slot Opening (Bar Spacing) = 105 μm (150 mesh) = 0.105 mm
The diameter of separation is the diameter of particles reporting 95% (or 50%) to the oversize fraction
  from Figure D.10:

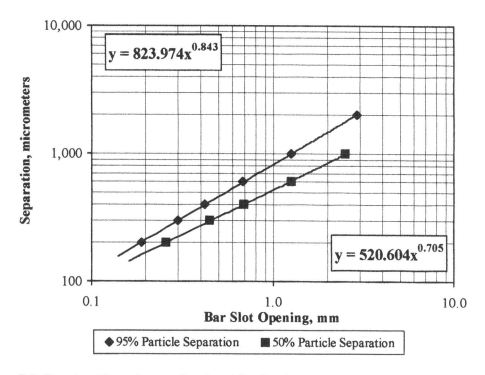

Figure D.10 Diameter of Separation as a Function of Slot Opening.

Percentage Particles in Oversize at 95%: Diameter of Particles Oversize = 123 μm
Percentage Particles in Oversize at 50%: Diameter of Particles Oversize = 106 μm
Capacity (C) from Figure D.11:
Length of Sieve Bend Screen Surface, mm = L
L = 800 mm: C = (108.26)(Slot Opening, mm) = 11.4 m³/hr per m (15.3 gpm per ft)
L = 1600 mm: C = (137.52)(Slot Opening, mm) = 14.4 m³/hr per m (19.4 gpm per ft)
−105 μm (−150 mesh) Soil in Feed to Screen = 0.0 kg/hr
Water in Feed Passing through Screen (Undersize) = 2920 kg/hr = 2920 L/hr (12.9 gpm)

$$\text{Minimum Width} = \frac{2920 \text{ L/hr}}{\dfrac{1000 \text{ L/m}^3}{11.4 \text{ m}^3/\text{hr per m}}} = 0.256 \text{ m or } 256 \text{ mm } (0.840 \text{ ft})$$

Radius of Sieve Bend Screen Surface, mm = R
Length of Sieve Bend Screen Surface, mm = L
Capacity of Non-Rapped Sieve Bend Screen (Table D.8) with L = 915 mm, R = 760 mm,
  Width = 610 mm, and Slot Opening = 0.105 mm:

$$\text{Capacity} = \frac{(\text{Available Width, mm})}{(\text{Minimum Width, mm})} \times (\text{ScreenUndersize, L/hr})$$

$$\text{Capacity} = \frac{(610 \text{ mm})}{(256 \text{ mm})} \times (2920 \text{ L/hr}) = 6960 \text{ L/hr} = 6.96 \text{ m}^3\text{/hr } (30.6 \text{ gpm})$$

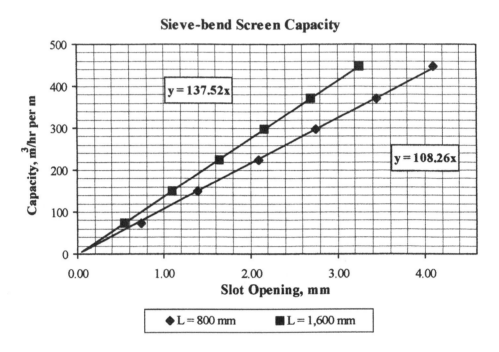

Figure D.11 Sieve Bend Screen Volumetric Capacity.

Summary of Design Results

	Unit Size	Number of Units	Screen Capacity m³/hr (gpm)
Demonstration Data	Non-Rapped Sieve Bend Screen R = 915 mm, L = 760 mm	1	7 (30)
Design Data	Rapped Sieve Bend Screen R = 1625 mm, L = 2390 mm	1	45 (195)

- Minimum Non-Rapped Sieve Bend Screen Size (Table D.8): Gravity Feed, Fixed Surface, 50°, 915 mm Surface Radius, 760 mm Surface Length, 610 mm Width
- The maximum capacity of the 915 mm by 760 mm sieve bend screen with a slot opening of 3.50 mm should be about 260 m³/hr/m (350 gpm/ft). With a screen width of 610 mm, the maximum capacity would be about 160 m³/hr (~700 gpm). However, the capacity of the sieve bend screen will be noticeably affected by much finer slot opening of 0.105 mm.
- Selection of the Optimum Rapped Sieve Bend Screen (Tables D.8 and D.9): Gravity Feed, Fixed Surface plus Rapped, 45°, 1625 mm Surface Radius, 2390 mm Surface Length, 610 mm Width. With a bar spacing of 0.300 mm, the maximum screen capacity should be about 68 m³/hr (300 gpm). With a bar spacing of 0.100 mm, the capacity should decrease to about 45 m³/hr (195 gpm).

Table D.8 Description of Sieve Bend Screens

Type Sieve Bend Screen	Screen Surface R mm	L mm	Available Widths mm	Slot Opening mm	Maximum Capacity (m³/hr)/m	Required Dimensions Height m	Floor Area m²/m
Gravity Feed, Fixed Surface, 50°	915 (36 in.)	760 (30 in.)	610 1220 1830	0.35 mm to 3.50 mm	260	1.829	6.4 10.4 14.9
Gravity Feed, Fixed Surface, 45°	2030 (80 in.)	1530 (60 in.)	610 1220 1830	0.35 mm to 3.50 mm	260	2.438	7.3 12.2 17.1
Gravity Feed, Reversible Surface, 60°	760 (30 in.)	760 (30 in.)	1220	0.35 mm to 3.50 mm	150	2.438	14.6
Pressure Feed, Reversible, 270°	510 (20 in.)	2390 (94 in.)	460	0.20 mm to 0.60 mm	225	2.134	21.9
Gravity Feed, Fixed Surface plus Rapped, 45°	1625 (64 in.)	2390 (94 in.)	610	0.07 mm to 0.30 mm	110	2.743	8.2

R – Radius of Sieve Bend Screen Surface, mm

L – Length of Sieve Bend Screen Surface, mm

Table D.9 Capacity of Rapped Sieve Bend Screen for Separations Finer than 300 μm

Slot Opening μm	Screen Bar Widths mm	Diameter of Separation μm	Screen Capacity m³/hr
100	1.5	95	44.3
150	1.5	135	52.2
200	1.5	190	60.2

Feed Falling Height of 102 cm (40 in.)

D.8 KNELSON CONCENTRATOR

Design Data

Mass Balance Data Feed	Feed Rate	kg/hr Solid	Water	% Distribution Solid	Water
Feed: −150 mesh Slurry	437.67 kg/hr	240.72	196.95	55.0	45.0
Feed: Fluidizing Water	6813.00 L/hr	0.00	6813.00	0.0	100.0
−105 μm , +37 μm Tailings (light fraction)		232.14	7006.27	3.2	96.8
−105 μm, +37 μm Concentrate (heavy fraction)		8.58	3.68	70.0	30.0
Totals Mass Input/Output		240.72	7009.95		

Demonstration Knelson Concentrator Specifications

- Size: 12 in. Knelson Concentrator
- Maximum Solid Feed Size: 2.0 mm (10 mesh)
- Minimum Solid Feed Size: ~37 μm (~400 mesh)
- Solid Feed Capacity: 0.0 kg/hr to 3600 kg/hr (0.0 lb/hr to 8000 lb/hr)

- Slurry Feed Capacity: 0 m³/hr to 11.4 m³/hr (0 gpm to 50 gpm)
- Feed Pulp Density: 0% to 75% (dwb)
- Fluidization Water: 6.78 m³/hr to 9.54 m³/hr (30 gpm to 42 gpm)
- Concentrate Volume: 2.72 L (0.72 gal)
- Concentrate Weight: 4 to 7 kg (9 to 15 lb)
- Concentrate Cycle Duration: 4 to 10 hr

Knelson Concentrator Design Calculations

Mass Balance Operational Data

Solid Feed Size = -105 μm by $+37$ μm (-150 mesh by $+400$ mesh)
Feed Pulp Density = %Solid-F = 55% (dwb)

$$\text{Feed \%Solids (by Volume)} = (F - V\%S) = \frac{100}{\left[1 + (S_g)_s \times \dfrac{(100 - (\%\text{Solids} - F))}{(\%\text{Solids} - F)}\right]}$$

$$\text{Feed \%Solids (by Volume)} = (F - V\%S) = \frac{100}{\left[1 + 2.60 \times \dfrac{(100 - (55.0))}{(55.0)}\right]} = 32.0 \text{ V\%}$$

$$\text{Feed Slurry Volumetric Rate} = \left[\frac{(\text{Fine} - F)}{(S_g)_s} + \frac{(\text{Fine} - F)(100 - (\%\text{Solids} - F))}{(\%\text{Solids} - F)}\right]$$

Solid Feed Capacity = (Fine-F) = 240.7 kg/hr (530.7 lb/hr)

$$\text{Feed Slurry Volumetric Rate} = \left[\frac{(240.7 \text{ kg/hr})}{(2.60 \text{ kg/L})} + \frac{(240.7 \text{ kg/hr})(100 - 55.0)}{(55.0)}\right]$$

Slurry Feed Capacity = 0.290 m³/hr (1.28 gpm)
Fluidization Water = 6813 L/hr = 6.183 m³/hr (30.0 gpm)
Maximum Concentrate Volume = 2.72 L (0.72 gallons)
Concentrate Solid = (Fine-C) = 8.58 kg/hr (18.9 lb/hr)
Concentrate Lead Content = 3.02% (dwb)
$(S_g)_{lead}$ = 11.34 kg/L
Average Density of Concentrate Soil Particle = $(S_g)_C$

$$(S_g)_C = \left[\frac{(2.60 \text{ kg/L})(100\% - 3.02\%)}{(100\%)} + \frac{(11.34 \text{ kg/L})(3.02\%)}{(100\%)}\right]$$

$(S_g)_C$ = 2.86 kg/L
At the end of each collection cycle, the concentrate was washed from the concentrator bowl
   with water diluting it to about 15% solids (dwb). It was assumed that the concentrate on the
   bowl had a pulp density of 70% solids (dwb).
Concentrate Pulp Density = %Solid-C = 70%

$$\text{Wet Concentrate Capacity Volumetric Rate} = \left[\frac{(\text{Fine} - C)}{(S_g)_C} + \frac{(\text{Fine} - C)(100 - (\%\text{Solids} - C))}{(\%\text{Solids} - C)}\right]$$

Solid Concentrate Capacity = (Fine-C) = 8.58 kg/hr (18.9 lb/hr)

$$\text{Wet Concentrate Capacity Volumetric Rate} = \left[\frac{(8.58 \text{ kg/hr})}{(2.86 \text{ kg/L})} + \frac{(8.58 \text{ kg/hr})(100 - 70.0)}{(70.0)}\right]$$

Wet Concentrate Volumetric Rate = 6.68 L/hr (0.030 gpm)
Concentrate Volume: 2.72 L (0.72 gal)

$$\text{Concentrate Cycle} = \frac{2.72 \text{ L}}{6.68 \text{ L/hr}} = 0.41 \text{ hr} (\sim 25 \text{ min})$$

Summary of Design Results

- The Knelson concentrator can be easily operated at the solid feed rate and pulp density of 240 kg/hr at 55% solids respectively.
- The maximum solid feed rate and pulp density would be 3600 kg/hr at ~75% solids respectively. A likely feed rate and pulp density could be 1800 kg/hr at ~40% solids, producing a slurry flow of 3.40 m³/hr.
- Fluidizing wash water flow must be at least 6.78 m³/hr (30 gpm).
- The cycle time for the collection of the concentrate containing 3% lead was quite short (~25 minutes). A higher lead content in the concentrate with a lower moisture content would increase the cycle time.

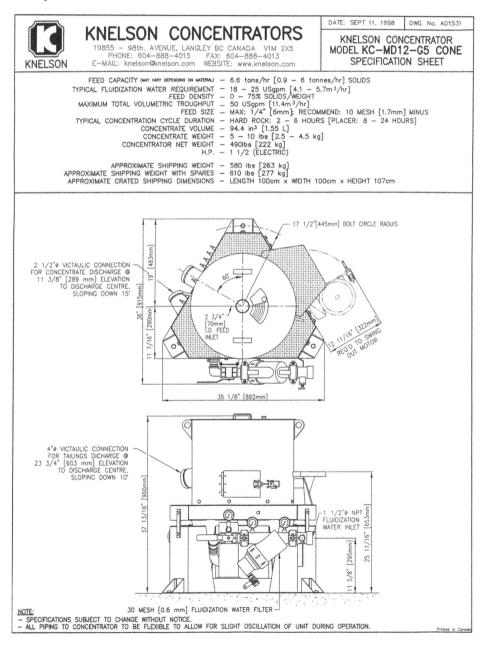

12" KNELSON CONCENTRATOR

With a rated throughput of up to 4tph of -6mm (1/4") solids, the 12" Knelson Concentrator has become an essential component of small-scale mechanized alluvial treatment plants and hardrock milling operations, as well as secondary and tertiary concentration circuits and gravity-based pilot plants.

The 12" Knelson Concentrator offers superior free-gold recovery and is ideal for small to medium-sized operations. The unit is backed by a full one year parts and labour warranty as well as an extended two year warranty on the inner cone.

When ordered with the optional stainless steel (S.S.) package, all critical wetted parts are of 303, 304, & 316 S.S. construction to resist deterioration in milling operations with corrosive environments.

Like other Knelson Concentrators, the 12" offers unsurpassed free-gold recoveries, ease of use, proven reliability and efficient operation.

✘ wear-resistant long-lasting polyurethane inner cone
✘ accepts a wide range of feed sizes and densities
✘ operation virtually unaffected by slimes and clays
✘ low water and power consumption
✘ quick and easy concentrate removal
✘ high throughput capacity
✘ low maintenance and operating costs
✘ replaceable inner linings
✘ security of concentrate is maintained during operation

 KNELSON GOLD CONCENTRATORS INC.

Pilot Plant Photographs

CONTENTS

Figure E.1 Soil Feed System and Gravity Separation Trailer.

Figure E.2 Coarse and Sand Gravity Separation Systems.

Figure E.3 Slimes Treatment System.

Figure E.4 Feed Hopper, Weigh Belt #1, Bucket Elevator, and Attrition Scrubber.

Figure E.5 Attrition Scrubber, Trommel, Mineral Jig, Weigh Belt #2, Sieve-bend Screen, and Reichert Spirals.

Figure E.6 Trommel, Mineral Jig, Weigh Belt #2, Sieve-bend Screen, and Reichert Spiral.

Figure E.7 Slimes Treatment System: Spiral Classifier, Hydrocyclone, and Knelson Bowl.

Figure E.8 Clean Water Tank, Dewatering Roll-Off Box, and Day Tank.

Figure E.9 Vat Leach System: Two Leach Tanks, Iron Precipitation and Ion-exchange Columns, and Solution Storage Tanks.

Figure E.10 Soil Hopper, Feed Belt, and Weigh Belt #1.

Figure E.11 Weigh Belt #1, Bucket Elevator, Attrition Scrubber, and Trommel.

Figure E.12 Trommel and Mineral Jig (Lower Right).

Figure E.13 Mineral Jig and Sweco Vibrating Screen #2.

Figure E.14A Sands System: Reichert Spiral Concentrators.

Figure E.14B Sands System: Reichert Spiral Concentrators.

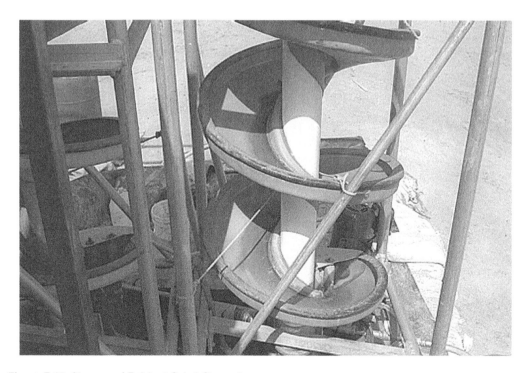

Figure E.15 Close-up of Reichert Spiral Channel.

Figure E.16 Slimes Treatment System Equipment Area (Hydrocyclone in Center).

Figure E.17 Hydrocyclone.

Figure E.18 Slimes Treatment: Feed Tank, Coriolis Flowmeters, and Knelson Bowl Concentrator.

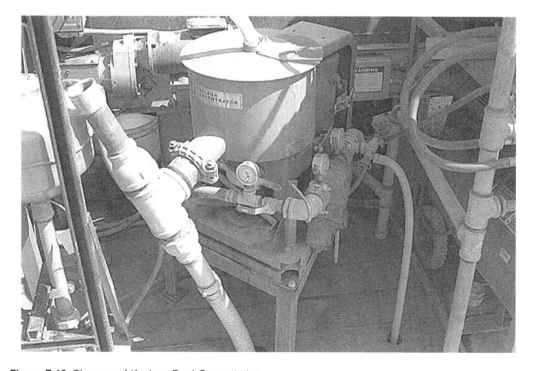

Figure E.19 Close-up of Knelson Bowl Concentrator.

Figure E.20 Slimes Sump and Hydrocyclone Feed Pump (Vertical Bowl Pump).

Figure E.21 Attrition Scrubber Auger (Upper Right), Sweco Screen #1, and Spiral Classifier Feed Point.

Figure E.22 Reichert Spiral System: Sweco Screen #3.

Figure E.23 Vertical Bowl (Sand) Pump.

Figure E.24 Wigwag Sampler.

Daily Log and Results of Feed Rate Tests

CONTENTS

F.1 INTRODUCTION

This appendix includes unedited daily project logs and feed rate observations for operations during the pilot demonstration project at NAS Miramar. The original objective of this pilot demonstration was to test the capability of the design of the Bureau of Reclamation (BOR) equipment to remove lead from small-arms firing range soils. However, BOR personnel opted to test the maximum capacity of the system.

Gravity separation technology was originally developed for the extraction of heavy metals such as gold or tin from placer deposits. This pilot test objective was to test this proven technology for its capability to remove lead from firing range soils. The plant was tested at the BOR facility in Salt Lake City and then operated as a pilot demonstration system at NAS Miramar.

At the direction of BOR personnel, the tests were conducted using a wide range of feed rates. The test rates were separated by about 113 kg/hr (250 lb/hr) increments below 1360 kg/hr (3000 lb/hr) and by 227 kg/hr (500 lb/hr) increments above 1360 kg/hr (3000 lb/hr). These tests indicated that the plant operated best at feed rates between 907 and 1360 kg/hr (2000 and 3000 lb/hr).

F.2 DAILY PROJECT LOG

During the tests, a daily log was maintained in the field to document field activities. This section summarizes the notes recorded in the field log by Dr. Jerry Montgomery of the Bureau of Reclamation (BOR).

F.2.1   First Field Demonstration: August 18 to August 30, 1996

Sunday, August 18th
   Mobilization to the project location.
Monday, August 19th
   Site preparation, including filling sandbags, laying out liner, and unloading trailers.
Tuesday, August 20th
   Pilot plant set-up.
Wednesday, August 21st
   The first shakedown tests were run, using water only.
Thursday, August 22nd
   Shakedown tests were run with soil for 2 hours. After 2 hours, a set of pumps failed and was unable to be repaired. Replacement bowl (vertical sand) pumps were located in Denver, Colorado, and were scheduled for delivery by Monday, August 26.
Friday, August 23rd
   More shakedown tests were run until problems were encountered with the SandPIPER pumps and some of the peristaltic pumps. The project team installed replacement pumps but the results were unsatisfactory. The team decided to suspend further tests until Monday, when the first new pumps were scheduled to arrive.
Saturday, August 24th
   Cleaned up and made changes in preparation for the new pump.
Sunday, August 25th
   No project activities were conducted on Sunday.
Monday, August 26th
   The replacement pumps did not arrive as scheduled. Therefore, the project team rigged four SandPIPER pumps to return the spiral concentrator water. They also rerouted the discharges to the second Sweco to divert bullets and rocks to a direction other than the −4.76 mm (−4 mesh) to +0.84 mm (20 mesh) fraction. Finally, they began changing the pumps on the hydrocyclone underflow circuit.
Tuesday, August 27th
   The new bowl (vertical sand) pumps were scheduled to arrive the following week. The team removed the thickeners and reconfigured the system so all the settling would occur in a roll-off box, making provisions for a second roll-off box. The hydrocyclone underflow pumps were exchanged and all temporary fixes were completed.

Wednesday, August 28th
    The pilot system was run all day, as summarized below:

7:00 a.m.	Started plant at about 227 kg/hr (500 lb/hr); ran for 2 hours.
9:00 a.m.	Increased feed to about 454 kg/hr (1000 lb/hr); ran for 1 hour.
10:00 a.m.	Increased feed to about 680 kg/hr (1500 lb/hr); ran for 1 hour.
11:00 a.m.	Increased feed to about 794 kg/hr (1750 lb/hr); ran for 1 hour.
12:00 p.m.	Increased feed to about 907 kg/hr (2000 lb/hr); ran for 1 hour.
1:00 p.m.	Increased feed to about 1021 kg/hr (2250 lb/hr); ran for 1 hour. Started having problems with SandPIPER pumps. A decision was made to order additional vertical pumps.
2:00 p.m.	Increased feed to about 1134 kg/hr (2500 lb/hr); ran for 1 hour.
3:00 p.m.	Increased feed to about 1247 kg/hr (2750 lb/hr); ran for 1 hour.
4:00 p.m.	Increased feed to about 1361 kg/hr (3000 lb/hr); ran for 1 hour.
5:00 p.m.	At 5:00, the feed on the plant was stopped. For a first test-run, the plant ran satisfactorily. The crew gathered information on how the various parts of the plant appeared to run at varying capacities. The pumps were causing problems.
6:00 p.m.	At 6:00, the plant was stopped for the day.

Thursday, August 29th

7:00 a.m.	Started plant at 1043 kg/hr (2300 lb/hr); ran for 6 hours.
1:00 p.m.	Increased the feed to 1270 kg/hr (2800 lb/hr); ran for 1/2 hour.
1:30 p.m.	Stopped the feed and initiated plant shutdown, taking care to flush clean all circuits.
3:30 p.m.	The plant was shut down. Collected the first test samples from all streams generated during the day's run. Cleaned up the site and pilot plant and prepared for travel. The pumps were still scheduled to arrive the following week.

Friday, August 30th
    Travel

F.2.2   Lead Values In Various Streams For First Field Trip

	Total Lead mg/kg
Gravel pit sand used in sandbags	**300**
Input soil or feed	**13,600**
Clean Streams	
Mineral Jig Tailings	1500
Spiral Tailings	900
Slimes	3700
Lead Removal Streams	
–4 mesh,+14 mesh from auger	54,400
Mineral jig concentrate	18,400
Reichert spiral concentrate	47,600
Knelson bowl concentrate	36,200

F.2.3   Second Field Demonstration: September 8 to September 25, 1996

Sunday, September 8th
    Travel
Monday, September 9th
    Commenced installing new bowl (vertical sand) pumps.

Tuesday, September 10th

Finished installing pumps and conducted shakedown test.

Wednesday, September 11th

7:00 a.m.	Started plant at approximately 227 kg/hr (500 lb/hr); ran about 1 hour.
8:00 a.m.	Stepped up feed to approximately 454 kg/hr (1000 lb/hr); ran about 2 hours.
10:00 a.m.	Stepped up feed to approximately 680 kg/hr (1500 lb/hr); ran the rest of day at this rate (4.5 hours).
2:30 p.m.	Started a shutdown procedure to determine the rate at which the plant would clean out if just the feed was stopped.
3:30 p.m.	Stopped the plant.

Thursday, September 12th

7:00 a.m.	Started plant at 3629 kg/hr (8000 lb/hr) to determine the maximum hopper feed rate and to load up the plant quickly.
7:30 a.m.	Dropped feed rate to approximately 680 kg/hr (1500 lb/hr) (680 kg); ran about 2.5 hours.
10:00 a.m.	Stepped up feed to approximately 907 kg/hr (2000 lb/hr) (907 kg); ran about 1/2 hour.
10:30 a.m.	Dropped feed to approximately 454 kg/hr (1000 lb/hr); ran for about 2 hour, until spiral recirculating lines and pumps became plugged.
11:00 a.m.	Stopped plant abruptly and plugged several lines. Repairs and lunch took about 2.5 hours.
1:30 p.m.	Started plant at approximately 454 kg/hr (1000 lb/hr); ran for 3 hours.

During this run, several plugged lines were experienced. Rather than shut the plant down, plugs were fixed during operations. It was concluded from this run that operating at a feed rate of 454 kg/hr allowed heavies to settle out, as there was not enough material flowing in the lines to keep them flushed and clean.

4:30 p.m.	Started orderly shutdown, requiring about 1 hour.
5:30 p.m.	Stopped the plant.

Friday, September 13th

7:00 a.m.	Started the plant at 680 kg/hr (1500 lb/hr); ran at this rate all day (7.75 hours). Some of the lines became plugged right after "Y" samplers were tripped to collect samples. Apparently, when the samplers were tripped, the surge in the line created a ram effect which settled out or stopped the heavies, resulting in a plug in the line. The "Y" samplers that caused the plugged lines were eliminated. As there were no large particles to separate in the slimes, the slime circuit never became plugged.
2:45 p.m.	The gear box on the feed hopper broke, so the plant was allowed to continue running, to clean the system.
4:00 p.m.	The plant was stopped for the day.

Saturday, September 14th

This day was used for general cleanup and maintenance. In addition, the gearbox representative visited the site and arranged for the delivery of a new gearbox on Monday.

Sunday, September 15th

No project activities were conducted on this date.

Monday, September 16th

On this day, the plant was fed by hand and ran most of the day, as follows.

8:00 a.m.	19-L (5-gal) plastic buckets of dirt were fed into the plant at the base of the bucket elevator. The buckets averaged about 27 kg (60 lb) and were fed into the plant about every 2 minutes. The estimated feed rate was between 680 and 907 kg/hr (1500 and 2000 lb/hr) with an average of 794 kg/hr (1750 lb/hr). The plant was run for about 4 hours before noon.

12:00 p.m.   Stopped the plant for lunch.
1:00 p.m.   The plant was started after lunch and ran for about two more hours at the average rate of 794 kg/hr (1750 lb/hr).
3:00 p.m.   Stopped feeding the plant.
3:45 p.m.   Stopped the plant.

Tuesday, September 17th
7:00 a.m.   Installed the new hopper gear and tested the system by varying the feed rate applied to the new hopper gear.
10:00 a.m.   Set the feed at 454 kg/hr (1000 lb/hr); ran for 1 hour.
11:00 a.m.   Stepped up feed to 794 kg/hr (1750 lb/hr); ran for 1 hour.
12:00 p.m.   Stepped down feed to 567 kg/hr (1250 lb/hr); ran for 1.3 hours.
1:20 p.m.   The crew was cleaning up spilled dirt around the project site and dumped two 19-L (5-gal) buckets of dirt straight onto the bucket elevator, causing an overload. This action stopped the plant abruptly. The bucket elevator was immediately cleaned out.

Wednesday, September 18th
7:00 a.m.   The plant was started, but sediment that had accumulated in the lines, because of the abrupt stop the day before, had to be cleaned out of the lines.
11:00 a.m.   With the lines now clean, the plant was started at about 907 kg/hr (2000 lb/hr) and ran for 5 hours. The plant ran very smoothly at this rate.
4:00 p.m.   Stopped feed to the plant.
4:30 p.m.   Stopped the plant but kept the spirals running. Although the majority of the plant is clear in 20 to 30 minutes, the spirals take about 15 to 20 minutes longer to clear since the spiral circuit contains rougher cleaner feedback loops.
4:45 p.m.   Stopped the spirals and shut the plant down.

Thursday, September 19th
7:00 a.m.   Started the plant at 340 kg/hr (750 lb/hr); and ran for 4 hours.
11:00 a.m.   Increased feed to 680 kg/hr (1500 lb/hr); and ran for 1 hour.
12:00 p.m.   Increased feed to 907 kg/hr (2000 lb/hr; and ran for 1.5 hours.
1:30 p.m.   Increased feed to 1134 kg/hr (2500 lb/hr); and ran for 0.50 hour.
2:00 p.m.   Increased feed to 1361 kg/hr (3000 lb/hr); and ran for 0.75 hour.
2:45 p.m.   Stopped feed to plant.
3:00 p.m.   Stopped washing and fines circuits.
3:15 p.m.   Stopped the rest of plant.

Friday, September 20th
On this day, the crew, at the direction of BOR personnel, ran a maximum capacity test:
7:00 a.m.   Started at 1361 kg/hr (3000 lb/hr); ran for 0.50 hour.
7:30 a.m.   Increased feed rate to 1588 kg/hr (3500 lb/hr); ran for 0.50 hour.
8:00 a.m.   Increased feed rate to 1814 kg/hr (4000 lb/hr); ran for 0.50 hour.
8:30 a.m.   Increased feed rate to 2041 kg/hr (4500 lb/hr); ran for 0.50 hour.
9:00 a.m.   Increased feed rate to 2268 kg/hr (5000 lb/hr); ran for 0.50 hour.
9:30 a.m.   Increased feed rate to 2495 kg/hr (5500 lb/hr); ran for 0.50 hour.
10:00 a.m.   Increased feed rate to 2722 kg/hr (6000 lb/hr); ran for 0.50 hour.
10:30 a.m.   Increased feed rate to 2948 kg/hr (6500 lb/hr); ran for 0.50 hour.
11:00 a.m.   Increased feed rate to 3175 kg/hr (7000 lb/hr); ran for 10 minutes.
11:10 a.m.   The plant abruptly stopped. The overrun was cleaned up, and the crew began to disassemble the plant.

Saturday, September 21st
The crew disassembled the plant and continued cleanup and packing.

Sunday, September 22nd
No project activities were conducted on this day.

Monday, September 23rd
Initiated plant packaging and site cleanup.
Tuesday, September 24th
Finished plant packaging and finished site cleanup, except for the leaching area. The first truck left the project site.
Wednesday, September 25th
The crew and the second truck, with the remainder of the plant, left the project site.

F.3 FEED RATES

The goal, outside the pilot demonstration objective, in varying feed rates was to determine the maximum feed rate at which clean product can be obtained. Only random samples were collected during this test, so no meaningful analysis of lead removal was possible. Throughout the various feed rates, the solids to water ratio was maintained between 35 and 50% by weight. Flow rates were measured assuming that water flow rates of 3.8 to 5.7 L/min (1.0 to 1.5 gpm) were needed for every 227 kg/hr (500 lb/hr) of material fed to the attrition scrubber.

A summary of pilot plant operation at each of the feed rates evaluated is included below.

F.3.1   227 kg/hr (500 lb/hr)

Two separate runs were conducted at 227 kg/hr (500 lb/hr), for a total of 3 hours. At this rate, it almost appeared as if nothing was being put into the plant. Material was coming out, but the individual streams contained so little material that the internal streams of the plant appeared to be empty except for the water colored by the clays in the slime stream.

F.3.2   340 kg/hr (750 lb/hr)

One run was conducted at 340 kg/hr (750 lb/hr), for a total of 4 hours. The results of this rate were slightly better than 227 kg/hr, and it almost appeared that lines were starting to sand. The plant was operating significantly under capacity at this level, and it was difficult to tell what was functioning. The cyclone was not functioning at all because the load was so small that there was no overflow and all the feed material was going to the underflow. The entire stream thus ended up in the Knelson bowl. The results of the analysis suggest that the high slime load resulted in hindered settling, reducing the efficiency of the Knelson bowl.

F.3.3   454 kg/hr (1000 lb/hr)

Six separate runs were conducted at 454 kg/hr (1000 lb/hr), for a total of 11 hours. Most of the test was within plus or minus 227 kg/hr of this feed rate. After running the plant at various levels around 454 kg/hr, it became obvious by watching the various process streams in the plant that at a rate of 454 kg/hr, process streams could suffer low capacity problems. During one run, the plant operated at the 454 kg/hr rate for about 3.5 hours and then began to develop obstructions in several of the process streams. Two lines eventually became completely plugged, and the plant had to be shut down. The shutdown caused three other lines to clog because they did not have time to clear. It was again found that at lower feed rates, the heavier particles were not swept along, and thus settled in low or flat spots which would eventually cause process lines to clog with a build-up of heavies.

F.3.4   567 kg/hr (1250 lb/hr)

One run was conducted at 567 kg/hr (1250 lb/hr), for a total of 1.5 hours. At this rate, the plant started to develop some of the same problems observed at the 454 kg/hr rate. Thus, this feed rate test was discontinued early. At this rate the −105 µm to + 44 µm (−150 mesh to +350 mesh) line was kept clean, but the −1.19 µm to + 105 µm (−16 mesh to +150 mesh) lines could not be kept clean.

F.3.5   680 kg/hr (1500 lb/hr)

Five separate runs were conducted at 680 kg/hr (1500 lb/hr), for a total of 17 hours (approximately 3.5 hours per run). Though the plant started to work as designed at this feed rate, the plant was still under product capacity and process streams were sluggish. The higher feed rate produced a corresponding higher feed rate to each stream, increasing the velocity of the slurry in that stream. At these higher velocities, the larger particles did not have time to settle out and clog the slurry pipe lines. Above this 794 kg/hr (1500 lb/hr) rate, we could keep the −1.19 mm to + 105 µm (−16 mesh to +150 mesh) lines clean.

F.3.6   794 kg/hr (1750 lb/hr)

Three separate runs were conducted at 794 kg/hr (1750 lb/hr), for a total of 8 hours (about 3 hours per run). At this rate, it appeared that the plant would run by itself. After the plant was started, the crew had only to add feed material.

F.3.7   907 kg/hr (2000 lb/hr)

Four separate runs were conducted at 907 kg/hr (2000 lb/hr), for a total of 8 hours (about 2 hours per run). At this feed rate, the plant seemed to be operating smoothly.

F.3.8   1021 kg/hr (2250 lb/hr)

Two separate runs were conducted at 1021 kg/hr (2250 lb/hr), for a total of 7 hours (about 3.5 hours per run). At this rate, the plant also seemed to be operating properly.

At the direction of BOR personnel, the test series designed to look at prolonged runs at feed rates higher than 907 kg/hr were then curtailed. However, a few quick tests were conducted with higher feed rates, but no sampling for lead removal analysis was conducted.

F.3.9   1134 kg/hr (2500 lb/hr)

Two separate runs were conducted at 1134 kg/hr (2500 lb/hr), for a total of about 2 hours (about 1 hour per run). The plant ran best within plus or minus 227 kg/hr (500 lb/hr) of this rate, or between 907 kg/hr (2000 lb/hr) and 1361 kg/hr (3000 lb/hr). At these rates, the plant could be started within 5 minutes, attain equilibrium within about 20 minutes, and continue to run smoothly until shut down. The plant would operate at this capacity with little attendance, other than to make certain that there was feed in the hopper. The plant operated between 907 kg/hr (2000 lb/hr) and 1361 kg/hr (3000 lb/hr) for several days in succession. During this portion of the test, the plant continued to operate very smoothly. An orderly shutdown was requisite to assure a smooth start up of the pilot plant on the following day.

F.3.10 1247 kg/hr (2750 lb/hr)

Two separate runs were conducted at 1247 kg/hr (2750 lb/hr), for 1.5 hours (about 3/4 hour per run). The plant continued to operate with no difficulties.

F.3.11 1361 kg/hr (3000 lb/hr)

Two separate runs were conducted at 1361 kg/hr (3000 lb/hr), for a total of 3 hours (about 1.5 hours per run). The plant continued to operate smoothly. At process rates between 907 kg/hr and 1361 kg/hr, the heavies were flushed and removed from the flat spots and no clogging occurred.

F.3.12 1588 kg/hr (3500 lb/hr)

One 0.50-hour run was conducted at 1588 kg/hr (3500 lb/hr). The system still performed well and none of the systems appeared to be overloaded.

F.3.13 1814 kg/hr (4000 lb/hr)

One 0.50-hour run was conducted at 1814 kg/hr (4000 lb/hr). At this flow rate, the heavies appeared to still be flushed and removed from the flat spots and no clogging occurred, but the lines appeared to be reaching capacity. The plant still ran well at this rate. None of the systems were overloaded, yet some appeared to be operating at full capacity.

F.3.14 2041 kg/hr (4500 lb/hr)

One 0.50-hour run was conducted at 2041 kg/hr (4500 lb/hr). This rate produced the first signs of system overload. The first function to fail was size classification in the spiral classifiers. They appeared to cease functioning properly at just over 1814 kg/hr. At the 2041 kg/hr rate, screening indicated that the smaller particles moving up the spiral conveyor were +149 μm (+100 mesh) rather than +105 μm (+150 mesh). At this point, the load on the Reichert spiral concentrator was constant and it did not appear overloaded. The rated maximum sand capacity of the Reichert concentrator used in this plant was 1350 kg/hr (2975 lb/hr). At the point that the classifiers failed, the spiral load was about 680 kg/hr (1500 lb/hr) to 907 kg/hr (2000 lb/hr). This estimate is based on the screening results that indicated between 30 and 40% of the soil reports to the Reichert concentrator circuit. The concentrators never overloaded, but the pumps supplying the spirals were working close to capacity at this point. The amount of material reporting to the mineral jig was minor, and it never came close to capacity. Although the spiral classifier circuit ceased to function at about 2041 kg/hr (4500 lb/hr), the feed rate was still increased in 227 kg/hr (500 lb/hr) increments to test the limits of the rest of the circuits. This continued until the rate of 3175 kg/hr (7000 lb/hr) was achieved.

F.3.15 2268 kg/hr (5000 lb/hr)

One 0.50-hour run was conducted at 2268 kg/hr (5000 lb/hr). Because larger particles were entering the slime circuit, the pieces of equipment designed to handle slime started to overload, and the finer particle circuits ceased functioning.

F.3.16 2495 kg/hr (5500 lb/hr)

One 0.50-hour run was conducted at 2495 kg/hr (5500 lb/hr). At this rate, the vertical pumps in the slimes circuit failed. They became overloaded with −177 μm (80 mesh) particles, combined with the total quantity of material now reporting to these pumps. Below the feed rate of 2495 kg/hr, the discharge from the attrition scrubber auger contained random clay balls, and above 2495 kg/hr, no clay balls were in the auger discharge. It thus appeared that a heavier load in the attrition scrubber resulted in better cleaning or scrubbing.

F.3.17 2722 kg/hr (6000 lb/hr)

One 0.50-hour run was conducted at 2722 kg/hr (6000 lb/hr). At rates between 2495 kg/hr and 2722 kg/hr, the rocks and bullets coming out of the attrition scrubber auger were completely clean. Before this feed rate, the rocks and bullet fragments still had clay or slimes attached. Other than this change in cleaning, there were no new observed changes in plant operation at this feed rate. The pumps were still overloaded and flooding the base.

F.3.18 2948 kg/hr (6500 lb/hr)

One 0.50-hour run was conducted at 2948 kg/hr (6500 lb/hr). There were no additional changes observed in plant operation at this feed rate.

F.3.19 3175 kg/hr (7000 lb/hr)

One 10-minute run was conducted at 3175 kg/hr (7000 lb/hr). At this point, the auger, which removed the bullets from the bottom of the attrition scrubber, failed and the plant was shut down. A large rock fell into the bucket elevator about 5 minutes before the auger jammed. It was unclear whether the jam was due to the rock or auger overload. Up to this point, the auger did not seem strained or overloaded. With the cessation of the auger function, however, the build-up of large and heavy particles quickly overburdened the attrition scrubber. In addition, the under portion of the plant was saturated with a pool of liquid, so the plant was shut down.

The feed hopper, weigh belt, bucket elevator, attrition scrubber, trommel, mineral jig, and vibrating Sweco screens, all along the primary circuit, never appeared to be strained. The feed hopper, weigh belt, and bucket elevator never exceeded their design capacity during this test. As the attrition scrubber was new and had an unspecified design capacity, this test explored its capabilities. The load coming up the auger, by all appearances, was approaching a maximum at the 3175-kg/hr (7000-lb/hr) level. However, the impellers in the main body of the scrubber did not show any signs of being impeded by the load they were receiving. The product coming out of the scrubber was actually better scrubbed at the higher load rates. The attrition scrubber needs to be disassembled to ascertain wear rates, durability of the lining, and the cause of the failure of the auger. The coarse fraction being discharged from the trommel never contained any −1.19 mm (16 mesh) particles. Thus, the trommel never overloaded or ceased to function. The load going to the mineral jig was relatively small even at the high feed rate, but the mineral jig was approaching maximum capacity at the 3175 kg/hr (7000 lb/hr) rate. In addition, the 45.7-cm (18-in.) vibrating Sweco screen used to dewater the mineral jig tailings appeared to be approaching its maximum capacity.

F.3.20 3629 kg/hr (8000 lb/hr)

Two separate 15-minute runs were conducted at 3629 kg/hr (8000 lb/hr), for a total period of 1/2 hour. On two occasions, the plant was started and ran for 15 minutes at 3629 kg/hr. This was the maximum feed rate of the hopper belt with the gate limit set at its 5.1-cm (2-in.) height level. The gate limit was set at about one-quarter capacity for the balance of the pilot demonstration. The hopper estimated feed capacity is about 13,608 kg/hr (30,000 lb/hr). Thus, the hopper is capable of easily feeding any amount of material that the pilot plant could possibly handle.

Analytical Discussion

CONTENTS

G.1 INTRODUCTION

Appendix G includes a synopsis of the analytical results for each size fraction, through each unit of equipment, and it includes an analysis of potential for rehabilitation of the treated soil.

G.2 ANALYTICAL DISCUSSION

G.2.1 Feed Soil
[Size −19.1 mm (−3/4 in.)]

Before this soil fraction was analyzed, all whole slugs were extracted. This provided an analysis without the bullet nugget effect. The bullet and nugget effect results from a single bullet, or nugget of lead, or a large enough concentration in a given volume of soil to act as if it were a single bullet or nugget. Removal of the individual bullets provided a more realistic view of how much lead will have to be removed from the soil besides the bullets. The sampling results range from a high of 16,050 mg/kg to a low of 3100 mg/kg. Even with the bullets removed, a nugget effect was observed, resulting in groupings of the samples in four value ranges. The groupings were: three samples in the 3000 to 3300 mg/kg range, one sample at 7050 mg/kg, six samples in the 11,200 to 13,600 mg/kg range, and one sample at 16,050 mg/kg range. The values in the 3000 to 3300 mg/kg range were the probable base level of fine lead contained in the berm soil. The steps above the average of 3170 mg/kg were about 4000 mg/kg apart, indicating that even without the bullets there was a substantial nugget effect. It also indicated that there was a predilection for a size of lead particle that caused the nugget effect. With the average dry weight of a sample running about 50 grams, it was estimated that the particles weighed about 0.2 grams or about 3 grains. The average 30-caliber bullet weighed about 150 grains, and the average 22-caliber bullet weighed about 80 grains. The average value of lead in the feed soil, not counting the whole bullets, was about 8115 mg/kg.

G.2.2 Screen #1 [Size –19.1 mm to +4.76 mm (–3/4 in. to +4 mesh)]

The 30 August sample consisted of ground-up bullets and rocks, explaining the 311 mg/kg lead in the sample. For the balance of the test, the bullets were extracted before the sample was analyzed. With the bullets removed, the average lead content of this fraction was low at 75 mg/kg. All that would be needed to return this fraction as a clean product is a step to separate the bullets from the rest of the +4.76 mm (+4 mesh) fraction. This additional separation process would result in the cleanest product from the plant without leaching.

G.2.3 Screen #2 [Size –4.76 mm to +1.19 mm (–4 mesh to +16 mesh)]

This fraction contained a large quantity of fine lead, and was the same size fraction that was discharged from the trommel and fed the mineral jig. Very little of this fraction actually reached the trommel and mineral jig because the material had previously settled in the attrition scrubber and was removed by the auger. The average lead content of this fraction was 24,955 mg/kg.

G.2.4 Mineral Jig Concentrate [Size –4.76 mm to +1.19 mm (–4 mesh to +16 mesh)]

The mineral jig received the coarse overflow fraction from the trommel scrubber. This product was a mixture of lead, copper, and heavy minerals. Once the mineral jig bed became established with the –4.79 mm to +1.19 mm lead particles, mineral jig efficiency increased. After the third day, the amount of lead in these concentrates ranged from 28,500 mg/kg (2.80%) to almost 575,000 mg/kg (57.5%). The mineral jig did an excellent job concentrating lead once the bed was well established with particles the same size as those it was being fed.

G.2.5 Mineral Jig Tailings [Size –4.76 mm to +1.19 mm (–4 mesh to +16 mesh)]

This was also described as weigh belt #2 product. This product also had the free lead removed. Chemical analysis of this fraction yielded lead values averaging about 797 mg/kg. The sample was not quite clean enough for industrial sites. Vat leaching might be required to decrease the lead content below 500 mg/kg.

G.2.6 Reichert Spiral Concentrator Feed [Size –1.19 mm to +105 μm (–16 mesh to +150 mesh)]

The Reichert spiral concentrator feed was pumped from the sump where the spiral classifier sands were mixed with dilution water. The lead content of this product averaged about 5900 mg/kg.

G.2.7 Reichert Spiral Concentrator Concentrate [Size −1.19 mm to +105 μm (−16 mesh to +150 mesh)]

This product had lead values that ranged from 13,500 mg/kg (1.35%) to 92,900 mg/kg (9.20%). The lead content averaged 47,045 mg/kg (4.70%). The product was very dark sand mixed with some light-grained sand.

G.2.8 Reichert Spiral Concentrator Tailings [Size −1.19 mm to +105 μm (−16 mesh to +150 mesh)]

The lead content of this product averaged about 757 mg/kg lead. The spiral was quite efficient in removing the free lead from this soil fraction. The lead content was about the same as the lead in the mineral jig tailings. Again, vat leaching of this fraction might be required to decrease the lead below 500 mg/kg.

G.2.9 Hydrocyclone Feed [Size −105 μm (−150 mesh)]

This feed was pumped from the sump containing the spiral classifier overflow. The lead content of this fraction averaged 5600 mg/kg. In the hydrocyclone, this fraction split into the two fractions described below.

G.2.10 Hydrocyclone Overflow [Size fraction −37 μm (−350 mesh)]

The lead content of this fraction averaged about 5380 mg/kg. This fraction was treated with a flocculent and pumped into the roll-off boxes to settle out the slimes fractions.

G.2.11 Hydrocyclone Underflow (Knelson Bowl Feed) [Size −105 μm to +37 μm (−150 mesh to +350 mesh)]

The lead content of this fraction averaged about 5160 mg/kg. The hydrocyclone underflow was pumped to the Knelson bowl concentrator as feed slurry.

G.2.12 Knelson Bowl Concentrator Concentrate [Size −105 μm to +37 μm (−150 mesh to +350 mesh)]

The lead content of these samples ranged from about 6400 mg/kg (0.64%) to 119,200 mg/kg (11.9%). The average lead content was 42,695 mg/kg (4.27%). The results had large variations because the grooves in the Knelson bowl were never full of concentrated product and there were considerable leftover slimes in the grooves. These slimes decreased the concentration of lead when the grooves were cleaned. It was concluded from the available data that the Knelson bowl was removing lead from the hydrocyclone underflow fraction.

G.2.13 Knelson Concentrator Tailings [Size −105 μm to +37 μm (−150 mesh to +350 mesh)]

The lead content of this fraction averaged 1740 mg/kg. There were problems with the Knelson bowl feed, and variable results were obtained for the concentrate lead values. The fact that lead was concentrated in the bowl and the tailings lead values were decreased, indicated that the Knelson bowl was working. However, the results did not indicate whether more lead could be removed by the Knelson bowl.

G.2.14 Mixed slimes [Size −105 μm (−150 mesh)]

The lead content of this fraction averaged 4525 mg/kg. Based on the mass balance, the mass of mixed slimes was about 40% of the original mass of the −19.0 mm (−3/4 in.) feed soil. The concentration of lead was about one-half the average lead content of the −19.0 mm feed samples. The lead content of mixed slimes fraction was also about 1100 mg/kg less than the material fed to the hydrocyclone and about 650 mg/kg less than the material fed to the Knelson bowl concentrator. This result indicated that although the Knelson bowl was only partially functioning, the lead content of the mixed slimes fraction was still decreased by at least 15%.

G.3 SOIL REUSE POTENTIAL

In this section, the potential for returning the cleaned soil fraction back to the berm or the potential for another use has been evaluated. The percentage data in the following table were from the BOR particle sizing tests. The lead concentrations were estimated from the results in the CMRI treatability tests and BOR leaching tests.

	Size Fractions Weight %		Estimated Lead Content mg/kg	
Soil Fraction Cleaned	Total	Cumulative	Before Leach	After Leach
Attrition Scrubber Auger Fractions	7%	7%	75	75
Mineral Jig Tailings	4%	11%	800	300
Reichert Spiral Tailings	39%	50%	800	300
Knelson Bowl Tailings	26%	76%	1700	500
Hydrocyclone Slimes	15%	91%	4500	660

The evaluation indicated that about 88% of the soil could be cleaned and mixed to form a single, clean product with the lead content less than 500 mg/kg. However, two options were available for mixing the soil fractions. The first was to mix the fractions without leaching, and the second was to perform a single-stage vat leach before mixing the soil fractions. The final mixture results for both options are given in the following table.

Fraction Percentage	Before Leach Lead Content mg/kg	After Leach Lead Content mg/kg	Estimated Mixture Results	
			Lead Content mg/kg Soil Unleached[1]	Lead Content mg/kg Soil Leached[1]
7	75	75	5	5
4	800	300	32	12
39	800	300	312	117
15	1700	500	255	75
Mixed Soil Fraction Total Lead = L			604	209
Final Lead in Remix = (100%/76%) × L, mg/kg			795	275

1. Example Calculation:
   Soil Mass for the Reichert Spiral Tailings at 39% ≡ (39 kg Tailings)/(100 kg Soil)
   Lead in Mineral Jig Tailings = ((39 kg)(800 mg/kg))/(100 kg Soil) = 312 mg/kg Soil

The remaining 15%, the hydrocyclone slimes, was very fine, and still contained about 4500 mg/kg. The total quantity of lead in the slime fraction was calculated to be:

$$\text{Slime Fraction Lead Content} = 15\% \equiv \frac{15 \text{ kg Slimes}}{100 \text{ kg Soil}}$$

$$\text{Lead in Slime Fraction} = \frac{(15 \text{ kg})(4500 \text{ mg/kg})}{(100 \text{ kg Soil})} = 675 \text{ mg/kg Soil}$$

The slimes would have to be treated further before they could be mixed with the remaining 76% of the soil. If the slimes could be leached to the same extent as the rest of the soil fraction, about 60% extraction, it would contain about 270 mg of lead/kg of soil. If combined and mixed with the 76% of the vat-leached soil, the lead concentration for the total soil would be about 545 mg/kg. Two-stage leaching of the sand and/or slimes fractions should decrease the lead concentration below 545 mg/kg in the recombined, rehwabilitated soil.

From the demonstration results, the average percentages of lead removed from the three primary fractions using the mineral jig, the Reichert spiral concentrator, and Knelson bowl are shown in the following table. The mixed slimes fractions were not treated in the demonstration.

Soil Fraction Cleaned	Feed (mg/kg)	Tailings (mg/kg)	% Lead Removed
Attrition Scrubber Auger Fractions	8115	75	99%
Mineral Jig	18,000	795	96%
Reichert Spiral Concentrator	5900	755	87%
Knelson Bowl Concentrator	5160	1740	66%
Mixed Slimes	2190	2190	0%

Process Control and Monitoring Data

CONTENTS

H.1 INTRODUCTION

Appendix H describes the I/O racks, wiring, signal processing, operations modes, and relays of the process control system used for the pilot plant demonstration at NAS Miramar.

H.2 PROCESS CONTROL PHYSICAL LAYOUT

H.2.1 I/O Racks

The AATDF system used a combination of three analog I/O racks for 48 I/O points, and four digital I/O racks for 64 digital I/O points. Any combination of analog and digital racks could be on the same serial communications link, but, for consistency, the analog modules were mounted on analog racks and the digital modules were mounted on digital racks. All of the I/O racks were mounted vertically on an aluminum panel assembly and were separated vertically and horizontally from adjacent racks by plastic Panduit™ wiring ducts.

H.2.2 Rack Identification

The three analog racks were mounted across the top of the panel and were addressed from right to left as "1," "2," and "3" as viewed from the front of the panel. Immediately below the analog racks were three of the four digital racks that were addressed from left to right as "4," "5," and "6." Immediately below rack 6 was the remaining digital rack — at address "7." Adjacent and to the left of digital rack 7 were the 24 volt and 5/15 volt power supplies that powered the racks and the rack analog modules.

H.2.3 Wiring

Field wiring for power and communications for each of the racks was fed through the wiring ducts separating the individual racks. Field wiring was fed out the bottom of the panel assembly and was routed through the wiring duct to the various sensors and equipment controllers on the gravity-separation-circuit trailer.

H.2.4 Signal Processing

The analog racks were loaded with a mixture of analog input and analog output modules. The analog input modules received and converted 4 to 20 milliamp (ma) signals from sensors, such as flow meters and mass meters, into digital signals that were processed by the computer. The analog output modules performed the reverse operation and converted digital signals from the computer into 4 to 20 ma signals. The controller program on the process equipment used these signals to control operation parameters, such as speed and actuator position, of the particular unit operation.

H.2.5 Auto/Manual Operation Modes

The digital racks were also loaded with 5 and 12 volt on-off relays to actuate larger 120 and 230 volt relays which turned the processing equipment "on" and "off." These large relays could be controlled from the computer-controlled OPTOMUX relays or from an array of two-position switches mounted on the door of the OPTOMUX enclosure. The array of two-position switches provided manual on-off control of the process equipment. A primary control switched control from auto (computer control: auto or manual) to panel manual control. When in the panel manual position, any on-off signal sent from the OPTOMUX relays were ignored, and on-off control of the equipment was entirely under the control of the manual switches on the panel. While this wiring scheme accomplished the intended goal of absolute manual control, it did not provide the OPTOMUX or the process control computer with the on-off status information of the process equipment. Thus, when running in manual control mode on the panel, the process-control computer was prevented from recording and archiving equipment on-off status. This defeated the two primary purposes of the process control computer: to advise operators of equipment operation status and to monitor and archive process operation data. This fault could be corrected by rewiring the system so that when in panel manual control mode, the OPTOMUX relays are biased and, in turn, would communicate the on-off status of the relay to the PCP.

H.2.6 Relays

The four digital OPTOMUX racks turned process equipment on the trailer on and off via OPTOMUX 5 volt relays. Two types of relays were in use: digital relays on rack number 4 and mechanical reed relays on racks 6 and 7. During shakedown at the U.S. Bureau of Mines Salt Lake City Research Center (SLRC), a problem was identified with the reed relays sticking in the "on" position when switched quickly and repeatedly from "on" to "off." This problem was acknowledged and could have been corrected by replacing the mechanical reed relays with digital relays, but this was not done due to time constraints. The digital relays, on the other hand, proved to be robust, and no problems were experienced with their operation. It is recommended that all mechanical reed relays be replaced with the solid-state digital relays.

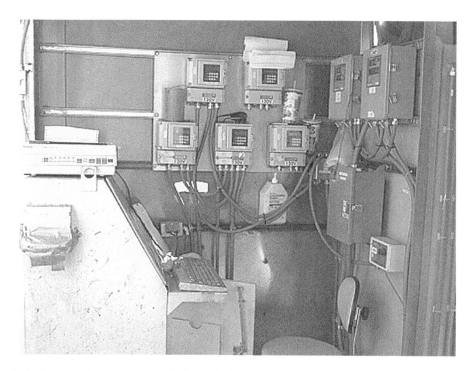

Figure H.1 Programming Computer and Electronic Controls.

Figure H.2 Electronic Control Systems and Main Power Panel.

APPENDIX I

Process Flowsheet and Mass Balance
for Full-Scale Plant

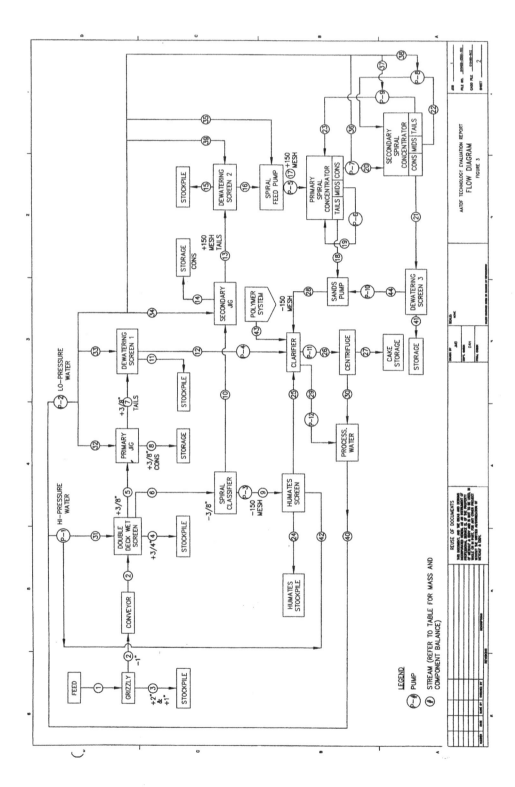

Figure I.1 Process Flowsheet for Full-scale Plant.

Table I.1 Treatability Results for the Soil at the Hypothetical Firing Range

Particle Size		Screen Analysis %dwb	Lead Analysis %dwb	Particle Size		Screen Analysis %dwb	Lead Analysis %dwb
mm	in./mesh			mm	in./mesh		
Oversize				**Middlings**			
+25.00	+1 in.	4.47	475	+0.59	+30 mesh	4.48	87
+19.00	+3/4 in.	2.69	325	+0.30	+50 mesh	5.11	95
				+0.11	+150 mesh	6.99	72
Humates				**Concentrate**			
+9.50	+3/8 in.	0.00	0	+9.50	+3/8 in.	3.98	140,530
				+4.76	+4 mesh	0.56	211,700
Tailings							
+9.50	+3/8 in.	3.80	525	+2.38	+8 mesh	0.24	382,708
+4.76	+4 mesh	3.17	73	+1.19	+16 mesh	0.56	20,958
+2.38	+8 mesh	5.51	180	+0.59	+30 mesh	1.03	12,982
+1.19	+16 mesh	12.61	129	+0.30	+50 mesh	1.15	13,536
+0.59	+30 mesh	8.50	104	+0.11	+150 mesh	1.82	2134
+0.30	+50 mesh	9.49	78	−0.11	−150 mesh	1.10	3111
+0.11	+150 mesh	15.49	84				
−0.11	−150 mesh	7.15	2509		**Total**	100.00	8495

Gravity Treatment

Humates − skimmed from +9.50 mm (+3/8 in.) tailings

+9.50 mm (+3/8 in.) − Jig tailings and concentrate

−9.50 mm by +4.76 mm (−3/8 in. by +4 mesh) — Jig tailings and concentrate

−4.76 mm by +1.19 mm (−4 mesh by +16 mesh) — Jig tailings and concentrate

−1.19 mm by +0.59 mm (−16 mesh by +30 mesh) — Shaking Table tailings, middlings, and concentrate

−0.59 mm by +0.297 mm (−30 mesh by +50 mesh) — Shaking Table tailings, middlings, and concentrate

−0.297 mm by +105 μm (−50 mesh by +150 mesh) — Shaking Table tailings, middlings, and concentrate

−105 μm (−150 mesh) — Shaking Table tailings, middlings, and concentrate

Table I.2 Mass Balance for Base Case Conceptual Design: Nominal Operating Rate at 13.60 t/hr (15 tons/hr)

Unit Operation	Flow	Description	Mass Flows Solid kg/hour	Mass Flows Liquid kg/hour	Distributions Solid wt%	Distributions Liquid wt%	Lead Content mg/kg	Comments
Double-Deck Grizzly Screen Openings: 50.8 mm (1 in.) 25.0 mm (2 in.)	1	Feed: Stockpiled Soil	14,242.1	1582.5	90.0	10.0	8495	From Feed Stockpile
	2	Oversize: +25.0 mm Soil	636.6	70.7	90.0	10.0	475	**To Clean Stockpile**
	3	Undersize: -25.0 mm Soil	13,605.4	1511.7	90.0	10.0	8871	To Double-Deck Wet Screen
Double-Deck Wet Screen	2	Feed: -25.0 mm Soil	13,605.4	1511.7	90.0	10.0	8871	From Grizzly
Screen Openings: 19.00 mm (3/4 in.) 9.50 mm (3/8 in.)	31	Feed: Screen Wash Water	0.0	25,206.3	0.0	100.0	0	High Pressure Water
	4	Oversize: +19.00 mm	366.0	91.5	80.0	20.0	325	**To Clean Stockpile**
	5	Oversize: -19.00 mm; +9.50 mm	1058.5	264.6	80.0	20.0	72,180	To Primary Jig
	6	Undersize: -9.50 mm	12,181.0	26,361.9	31.6	68.4	3626	To Spiral Classifier
Primary Jig Jig Cut Size: 9.50 mm (3/8 in.)	5	Feed: -19.05 mm; +9.50 mm	1058.5	264.6	80.0	20.0	72,180	From Double-Deck Wet Screen
	32	Feed: Jig Wash Water	0.0	1060.1	0.0	100.0	0	Low Pressure Water
	8	Concentrates	541.7	232.2	70.0	30.0	140,530	**Lead to Storage**
	7	Tailings	516.8	1092.6	32.1	67.9	524	To Dewatering Screen #1
Dewatering Screen #1 Screen Opening: 2.38 mm (8 mesh)	7	Feed: Tailings Primary Jig	516.8	1092.6	32.1	67.9	524	From Primary Jig
	33	Feed: Screen Wash Water	0.0	287.3	0.0	100.0	0	Low Pressure Water
	11	Oversize: +2.38 mm Soil	490.9	163.6	75.0	25.0	524	**To Clean Stockpile**
	12	Undersize: -2.38 mm Soil	25.8	1216.2	2.1	97.9	524	To Clarifier
			0.0	0.0	1609.3		0.00	
Spiral Classifier	6	Feed: -9.50 mm	12,181.0	26,361.9	31.6	68.4	3626	From Double-Deck Wet Screen
	10	Sand Fraction: +105 micron	10,999.5	5922.8	65.0	35.0	3737	To Secondary Jig
	9	Slime Fraction: -105 micron; humates	1181.4	20,439.1	5.5	94.5	2589	To Humates Screen
Secondary Jig Jig Cut Size: 1.19 mm (16 mesh)	10	Feed: +105 micron Sand Fraction	10,999.5	5922.8	65.0	35.0	3737	From Spiral Classifier
	34	Feed: Jig Wash Water	0.0	11,016.1	0.0	100.0	0	Low Pressure Water
	14	Concentrates	195.1	83.6	70.0	30.0	164,079	**Lead To Storage**
	13	Tailings	10,804.4	16,855.3	39.1	60.9	842	To Dewatering Screen #2
Dewatering Screen #2 Screen Opening: 1.19 mm (16 mesh)	13	Feed: Tailings Secondary Jig	10,804.4	16,855.3	70.0	30.0	842	From Secondary Jig
	39	Feed: Screen Wash Water	0.0	3063.7	0.0	100.0	0	Low Pressure Water
	15	Oversize: +1.19 mm	3186.0	1715.6	65.0	35.0	134	**To Clean Stockpile**
	16	Undersize: -1.19 mm; +105 micron	7618.4	18,203.4	29.5	70.5	1138	To Spiral Feed Pump

Equipment	Stream	Description						Notes
Spiral Feed Pump	16	Feed: −1.19 mm; +105 micron	7618.4	18,203.4	29.5	70.5	1138	From Dewatering Screen #2
	35	Feed: Dilution Water Feed	0.0	0.0	0.0	100.0	0	Low Pressure Water
	17	Feed: Spiral Concentrators	7618.4	18,203.4	29.5	70.5	1138	To Primary Spiral Concentrator
Primary Spiral Concentrator	17	Feed: From Spiral Feed Pump	7618.4	18,203.4	29.5	70.5	1138	Total Feed at 30% Solids
	19	Feed: Primary Middlings	1115.7	1673.6	40.0	60.0		
	23	Feed: Secondary Tailings	552.3	2668.1	17.1	82.9		
	37	Feed: Dilution Water Tailings	0.0	0.0	0.0	100.0	0	Low Pressure Water
	18	Tailings	7054.9	19,755.8	26.3	73.7	573	To Sands Pump
	19	Middlings	1115.7	1673.6	40.0	60.0		To Primary Spiral Concentrator
	20	Concentrate	1115.7	1115.7	50.0	50.0		To Secondary Spiral Concentrator
Secondary Spiral Concentrator	20	Feed: Primary Spiral Concentrate	1115.7	1115.7	50.0	50.0		Total Feed at 30% Solids
	36	Feed: Dilution Water Concentrate	0.0	1593.0	0.0	100.0		Low Pressure Water
	22	Feed: Secondary Middlings	563.5	845.3	40.0	60.0		
	38	Feed: Dilution Water Middlings	0.0	522.8	0.0	100.0		Low Pressure Water
	23	Tailings	552.3	2668.1	17.1	82.9		To Primary Spiral Concentrator
	22	Middlings	563.5	845.2	40.0	60.0		To Secondary Spiral Concentrator
	21	Concentrate	563.5	563.5	50.0	50.0	8215	To Dewatering Screen #3
Dewatering Screen #3 Screen Opening: 350 micron (45 mesh)	21	Feed: Secondary Spiral Concentrate	563.5	563.5	50.0	50.0	8215	To Sands Pump
	44	Undersize: Water	0.0	464.0	0.0	100.0	0	
	41	Oversize: Concentrate	563.5	99.4	85.0	15.0	8215	**Lead Storage**
Sands Pump	18	Feed: Primary Spiral Tailings	7054.9	19,755.8	26.3	73.7	573	From Primary Spiral Concentrator
	44	Feed: Undersize Water	0.0	464.0	0.0	100.0	0	From Screen #3
	28	To Clarifier	7054.9	20,219.8	25.9	74.1	573	
Humates Screen Screen Opening: 2.00 mm (10 mesh)	9	Feed: Classifier Slime Fraction	1181.4	20,439.1	5.5	94.5	2589	From Spiral Classifier
	42	Feed: Screen Wash Water	0.0	656.7	0.0	100.0	0	High Pressure Water
	24	Oversize: +2.00 mm Fraction	23.6	23.6	50.0	50.0	5178	**To Humates Stockpile**
	25	Undersize: −2.00 mm Fraction	1157.8	21,072.2	5.2	94.8	2536	To Clarifier

continued

Table I.2 (continued) Mass Balance for Base Case Conceptual Design: Nominal Operating Rate at 13.60 t/hr (15 tons/hr)

Unit Operation	Flow	Description	Mass Flows		Distributions		Lead Content	Comments
			Solid kg/hour	Liquid kg/hour	Solid wt %	Liquid wt%	mg/kg	
Clarifier	12	Feed: -2.38 mm Fraction	25.8	1216.2	2.1	97.9	524	From Dewatering Screen #1
	25	Feed: -2.00 mm Fraction	1157.8	21,072.2	5.2	94.8	2536	From Humates Screen
	28	Feed: -1.19 mm Fraction	7054.9	20,219.8	25.9	74.1	573	From Sands Pump
	43	Feed: Polymer	0.0	0.0	100.0	100.0	0	
	26	Underflow: -2.38 mm Soil	8238.6	32,954.3	20.0	80.0	848	To Centrifuge
	29	Overflow: Clean Water	0.0	9553.9	0.0	100.0	0	To Process Water Storage
Centrifuge	26	Feed: -2.38 mm Soil	8238.6	32,954.3	20.0	80.0	848	From Clarifier
	30	Overflow: Clean Water	0.0	27,462.0	0.0	100.0	0	To Process Water Storage
	27	Underflow: -2.38 mm Fraction	8238.6	5492.4	60.0	40.0	848	To Cake Storage
Process Water Storage	29	Feed: Clean Water from Clarifier	0.0	9553.9	0.0	100.0	0	From Clarifier
	30	Feed: Clean Water from Centrifuge	0.0	27,462.0	0.0	100.0	0	From Centrifuge
	40	Recycle Process Water	0.0	37,015.8	0.0	100.0	0	
Solid and Water Balance		Inflow Wet Contaminated Soil	14,242.1	1582.5	90.0	10.0	8495	
		Inflow Process Water	0.0	43,406.0	0.0	100.0	0	
		Outflow Washed Soil & Concentrates	14,242.1	7972.7	64.1	35.9	8495	
		Outflow Recycle Process Water	0.0	37,015.8	0.0	100.0	0	
		Make-up Process Water	0.0	6390.2	0.0	100.0	0	

Equipment Calculations for Full-Scale Plant

CONTENTS

Table J.1 Mass Balance for Full-Scale Plant Design: Maximum Operating Rate at 18.14 t/hr (20 t/hr)

Unit Operation	Flow	Description	Mass Flows Solid kg/hour	Mass Flows Liquid kg/hour	Distributions Solid wt %	Distributions Liquid wt%	Lead Content mg/kg	Comments
Double-Deck Grizzly								
Screen Openings:								
50.8 mm (1 in)								
25.0 mm (2 in)								
	1	Feed: Stockpiled Soil	18,989.4	2109.9	90.0	10.0	8495	From Feed Stockpile
	2	Oversize: +25.0 mm Soil	848.8	94.3	90.0	10.0	475	**To Clean Stockpile**
	3	Undersize: –25.0 mm Soil	18,140.6	2015.6	90.0	10.0	8871	To Double-Deck Wet Screen
Double-Deck Wet Screen								
Screen Openings:								
19.00 mm (3/4 in)								
9.50 mm (3/8 in)								
	2	Feed: –25.0 mm Soil	18,140.6	2015.6	90.0	10.0	8871	From Grizzly
	31	Feed: Screen Wash Water	0.0	33,608.4	0.0	100.0	0	High Pressure Water
	4	Oversize: +19.00 mm	488.0	122.0	80.0	20.0	325	**To Clean Stockpile**
	5	Oversize: –19.00 mm; +9.50 mm	1411.3	352.8	80.0	20.0	72,180	To Primary Jig
	6	Undersize: –9.53 mm	16,241.3	35,149.2	31.6	68.4	3626	To Spiral Classifier
Primary Jig								
Jig Cut Size:								
9.50 mm (3/8 in)								
	5	Feed: –19.05 mm; +9.50 mm	1411.3	352.8	80.0	20.0	72,180	From Double-Deck Wet Screen
	32	Feed: Jig Wash Water	0.0	1413.5	0.0	100.0	0	Low Pressure Water
	8	Concentrates	722.3	309.6	70.0	30.0	140,530	**Lead to Storage**
	7	Tailings	689.0	1456.7	32.1	67.9	524	To Dewatering Screen #1
Dewatering Screen #1								
Screen Opening:								
2.38 mm (8 mesh)								
	7	Feed: Tailings Primary Jig	689.0	1456.7	32.1	67.9	524	From Primary Jig
	33	Feed: Screen Wash Water	0.0	383.0	0.0	100.0	0	Low Pressure Water
	11	Oversize: +2.38 mm Soil	654.6	218.2	75.0	25.0	524	**To Clean Stockpile**
	12	Undersize: –2.38 mm Soil	34.5	1621.6	2.1	97.9	524	To Clarifier
Spiral Classifier								
	6	Feed: –9.50 mm	16,241.3	35,149.2	31.6	68.4	3626	From Double-Deck Wet Screen
	10	Sand Fraction: +105 micron	14,666.0	7897.1	65.0	35.0	3737	To Secondary Jig
	9	Slime Fraction:–105 micron; humates	1,575.2	27,252.2	5.5	94.5	2589	To Humates Screen
Secondary Jig								
Jig Cut Size:								
1.19 mm (16 mesh)								
	10	Feed: +105 micron Sand Fraction	14,666.0	7897.1	65.0	35.0	3737	From Spiral Classifier
	34	Feed: Jig Wash Water	0.0	14,688.2	0.0	100.0	0	Low Pressure Water
	14	Concentrates	260.2	111.5	70.0	30.0	164,079	**Lead To Storage**
	13	Tailings	14,405.9	22,473.8	39.1	60.9	842	To Dewatering Screen #2
Dewatering Screen #2								
Screen Opening:								
1.19 mm (16 mesh)								
	13	Feed: Tailings Secondary Jig	14,405.9	22,473.8	70.0	30.0	842	From Double-Deck Wet Screen
	39	Feed: Screen Wash Water	0.0	4084.9	0.0	100.0	0	Low Pressure Water
	15	Oversize: +1.19 mm	4248.1	2287.4	65.0	35.0	134	**To Clean Stockpile**
	16	Undersize: –1.19 mm; +105 micron	10,157.8	24,271.2	29.5	70.5	1138	To Spiral Feed Pump

Equipment	Stream	Description						Destination
Spiral Feed Pump	16	Feed: −1.19 mm; +105 micron	10,157.8	24,271.2	29.5	70.5	1138	From Dewatering Screen #2
	35	Feed: Dilution Water Feed	0.0	0.0	0.0	100.0	0	Low Pressure Water
	17	Feed: Spiral Concentrators	10,157.8	24,271.2	29.5	70.5	1138	To Primary Spiral Concentrator
Primary Spiral Concentrator	17	Feed: From Spiral Feed Pump	10,157.8	24,271.2	29.5	70.5	1138	Total Feed at ~29% Solids
	19	Feed: Primary Middlings	1487.6	2231.4	40.0	60.0		
	23	Feed: Secondary Tailings	736.4	3557.3	17.1	82.9		
	37	Feed: Dilution Water Tailings	0.0	0.0	0.0	100.0	0	Low Pressure Water
	18	Tailings	9406.6	26,341.0	26.3	73.7	573	To Sands Pump
	19	Middlings	1487.6	2231.4	40.0	60.0		To Primary Spiral Concentrator
	20	Concentrate	1487.6	1487.6	50.0	50.0		To Secondary Spiral Concentrator
Secondary Spiral Concentrator	20	Feed: Primary Spiral Concentrate	1487.6	1487.6	50.0	50.0		Total Feed at 30% Solids
	36	Feed: Dilution Water Concentrate	0.0	2124.0	0.0	100.0		Low Pressure Water
	22	Feed: Secondary Middlings	751.3	1127.0	40.0	60.0		
	38	Feed: Dilution Water Middlings	0.0	697.0	0.0	100.0		Low Pressure Water
	23	Tailings	736.4	3557.3	17.1	82.9		To Primary Spiral Concentrator
	22	Middlings	751.3	1126.9	40.0	60.0		To Secondary Spiral Concentrator
	21	Concentrate	751.3	751.3	50.0	50.0	8,215	To Dewatering Screen #3
Dewatering Screen #3 Screen Opening: 350 micron (45 mesh)	21	Feed: Secondary Spiral Concentrate	751.3	751.3	50.0	50.0	8215	To Sands Pump
	44	Undersize: Water	0.0	618.7	0.0	100.0	0	
	41	Oversize: Concentrate	751.3	132.6	85.0	15.0	8215	**Lead Storage**
Sands Pump	18	Feed: Primary Spiral Tailings	9406.6	26,341.0	26.3	73.7	573	From Primary Spiral Concentrator
	44	Feed: Undersize Water	0.0	618.7	0.0	100.0	0	From Screen #3
	28	To Clarifier	9406.6	26,959.7	25.9	74.1	573	
Humates Screen Screen Opening: 2.00 mm (10 mesh)	9	Feed: Classifier Slime Fraction	1575.2	27,252.2	5.5	94.5	2589	From Spiral Classifier
	42	Feed: Screen Wash Water	0.0	875.6	0.0	100.0	0	High Pressure Water
	24	Oversize: +2.00 mm Fraction	31.5	31.5	50.0	50.0	5178	**To Humates Stockpile**
	25	Undersize: −2.00 mm Fraction	1543.7	28,096.3	5.2	94.8	2536	To Clarifier

continued

Table J.1 (continued) Mass Balance for Full-Scale Plant Design: Maximum Operating Rate at 18.14 t/hr (20 t/hr)

Unit Operation	Flow	Description	Mass Flows		Distributions		Lead	Comments
			Solid kg/hour	Liquid kg/hour	Solid wt %	Liquid wt%	Content mg/kg	
Clarifier	12	Feed: −2.38 mm Fraction	34.5	1621.6	2.1	97.9	524	From Dewatering Screen #1
	25	Feed: −2.00 mm Fraction	1543.7	28,096.3	5.2	94.8	2536	From Humates Screen
	28	Feed: −1.19 mm Fraction	9406.6	26,959.7	25.9	74.1	573	From Sands Pump
	43	Feed: Polymer	0.0	0.0	100.0	100.0	0	
	26	Underflow: −2.38 mm Soil	10,984.8	43,939.0	20.0	80.0	848	To Centrifuge
	29	Overflow: Clean Water	0.0	12,738.5	0.0	100.0	0	To Process Water Storage
Centrifuge	26	Feed: −2.38 mm Soil	10,984.8	43,939.0	20.0	80.0	848	From Clarifier
	30	Overflow: Clean Water	0.0	36,615.8	0.0	100.0	0	To Process Water Storage
	27	Underflow: −2.38 mm Fraction	10,984.8	7323.2	60.0	40.0	848	**To Cake Storage**
Process Water Storage	29	Feed: Clean Water from Clarifier	0.0	12,738.5	0.0	100.0	0	From Clarifier
	30	Feed: Clean Water from Centrifuge	0.0	36,615.8	0.0	100.0	0	From Centrifuge
	40	Recycle Process Water	0.0	49,354.4	0.0	100.0	0	
Solid and Water Balance		Inflow Wet Contaminated Soil	18,989.4	2109.9	90.0	10.0	8495	
		Inflow Process Water	0.0	57,874.6	0.0	100.0	0	
		Outflow Washed Soil & Concentrates	18,989.4	10,630.2	64.1	35.9	8495	
		Outflow Recycle Process Water	0.0	49,354.4	0.0	100.0	0	
		Make-up Process Water	0.0	8520.3	0.0	100.0	0	

J.1 GRIZZLY SCREEN

Design Data

Design Balance Data: Table J.1

Description of Flow	kg/hr Solid	kg/hr Water	% Distribution Solid	% Distribution Water
Feed: Stockpiled Soil	18,989.4	2109.9	90.0	10.0
+25.0 mm Soil	848.8	94.3	90.0	10.0
−25.0 mm Soil	18,140.6	2015.6	90.0	10.0
Totals Mass Input/Output	18,989.4	2109.9		

Grizzly Screen Specifications

- Type – Vibrating Screen
- Grizzly Opening 50.8 mm (2 in.)
- Screen Opening 25.0 mm (1 in.)
- Mass Balance Grizzly Feed = 18,989 kg/hr (41,871 lb/hr)

Grizzly Screen Design Calculations

Reference: Weiss, 1985, pp. 3E-1 to 3E-19

Total Screen Area (A)

$$A = \frac{\text{kg/hr of undersize in feed}}{C \times B \times O \times F \times E \times S \times W \times D}$$

where:

A = Total Screen Area, m^2

C = Basic Screen Capacity, kg/hr per m^2 (Figure J.1)

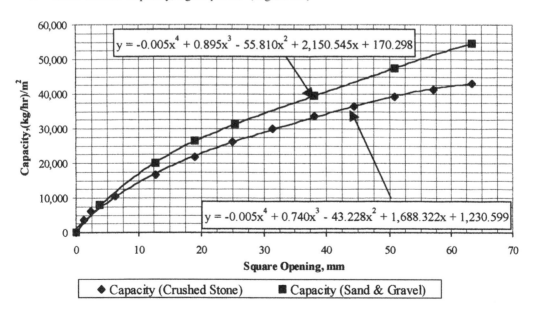

$$y = -0.005x^4 + 0.895x^3 - 55.810x^2 + 2{,}150.545x + 170.298$$

$$y = -0.005x^4 + 0.740x^3 - 43.228x^2 + 1{,}688.322x + 1{,}230.599$$

◆ Capacity (Crushed Stone)   ■ Capacity (Sand & Gravel)

Figure J.1 Screen Capacity (C).

Note: Basic Screen Capacity for soil having 1602 kg/m^3 (100 lb/ft^3) bulk density and at a screen angle of 18 degrees.

$$B = \text{Bulk Density Factor} = \frac{\text{Soil Bulk Density, kg/m}^3}{1602 \text{ kg/m}^3}$$

$$O = \text{Open Area Factor} = \frac{\text{Actual Screen \% Open Area}}{\text{\% Open Area (Table Value)}} \text{ (Table J.2)}$$

F = Fines Factor (Table J.3)
E = Efficiency Factor (Table J.4)
S = Slot Factor = 1.00 (for square opening screens)
W = Wet Screening Factor (Table J.4)
D = Deck Factor (Table J.5)

Grizzly Screen Capacities

Bulk Density of Soil = 1780 kg/m³ (111 lb/ft³)
Capacity for Crushed Stone from Figure J.1:
C (50.8 mm Grizzly) = 39,300 kg/hr per m²
C (25.0 mm Screen) = 26,000 kg/hr per m²
25.0 mm Screen Size
−25.0 mm Soil in Feed = 18,140 kg/hr
C (25.0 mm Screen) = 26,000 kg/hr per m²

Table J.2 % Open Area (%O.A.) for Square Screen Openings

Size mm	1.19	2.38	6.3	12.5	19.0	25.0	38.1	50.8	63.5
Size in. (mesh)	(16)	(8)	1/4	1/2	3/4	1	1 1/2	2	2 1/2
%O. A.	31	35	42	53	56	58	64	67	69

Table J.3 Fines (F) and Efficiency (E) Factors

%	0	10	20	30	40	50	60	70	80	85	90	95
F Factor	0.44	0.55	0.70	0.80	1.00	1.20	1.40	1.80	2.20	2.50	3.00	3.75
E Factor								2.25	1.75	1.50	1.25	1.00

Notes:
1. % is the percentage of fines in the feed to the screen or the efficiency of separation.
2. **F**, the fines factor, is a measure of the amount of material, in the feed to a screen deck, which is less than one-half the size of the opening in the deck. It is the measure of the difficulty of screening when compared to 40% fines in the feed (F = 1.00).
3. **E**, the efficiency of separation, is the ratio in % of the amount of material that actually passes the screen opening divided by the amount of material that should pass based on screen analysis.

Table J.4 Wet Screening (W) Factor

Size Opening, mm[1]	<0.8	1.6	3.2	4.8	8.0	9.5	12.5	19.0	25.0	50.8
Size Opening, in.	<1/32	1/16	1/8	3/16	5/16	3/8	1/2	3/4	1	2
W Factor	1.25	3.00	3.50	3.50	3.00	2.50	1.75	1.35	1.25	1.00

Note:
1. Square Opening

Table J.5 Deck (D) Factors

Deck	Top	2nd	3rd
D Factor	1.00	0.90	0.80

$$B = \frac{1780 \text{ kg/m}^3}{1602 \text{ kg/m}^3} = 1.11$$

Actual Screen %O.A. = 58%

$$O = \text{Open Area Factor} = \frac{58\%}{58\%} = 1.00$$

F = 1.00 (For −12.5-mm fines at ~ 94%, F = 3.00; used a conservative value)
E = 1.00 (% Efficiency = 95)
S = 1.00 (for square opening screens)
W = 1.00 (dry screening)
D = 0.90 (2nd deck)

$$A = \frac{18,140 \text{ kg/hr}}{26,000 \text{ kg/hr/m}^2 \times 1.00 \times 1.11 \times 1.00 \times 1.00 \times 1.00 \times 1.00 \times 0.90}$$

A = 0.698 m² (7.52 ft²)

Summary of Design Results:

- Vibrating Screen Size: Width = 0.91 m (3 ft); Length = 1.83 m (6 ft) with square screen opening of 25.0 mm (Table J.6)
- Actual Top Deck Area (Grizzly Screen Opening) = 1.44 m² (15.5 ft²)
- Actual Bottom Deck Area (25 mm Screen Opening) = 1.21 m² (13.0 ft²)
- Bulk Density Factor B = 1.11 and Deck Factor D = 0.90
- Maximum Solid Capacity = (1.21 m²)(26,000 kg/hr-m²)(1.11)(0.90) = 31,428 kg/hr

Maximum Depth of Bed

Maximum Depth = (4)(Screen Opening) = (4)(25.0 mm) = 100.0 mm or 10.0 cm
Width of Screen = 0.91 m
Bulk Density of Soil = 1780 kg/m³ (111 lb/ft³)
Figure J.2: Depth of Bed Factor = ~18,000 kg/hr per cm

Table J.6 Relationship Between Screen Size and Screen Area

Screen Size, m (ft)	Actual Screen Area Top Deck m² (ft²)	Actual Screen Area Bottom Deck m² (ft²)
0.91 × 1.83 (3 × 6)	1.44 (15.5)	1.21 (13.0)
0.91 × 2.44 (3 × 8)	1.90 (20.5)	1.58 (17.0)
0.91 × 3.05 (3 × 10)	2.42 (26.0)	2.00 (21.5)
1.22 × 2.44 (4 × 8)	2.65 (28.5)	2.23 (24.0)
1.22 × 3.05 (4 × 10)	3.34 (36.0)	2.79 (30.0)
1.22 × 3.66 (4 × 12))	4.00 (43.0)	3.34 (36.0)
1.22 × 4.27 (4 × 14)	4.65 (50.0)	3.90 (42.0)
1.52 × 2.44 (5 × 8)	3.30 (35.5)	6.69 (30.0)
1.52 × 3.05 (5 × 10)	4.27 (46.0)	3.06 (33.0)
1.52 × 3.66 (5 × 12)	5.11 (55.0)	4.27 (46.0)

$$\text{Feed Passing Over 25.0 mm Deck} = \left(\frac{4.47\%}{100}\right)(18,989 \text{ kg/hr}) = 849 \text{ kg/hr}$$

$$\text{Depth of Bed} = \frac{849 \text{ kg/hr}}{18,000 \text{ kg/hr per cm}} \approx 0.047 \text{ cm}$$

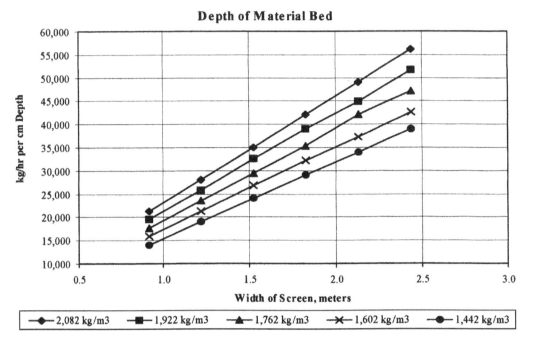

Figure J.2 Depth of Material Bed.
Notes:

1. The five curves are for soil bulk densities that increase from 1442 to 2082 kg/m³ of soil.
2. Depth of material bed is based on a soil flow rate of 18.29 m/min and a screen slope of 18 degrees.
3. The depth of bed at the discharge end must not be greater than four times the size of opening in the screen deck.

J.2 DOUBLE-DECK WET SCREEN

Design Data

Design Balance Data: Table J.1

Description of Flow	kg/hr		% Distribution	
	Solid	Water	Solid	Water
Feed: −25.0 mm Soil	18,140.6	2015.6	90.0	10.0
Feed: Wash Water	0.0	33,608.4	0.0	100.0
+19.0 mm Soil	488.0	122.0	80.0	20.0
−19.0 mm by +9.5-mm Soil	1411.3	352.8	80.0	20.0
−9.5 mm Soil	16,241.3	35,149.2	31.6	68.4
Totals Mass Input/Output	18,140.6	35,624.0		

Double-Deck Wet Screen Specifications

- Type – Vibrating Screen
- Screen Openings 19.0 mm (3/4 in.) and 9.5 mm (3/8 in.)
- Mass Balance Wet Screen Feed = 18,140 kg/hr (40,000 lb/hr)
- High Pressure Wash Water Requirement = 24.7 L/min per m³ (5 gpm per yd³)

Double-Deck Wet Screen Design Calculations

Reference: Weiss, 1985, pp. 3E-1 to 3E-19

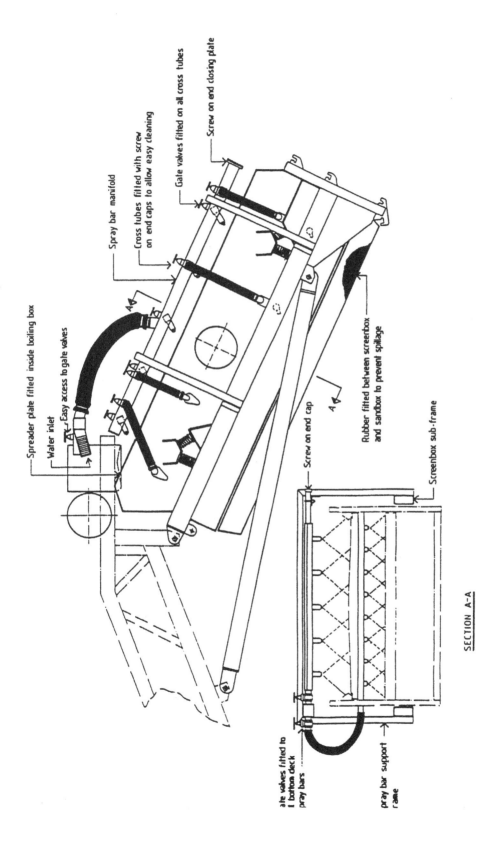

SECTION A–A

Finlay Double-deck hydrascreen.

Wet Screen Capacity

Bulk Density of Soil = 1780 kg/m³ (111 lb/ft³)
Capacity for Crushed Stone (C) from Figure J.1:
C (–19.0 mm) = 22,000 kg/hr per m²
C (–9.5 mm) = 13,500 kg/hr per m²

19.0 mm Screen Size

–19.0-mm (–3/4-in.) Soil in Feed = 17,653 kg/hr
C (19.0 mm) = 22,000 kg/hr per m²

$$B = \frac{1780 \text{ kg/m}^3}{1602 \text{ kg/m}^3} = 1.11$$

%O.A. = ~56% (Table J.2)

$$O = \text{Open Area Factor} = \frac{56\%}{56\%} = 1.00$$

F = 1.00 (for –9.5 mm fines at ~84%, F = 2.40, Table J.3; used a conservative value)
E = 1.00 (% Efficiency = 95, Table J.3)
S = 1.00 (for square opening screens)
W = 1.00 (for 19.0 mm opening, W = 1.35, Table J.4; used F = 1.00 assuming dry screening)
D = 1.00 (top deck, Table J.5)

$$A = \frac{17,653 \text{ kg/hr}}{22,000 \text{ kg/hr/m}^2 \times 1.11 \times 1.00 \times 1.00 \times 1.00 \times 1.00 \times 1.00 \times 1.00}$$

A = 0.723 m² (7.78 ft²)

Summary of Design Results

- Vibrating Screen Size: Width = 0.91 m (3 ft); Length = 1.83 m (6 ft) with square screen opening of 19.0 mm (Table J.6)
- Actual Top Deck Area = 1.44 m² (15.5 ft²)
- Bulk Density Factor B = 1.11
- Maximum Solid Capacity = (1.44 m²)(22,000 kg/hr per m²)(1.11) = 35,164 kg/hr

Maximum Depth of Bed

Maximum Depth Top Screen = (4)(Screen Opening) = (4)(19.0 mm) = 76 mm or 7.6 cm
Width of Screen = 0.91 m
Figure J.2: Depth of Bed Factor = ~ 18,000 kg/hr per cm
Feed Passing Over 19.0 mm Deck = 488 kg/hr

$$\text{Depth of Bed} = \frac{488 \text{ kg/hr}}{18,000 \text{ kg/hr per cm}} \approx 0.027 \text{ cm}$$

9.5 mm Screen Size

–9.5-mm (–3/8-in.) Soil in Feed = 16,241 kg/hr
C (9.5 mm) = 13,500 kg/hr per m²

$$B = \frac{1780 \text{ kg/m}^3}{1602 \text{ kg/m}^3} = 1.11$$

%O.A. = 48% (Table J.2)

O = Open Area Factor = $\dfrac{48\%}{48\%}$ = 1.00

F = 1.00 (for minus 4.7 mm fines at ~ 79%, F = 2.20, Table J.3; used a conservative value of 1.00)

E = 1.00 (% Efficiency = 95, Table J.3)

S = 1.00 (for square opening screens)

W = 1.00 (for 9.5 mm opening, W = 2.50, Table J.4; used a value of 1.00 assuming dry screening)

D = 0.90 (2nd deck, Table J.5)

$$A = \frac{16{,}241 \text{ kg/hr}}{13{,}500 \text{ kg/hr/m}^2 \times 1.11 \times 1.00 \times 1.00 \times 1.00 \times 1.00 \times 1.00 \times 0.90}$$

A = 1.204 m² (13.0 ft²)

Summary of Design Results

- Vibrating Screen Size: Width = 0.91 m (3 ft); Length = 1.83 m (6 ft) with square screen opening of 9.50 mm (Table J.6)
- The lower 9.5 mm screen governed the double-deck screen capacity
- Actual Bottom Deck Area = 1.21 m² (13.0 ft²)
- Bulk Density Factor B =1.11 and Deck Factor D = 0.9
- Maximum Solid Capacity = (1.21 m²)(13,500 kg/hr-m²)(B=1.11)(D=0.9) = 16,319 kg/hr

Maximum Depth of Bed

Maximum Depth Top Screen = (4)(Screen Opening) = (4)(9.5 mm) = 38 mm or 3.8 cm

Width of Screen = 0.91 m

Figure J.2: Depth of Bed Factor = ~18,000 kg/hr per cm

Feed Passing Over 9.5-mm Deck = 1411 kg/hr

Depth of Bed = $\dfrac{1411 \text{ kg/hr}}{18{,}000 \text{ kg/hr per cm}} \approx 0.078$ cm

J.3 PRIMARY MINERAL JIG

Design Data

Design Balance Data: Table J.1

Description of Flow	kg/hr		% Distribution	
	Solid	Water	Solid	Water
Feed: −19.0 mm by +9.5 mm Soil	1411.3	352.8	80.0	20.0
Feed: Dilution Water	0.0	1413.5	0.0	100.0
Concentrate	722.3	309.6	70.0	30.0
Tailings	689.0	1456.7	32.1	67.9
Totals Mass Input/Output	1411.3	1766.3		

Mineral Jig Specifications

- Type – Simplex
- Mass Balance Jig Feed = 1411 kg/hr (2460 lb/hr)
- Mass Balance Media Flow = (1.414 + 0.353 m³/hr) = 1.767 m³/hr (7.78 gpm)

Mineral Jig Design Calculations

Reference: Svedala, 1996, pp. 4:11 to 4:13

Mineral Jig Capacity

Design Rate for Jig Feed (–19.0 mm by +9.5 mm) Fraction = 1411 kg/hr
Capacity from Figure J.3
Specific Gravity of Lead Metal = 11.4 kg/L
Estimated Mineral Jig Capacity (Lower Limit) = 7500 kg/hr/m²
Estimated Mineral Jig Capacity (Upper Limit) = 12,800 kg/hr/m²

$$\text{Jig Bed Area (Lower Limit)} = \frac{\text{Mineral Jig Feed, kg/hr}}{\text{Capacity, kg/hr} - \text{m}^2} = \frac{1411 \text{ kg/hr}}{7500 \text{ kg/hr} - \text{m}^2}$$

Mineral Jig Bed Area (Lower Limit) = 0.188 m² (2.03 ft²)
Jig Bed Area (Upper Limit) = (1411 kg/hr)/(12,800 kg/hr/m²) = 0.110 m² (1.19 ft²))

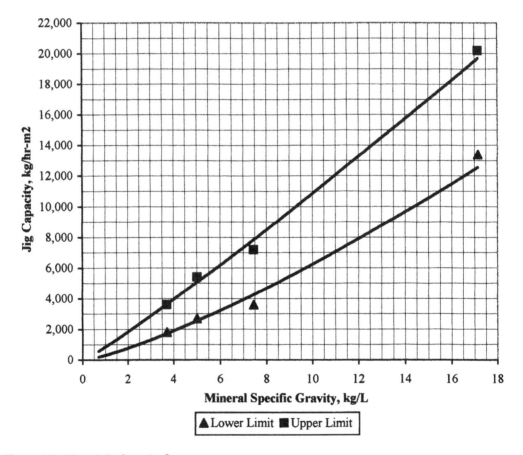

Figure J.3   Mineral Jig Capacity Curves.

Table J.7   Gravity Separation Mineral Jig, Sizing

Mineral Jig Type[1]	Bed Area		Media Flow (water)	
	m²	ft²	m³/hr	gpm
Simplex 12 × 18	0.14	1.50	1.3 to 2.0	5.7 to 8.7
Simplex 16 × 24	0.25	2.67	1.5 to 3.0	6.6 to 13.2
Simplex 24 × 36	0.56	6.00	5.0 to 7.0	22.0 to 31.0
Duplex 12 × 18	0.28	3.00	2.0 to 4.0	8.8 to 17.6
Duplex 16 × 24	0.50	5.33	3.0 to 5.0	13.2 to 22.0
Duplex 24 × 36	1.12	12.00	9.0 to 14.0	40.0 to 61.0

1.Type number refers to width and length of particle bed area (in.).

Summary of Design Results

- Selected Mineral Jig Size (Table J.7): Simplex 16×24
- Total Bed Area per Jig = 0.25 m² (2.67 ft²)
- Media Flow Range per Jig = 1.5 to 3.0 m³/hr (6.6 to 13.2 gpm)
- Number of Mineral Jigs Required (Lower and Upper Limits) = 1
- Solid Feed Rate (Lower Limit) = 0.25 m² × 7500 kg/hr/m² = 1875 kg/hr
- Solid Feed Rate (Upper Limit) = 0.25 m² × 12,800 kg/hr/m² = 3200 kg/hr
- Media Flow (Lower Limit) = 1.767 m³/hr × (1857/1411) = 2.35 m³/hr (10.3 gpm)
- Media Flow (Upper Limit) = 1.767 m³/hr × (3200/1411) = 4.00 m³/hr (17.6 gpm)
- The Simplex 16×24 jig had excess capacity.

J.4 DEWATERING SCREEN #1

Design Data

Design Balance Data: Table J.1

Description of Flow	kg/hr		% Distribution	
	Solid	Water	Solid	Water
Feed: Jig Tailings	689.0	1456.7	32.1	67.9
Feed: Fresh Water	0.0	383.0	0.0	100.0
Oversize: + 2.38 mm (8 mesh)	654.6	218.2	75.0	25.0
Undersize: - 2.38 mm (8 mesh)	34.5	1621.6	2.1	97.9
Totals Mass Input/Output	689.0	1839.7		

Dewatering Screen Specifications

- Type – Vibrating Screen
- Screen Openings = 2.38 mm (8 mesh)
- Mass Balance Wet Screen Feed = 689 kg/hr (1519 lb/hr)
- Low Pressure Wash Water Requirement = 14.8 L/min per m³ (3.0 gpm per yd³)

Dewatering Screen Design Calculations

Reference: Weiss, 1985, pp. 3E-1 to 3E-19

Wet Screen Capacity

Bulk Density of Soil = 1780 kg/m³ (111 lb/ft³)
Capacity from Figure J.1:
C (2.38 mm) = 5000 kg/hr per m²

2.38 mm Screen Size

–2.38 mm (–8 mesh) Soil in Feed to Screen = 35 kg/hr (76 lb/hr)

$$B = \frac{1780 \text{ kg/m}^3}{1602 \text{ kg/m}^3}$$

Table J.2:
%O.A. = ~36%

$$O = \text{Open Area Factor} = \frac{36\%}{36\%} = 1.00$$

Table J.3:

F = 1.00 (for −1.19 mm fines at ~70%, F = 1.80, used a conservative value)

E = 1.00 (% Efficiency = 95)

S = 1.00 (for square opening screens)

Table J.4:

W = 1.00 (for 2.38 mm opening, W = ~3.25, used a conservative value)

$$A = \frac{35 \text{ kg/hr}}{5000 \text{ kg/hr/m}^2 \times 1.11 \times 1.00 \times 1.00 \times 1.00 \times 1.00 \times 1.00 \times 1.00}$$

$A = 0.00631 \text{ m}^2 \ (0.0679 \text{ ft}^2)$

Summary of Design Results

- Vibrating Screen Size: Width = 0.91 m (3 ft); Length = 1.83 m (6 ft) with square screen opening of 2.38 mm (Table J.6)
- Single Deck Area = 1.44 m² (15.5 ft²)
- Bulk Density Factor B = 1.11
- Maximum Solid Capacity = (1.44 m²)(5000 kg/hr-m²)(1.11) = 7992 kg/hr

J.5 SPIRAL CLASSIFIER

Design Data

Design Balance Data: Table J.1	kg/hr		% Distribution	
Description of Flow	Solid	Water	Solid	Water
Feed: −9.5 mm Soil	16,241.3	35,149.2	31.6	68.4
Feed: Wash Water	0.0	0.0	0.0	100.0
−9.5 mm, +105 μm (Sand) Product	14,666.0	7897.1	65.0	35.0
−105 μm Slimes (Overflow) Product	1575.2	27,252.2	5.5	94.5
Totals Mass Input/Output	16,241.3	35,149.2		

Spiral Classifier Specifications

- Particle Size of Separation = 105 μm (150 mesh)
- Slurry Feed Rate = 16,241 kg solids/hr (17.91 ton/hr)
- O/F: C = Overflow Capacity = 1575 kg solids/hr (1.74 ton/hr)
- O/F: %S = Overflow Pulp Density = 5.50% solids
- $(S_g)_s$ = Specific Gravity of Solids = 2.6 kg/L

Calculation of Overflow (O/F) Pool Area

Reference Appendix D, pp. 286–288: Svedala, 1996, pp. 3:11–3:13

O/F Volume Rate (O/F: V)

$$O/F: V = \frac{(O/F:C)}{(S_g)_s} + \frac{(O/F:C)(100\% - O/F:\%S)}{(O/F:\%S)}$$

$$O/F: V = \frac{(1575 \text{ kg/hr})}{2.6 \text{ kg/L}} + \frac{(1575 \text{ kg/hr})(100\% - 5.50\%)}{(5.50\%)} = 27,667 \text{ L/hr}$$

$$O/F: V = \frac{27,667 \text{ L/hr}}{1000 \text{ L/m}^3} = 27.67 \text{ m}^3/\text{hr} \ (121.8 \text{ gpm})$$

O/F Volume % Solids (O/F V: %S)

$$O/F \text{ V:}\%S = \cfrac{100}{\left[1 + \cfrac{(S_g)_s \times (100\% - O/F\text{:}\%\text{Solids})}{O/F\text{:}\%\text{Solids}}\right]}$$

$$O/F \text{ V:}\%S = \cfrac{100}{\left[1 + \cfrac{(2.60) \times (100\% - 5.50\%)}{5.50\%}\right]} = 2.19\%$$

From the figure on page 241 in Appendix D, the settling rate of 105 μm (150 mesh) particles at 2.19% O/F V: %S was about 30 m/hr (1.6 ft/min).

O/F Pool Area (O/F: PA)

Specific Gravity Correction:

$$CSR = \text{Corrected Settling Rate} = (m/hr) \times \left[\frac{(S_g)_s - 1}{1.65}\right]^{0.5}$$

$$CSR = (30 \text{ m/hr}) \times \left[\frac{2.6 - 1}{1.65}\right]^{0.5} = 29.5 \text{ m/hr}$$

The factor, 0.7, is for the disturbance in the settling pool caused by the rotating spiral.

$$O/F\text{:PA} = \frac{O/F\text{:V}}{(0.70)(CSR)}$$

$$O/F\text{:PA} = \frac{27.57 \text{ m}^3/\text{hr}}{(0.70)(29.5 \text{ m/hr})} = 1.335 \text{ m}^2 \ (14.37 \text{ ft}^2)$$

Summary of Design Results for Overflow Pool Area

- Classifier size based on maximum pool area of 1.34 m² (Table J.8).
- 600 mm (24 in.), Simplex SP, 100 MF with pool area of 1.07 m² to 1.55 m².
- With a pool area of 1.55 m², the slurry capacity was 32.0 m³/hr. At a solids content of 5.5%, the overflow solid capacity was about 1800 kg/hr. Accordingly, the solid feed capacity was 21,700 kg/hr.

Calculation of Sand Raking Capacity

Sand Raking Capacity (Sand: V)

Sand: V = Sand Raking Volume = m³/hr
Sand: C = Sands Capacity, kg/hr = 14,743 kg/hr
Sand: %S = Sands % Solids = 65.0%

$$\text{Sand: V} = \cfrac{\left[\cfrac{(\text{Sand:C})}{(S_g)_s} + \cfrac{(\text{Sand:C})(100\% - \text{Sand }\%S)}{\text{Sand }\%S}\right]}{1000 \text{ L/m}^3}$$

$$\text{Sand: V} = \cfrac{\left[\cfrac{(14{,}666 \text{ kg/hr})}{2.6 \text{kg/L}} + \cfrac{(14{,}666 \text{ kg/hr})(100\% - 65\%)}{65\%}\right]}{1000 \text{ L/m}^3} = 13.54 \text{ m}^3/\text{hr} \ (59.6 \text{ gpm})$$

Table J.8 Spiral Classifier – Pool Area and Raking Capacity

Diameter Spiral		Classifier Design		Spiral rpm	Rake Capacity m³/hr		Tank	Available Pool Area m²	
in.	mm	type	spiral	min/max	min	max	Model	min	max
24	600	Simplex	SP	6/16	3.6	9.6	100	1.07	1.55
			DP	6/16	6.6	17.6	125	1.54	2.08
							150	2.05	2.64
30	750	Simplex	SP	5/14	5.0	14.0	100	1.76	2.37
			DP	5/14	9.5	26.6	125	2.52	3.21
36	900	Simplex	SP	4/11	8.0	22.0	100	2.41	3.36
			DP	4/11	16.0	44.0	125	3.49	4.53
							150	4.66	5.81
42	1000	Simplex	SP	3.5/10	9.4	27.0	100	3.50	4.60
			DP	3.5/10	19.2	55.0	125	4.94	6.17
48	1200	Simplex	SP	3/9	15.0	45.0	100	4.10	5.95
			DP	3/9	29.7	89.1	125	5.95	7.99
							150	7.99	10.20
60	1500	Simplex	SP	3/7	29.4	68.6	100	6.35	9.28
			DP	3/7	59.1	137.9	125	9.26	12.43
							150	12.41	15.88
72	1800	Simplex	SP	2/6	26.0	78.0	100	9.12	13.24
			DP	2/6	63.2	189.6	125	13.24	17.78
							150	17.78	22.61

All classifiers have "Modified Flair Tanks" and maximum slope of 29.1 cm per m (3.5-in. per ft)

Simplex, Single spiral in tank

SP – Single-pitch spirals

DP – Double-pitch spirals

Model 100 – 100% Spiral submergence in hindered settling zone

Model 125 – 125% Spiral submergence in hindered settling zone

Model 150 – 150% Spiral submergence in hindered settling zone

Summary of Design Results for Sand Raking Capacity

- The required Classifier raking capacity was 13.5 m³/hr.
- Classifier Size: 750 mm (30 in.), Simplex SP, 100 MF with raking capacity of 5.0 m³/hr to 14.0 m³/hr (Table J.8).
- With a raking capacity of 14.0 m³/hr, the sand capacity was 15,200 kg/hr. Consequently, the solid feed capacity was 16,800 kg/hr.

Calculation of Sand Compression Pool Area

Sand Compression Pool Area (Sand: CPA)

From the Figure on page 241 in Appendix D, settling rate at 40-volume % for 105 μm (150 mesh) particles was about 1.8 m/hr.

Specific Gravity Correction:

$$CSR = \text{Corrected Settling Rate} = (m/hr) \times \left[\frac{(S_g)_s - 1}{1.65}\right]^{0.5}$$

$$CSR = (1.80 \text{ m/hr}) \times \left[\frac{2.60 - 1}{1.65}\right]^{0.5} = 1.77 \text{ m/hr } (0.097 \text{ ft/min})$$

$$CZV = \text{Compression Zone Volume} = \frac{\dfrac{(\text{Sand:C, kg/hr})}{1000 \text{ kg/t}}}{\dfrac{0.70 \times 0.80 \times (S_g)_s \times 40\%}{100\%}}$$

The factor, 0.7, is for the disturbance in the settling pool caused by the rotating spiral. The factor, 0.8, refers to the smaller pool area available at the compression level.

$$CZV = \frac{\dfrac{(14{,}666 \text{ kg/hr})}{1000 \text{ kg/t}}}{\dfrac{0.70 \times 0.80 \times 2.60 \times 40\%}{100\%}} = 25.18 \text{ m}^3/\text{hr}$$

$$\text{Sand:CPA} = \frac{CZV \text{ m}^3/\text{hr}}{CSR \text{ m}/\text{hr}} = \frac{25.18 \text{ m}^3/\text{hr}}{1.77 \text{ m}/\text{hr}} = 14.23 \text{ m}^2 \ (153.1 \text{ ft}^2)$$

Summary of Design Results for Sand Compression Pool Area

- Classifier size based on a sand compression area of 14.23 m² (Table J.8).
- 1500 mm (60 in.), Simplex SP, 150 MF with pool area of 12.4 m² to 15.9 m².
- With a pool area of 15.9 m², the wet sand capacity was 28.1 m³/hr. The sands solid capacity was 16,400 kg/hr. Accordingly, the maximum solid feed rate to the classifier was 18,200 kg/hr.

Alternate Calculation of Spiral Classifier Capacity

Reference: Weiss, 1985, pp. 3D-46 to 3D-52

Classifier Pool Area (Size) Required

Using Figure J.4:
Cut point = 105 μm (150 mesh)
Overflow Capacity per unit Pool Area = 27,900 kg/24 hr/m²
Overflow Capacity per unit Pool Area = (27,900 kg/24 hr/m²) = 1162 kg/hr/m²
Mass Balance Overflow Solid Rate = 1575 kg/hr (3473 lb/hr)

$$\text{Pool Area} = \frac{\text{Mass Balance Rate}}{\text{Overflow Capacity}} = \frac{1575 \text{ kg/hr}}{1162 \text{ kg/hr/m}^2} = 1.36 \text{ m}^2 \ (14.6 \text{ ft}^2)$$

Summary of Design Results for Alternate Capacity Calculations

- Classifier size based on a pool area of 1.36 m² (Table J.8).
- 600 mm (24 in.), Simplex SP, 100 MF with pool area of 1.54 m² to 2.08 m².
- For an average pool area of 1.81 m², the slurry capacity was 31.6 m³/hr. The overflow solid capacity was 2100 kg/hr at 5.5% solids. The solid feed capacity was 18,561 kg/hr.

Conclusions

- Sand compression pool area governed in the selection of the classifier.
- Classifier Size Required: 1500 mm (60 in.), Simplex 150 MF.
- The maximum flow of sand product was 28.1 m³/hr (123 gpm) at 80% solids.
- The maximum overflow rate was 328 m³/hr (1440 gpm) at 5.5% solids.
- The sand solid capacity based on the sand compression pool area was 16,400 kg/hr.
- The overflow solid capacity based on the sand compression pool area was 18,700 kg/hr.

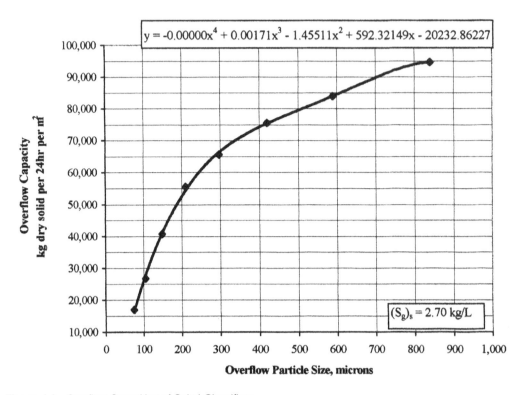

$$y = -0.00000x^4 + 0.00171x^3 - 1.45511x^2 + 592.32149x - 20232.86227$$

$(S_g)_s = 2.70 \text{ kg/L}$

Figure J.4   Overflow Capacities of Spiral Classifiers.

J.6 SECONDARY MINERAL JIG

Design Data

Design Balance Data: Table J.1	kg/hr		% Distribution	
Description of Flow	Solid	Water	Solid	Water
Feed: +105 μm Sand Fraction	14,666.0	7897.1	65.0	35.0
Feed: Dilution Water	0.0	14,688.2	0.0	100.0
Concentrate	260.2	111.5	70.0	30.0
Tailings	14,405.9	22,473.8	39.1	60.9
Totals Mass Input/Output	14,666.0	22,585.3		

Mineral Jig Specifications

- Type: Duplex
- Mass Balance Mineral Jig Feed = 14,666 kg/hr (32,339 lb/hr)
- Mass Balance Media Flow = (7.90 m³/hr + 14.67 m³/hr) = 22.57 m³/hr (99.4 gpm)

Mineral Jig Design Calculations

Reference: Svedala, 1996, pp. 4:11 to 4:13

Mineral Jig Capacity

Design Rate for Mineral Jig Feed (+105 μm) Fraction = 14,666 kg/hr
Capacity from Figure J.3

Specific Gravity of Lead Metal = 11.4 kg/L
Estimated Mineral Jig Capacity (Lower Limit) = 7500 kg/hr/m^2
Estimated Mineral Jig Capacity (Upper Limit) = 12,800 kg/hr/m^2

$$\text{Mineral Jig Bed Area (Lower Limit)} = \frac{\text{Jig Feed, kg/hr}}{\text{Capacity, kg/hr} - \text{m}^2} = \frac{14,666 \text{ kg/hr}}{7500 \text{ kg/hr} - \text{m}^2}$$

Mineral Jig Bed Area (Lower Limit) = 1.955 m^2 (21.1 ft^2)
Mineral Jig Bed Area (Upper Limit) = (14,666 kg/hr)/(12,800 kg/hr/m^2) = 1.145 m^2 (12.3 ft^2)

Summary of Design Results

- Selected Mineral Jig (Table J.7): Duplex 24 × 36
- Bed Area per Duplex Jig = 1.12 m^2 (12.0 ft^2)
- Media Flow Range per Duplex Jig = 9.0 to 14.0 m^3/hr (40 to 61 gpm)
- Number of Mineral Jigs Required (Lower Limit) = 2
- Number of Mineral Jigs Required (Upper Limit) = 1
- Total Bed Area (Lower Limit) 2 units × 1.12 m^2 = 2.24 m^2 (24.0 ft^2)
- Total Bed Area (Upper Limit) 1 unit × 1.12 m^2 = 1.12 m^2 (12.0 ft^2)
- Solid Feed Rate (Lower Limit) = 2.24 m^2 × 7500 kg/hr/m^2 = 16,800 kg/hr
- Solid Feed Rate (Upper Limit) = 1.12 m^2 × 12,800 kg/hr/m^2 = 14,350 kg/hr
- Media Flow (Lower Limit) = 22.57 m^3/hr × (16,800/14,666)/2 = 12.9 m^3/hr (56.9 gpm)
- Media Flow (Upper Limit) = 22.57 m^3/hr × (14,350/14,666)/1= 22.1 m^3/hr (97.3 gpm)
- Used one Duplex 24 × 36 jig in treatment plant with increased media flow.

J.7 DEWATERING SCREEN #2

Design Data

Design Balance Data: Table J.1

Description of Flow	kg/hr		% Distribution	
	Solid	Water	Solid	Water
Feed: Secondary Jig Tailings	14,405.9	22,473.8	70.0	30.0
Feed: Fresh Water	0.0	4084.9	0.0	100.0
Oversize: + 1.19 mm (16 mesh)	4248.1	2287.4	65.0	35.0
Undersize: −1.19 mm (16 mesh)	10,157.8	24,271.2	29.5	70.5
Totals Mass Input/Output	14,405.9	26,558.7		

Dewatering Screen Specifications

- Type – Vibrating Screen
- Screen Openings =1.19 mm (16 mesh)
- Mass Balance Wet Screen Feed = 14,406 kg/hr (31,765 lb/hr)
- Low Pressure Wash Water Requirement = 7.57 L/min per m^3 (1.53 gpm per yd^3)

Dewatering Screen Design Calculations

Reference: Weiss, 1985, pp. 3E-1 to 3E-19

Wet Screen Capacity

Bulk Density of Soil = 1780 kg/m^3 (111 lb/ft^3)
Capacity for Sand and Gravel from Figure J.1:
C (1.19 mm) = 3180 kg/hr per m^2

1.2 mm Screen Size

−1.19 mm (−16 mesh) Soil in Feed to Screen = 10,158 kg/hr (22,398 lb/hr)

$$B = \frac{1780 \text{ kg/m}^3}{1602 \text{ kg/m}^3} = 1.11$$

%O.A. = ~31% (Table J.2)

$$O = \text{Open Area Factor} = \frac{31\%}{31\%} = 1.00$$

F = 1.20 (for −0.59 mm fines at ~48%, F = 1.20, Table J.3)
E = 1.00 (% Efficiency = 95, Table J.4)
S = 1.00 (for square opening screens)
W = 2.00 (for 1.2 mm opening, W = ~2.0, Figure J.4)

$$A = \frac{10,158 \text{ kg/hr}}{3180 \text{ kg/hr/m}^2 \times 1.11 \times 1.00 \times 1.20 \times 1.00 \times 1.00 \times 2.00 \times 1.00}$$

A = 1.20 m² (12.9 ft²)

Summary of Design Results

- Vibrating Screen Size: Width = 0.91 m (3 ft); Length = 2.43 m (8 ft) with square screen opening of 1.19 mm (Table J.6)
- Single Deck Area = 1.90 m² (20.5 ft²)
- Bulk Density Factor B = 1.11, Fines Factor F = 1.2, and Wet Screen Factor F = 2.0
- Maximum Solid Capacity = (1.90 m²)(3180 kg/hr-m²)(1.11)(1.2)(2.0) = 16,100 kg/hr

J.8 PRIMARY MD LG7 SPIRAL CONCENTRATOR

Design Data

Design Balance Data: Table J.1

Description of Flow	kg/hr		% Distribution	
	Solid	Water	Solid	Water
Feed: −1.19 mm; +105 μm	10,157.8	24,271.2	29.5	29.5
Feed: Primary Middlings	1487.6	2231.4	40.0	60.0
Feed: Secondary Tailings	736.4	3557.0	17.1	82.9
Feed: Dilution Water Tailings	0.0	0.0	0.0	100.0
−1.19 mm Tailings (light fraction)	9406.6	26,341.0	26.3	73.7
−1.19 mm Middlings	1487.6	2231.4	40.0	60.0
−1.19 mm Concentrate (heavy fraction)	1487.6	1487.6	50.0	50.0
Totals Mass Input/Output	12,381.8	30,060.0		

Spiral Concentrator Specifications

Reference: LG7 Spiral Concentrator Brochure, MD Mineral Technologies, Orange Park, FL.

Optimum Operational Requirements

- Model: MD LG7 Spiral Concentrator
  - Single Start – 1 spiral per column
  - Twin Start – 2 spirals per column
  - Triple Start – 3 spirals per column

- Feed Particle Size Range: 37 μm (400 mesh) to 2.0 mm (10 mesh)
- Slurry % Solids (w/w): 20% and 40% (good operation between 15% and 60%)
- No wash water required
- Head Feed (Per Start)
  - Capacity: 500 kg/hr to 2000 kg solid per hour
  - Pulp Volume: 3.4 m^3/hr to 5.0 m^3/hr (15.4 gpm to 22.0 gpm)
- Concentrate Removal (Per Start)
  - Rate: up to 300 kg solid per hour
  - Pulp Density (w/w): 30% to 60% solids

Spiral Concentrator Design Calculations

Reference: Weiss, 1985, pp. 3D-46 to 3D-52

Full-Scale Operational Requirements

Sand: F = Total Spiral Solid Feed = 12,382 kg/hr (27,302 lb/hr)
Slurry: F = Total Spiral Slurry Feed = 42,442 kg/hr (93,585 lb/hr)
%Solids: F = % Solids of Total Feed = 29.17%
Total Concentrate = 1488 kg/hr (3273 lb/hr)
%SF: V = %Solids of Total Feed (by Volume)

$$\%SF{:}V = \frac{100}{\left[1 + 2.6 \times \frac{(100 - \%Solids{:}F)}{\%Solids{:}F}\right]}$$

$$\%SF{:}V = \frac{100}{\left[1 + 2.6 \times \frac{(100 - 29.17)}{29.17}\right]} = 13.67\%$$

Slurry Volumetric Rate = SVR

$$SVR = \left[\frac{Sand{:}F}{(S_g)_s} + \frac{(Sand{:}F)(100\% - \%Solids{:}F)}{\%Solids{:}F}\right]$$

$$SVR = \left[\frac{12,382 \text{ kg/hr}}{2.60 \text{ kg/L}} + \frac{(12,382 \text{ kg/hr})(100\% - 29.17\%)}{29.17\%}\right] = 34,828 \text{ L/hr} = 34.8 \text{ m}^3/\text{hr} \ (153 \text{ gpm})$$

Nominal SVR per Spiral (Triple Start) = (3)(4.2 m^3/hr) = 12.6 m^3/hr (66 gpm)
Maximum Concentrate per Spiral (Triple Start) = (3)(300 kg/hr) = 900 kg/hr

Summary of Design Results

- Model: MD LG7, Triple Start
- Number of Spiral Concentrator Units Required = 3
- SVR (Triple Start) = (3 units)(12.6 m^3/hr) = 37.8 m^3/hr (132 gpm)
- Concentrate (Triple Start) = (3 units)(900 kg/hr) = 2700 kg/hr
- The three spiral concentrators were acceptable for a soil feed rate of 13.91 t/hr (15.3 tons/hr) at 30% solids.

LG7 Spiral Concentrator

Manufacturer Contact

- MD Mineral Technologies, Inc., 1857 Wells Street, Orange Park, FL 32073
    Phone: 904-215-9006
    Web Site: www.mdmintec.com.au
    E-mail: mdmtusa@aol.com

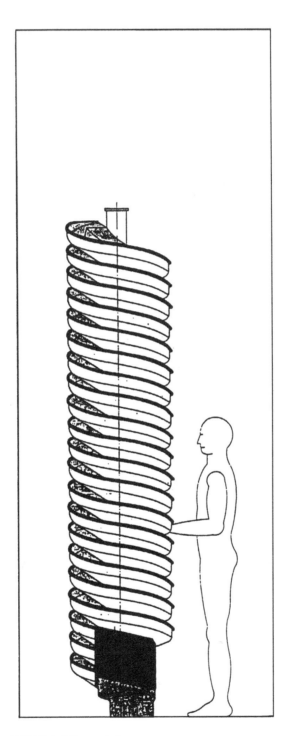

Figure J.4A Schematic of LG7 Spiral Concentrator

MDLG7 Spiral Concentrator

The LG7 spiral concentrator is comprised of one or multiple spiral troughs, center column, feed box, and product box.

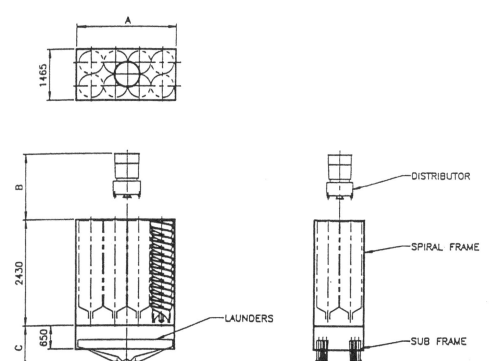

Figure J.4B LG7 Spiral Concentrator: Sprial Banks, Double Row.

Starts

The LG7 spiral concentrator can be supplied in three forms:

- Single Start – 1 spiral trough per column.
- Double Start – 2 spiral troughs per column.
- Triple Start – 3 spiral troughs per column.

Turns

Each trough has six turns, beginning at the feed box and ending at the product box.

Feed Box

The entry point at the top of the spiral for the slurry feed.

Product Box

Located at the base of the spiral to divide the products generated by the product splitters.

Product Splitters

Device located in the product box to separate the mineral flows into four products referred to as concentrate, middlings, high density tailings, and low density tailings.

Spiral Banks

The LG7 spirals are assembled into banks of spirals. Banks are supplied with feed distributor, spiral sub-frame, and product launders.

Design Data

- Simple operation with one set of splitters per spiral. Multiple starts have ganged splitters.
- Product outlets are also ganged for ease of operation.
- No wash water required.
- The LG7 spiral is constructed in strong lightweight fiberglass with polyurethane covering and castings for corrosion and abrasion resistance.
- Capacity (Per Start): up to 2000 kg/hr (2.21 t/hr) depending on application.
- Size Range: 30 μm (~400 mesh) to 2 mm (10 mesh).
- Pulp Density (w/w): up to 60% solids.
- Pulp Volume (maximum): 5.0 m³/hr.
- Concentrate Removal (Per Start):
- Rate: up to 0.30 t/hr.
- Pulp Density (w/w): 30 to 60% solids.

Number of Spirals	Number of Starts	Dimensions (mm)			Installed Mass[1] (tonnes)	
		A	B	C	Top Fed	Bottom Fed
2 TWIN	4	795	1450		0.32	
4 TWIN	8	1465	1550	1065	0.67	0.70
6 TWIN	12	2135	1550	1190	0.90	0.96
8 TWIN	16	2805	1716	1190	1.20	1.25
10 TWIN	20	3475	1816	1370	1.55	1.60
12 TWIN	24	4280	1816	1640	1.80	
12 TWIN	24	4570	1816	1640		1.92

1. For Operating Mass: add 10 kg to 20 kg per start per 1000 kg/hr of feed slurry.

J.9 SECONDARY MD LG7 SPIRAL CONCENTRATOR

Design Data

Design Balance Data: Table J.1	kg/hr		% Distribution	
Description of Flow	Solid	Water	Solid	Water
Feed: Primary Spiral Concentrate	1487.6	1487.6	50.0	50.0
Feed: Dilution Water Concentrate	0.0	2124.0	0.0	100.0
Feed: Secondary Middlings	751.3	1127.0	40.0	60.0
Feed: Dilution Water Middlings	0.0	697.0	0.0	100.0
Secondary Tailings (light fraction)	736.4	3557.3	17.1	82.9
Secondary Middlings	751.3	1126.9	40.0	60.0
Secondary Concentrate (heavy fraction)	751.3	751.3	50.0	50.0
Totals Mass Input/Output	2238.9	5435.5		

Spiral Concentrator Specifications

Reference: LG7 Spiral Concentrator Brochure, MD Mineral Technologies, Orange Park, FL

Optimum Operational Requirements

- Model: MD LG7 Spiral Concentrator
  - Single Start – 1 spiral per column
  - Twin Start – 2 spirals per column
  - Triple Start – 3 spirals per column
- Feed Particle Size Range: 37 μm (400 mesh) to 2.0 mm (10 mesh)
- Slurry %Solids (w/w): 20% and 40% (good operation between 15% and 60%)
- No Wash Water Required
- Head Feed (Per Start)
  - Capacity: 500 kg/hr to 2000 kg solid per hour
  - Pulp Volume: 3.4 m³/hr to 5.0 m³/hr (15.4 gpm to 22.0 gpm)
- Concentrate Removal (Per Start)
  - Rate: up to 300 kg solid per hour
  - Pulp Density (w/w): 30% to 60% solids

Spiral Concentrator Design Calculations

Reference: Weiss, 1985, pp. 3D-46 to 3D-52

Full-Scale Operational Requirements

Sand: F = Total Spiral Solid Feed = 2239 kg/hr (4937 lb/hr)
Slurry: F = Total Spiral Slurry Feed = 7674 kg/hr (16,921 lb/hr)
%Solids: F = % Solids of Total Feed = 29.17%
Total Concentrate = 752 kg/hr (1657 lb/hr)
%SF: V = % Solids of Total Feed (by Volume)

$$\%SF: V = \frac{100}{\left[1 + 2.6 \times \dfrac{(100 - \%Solids: F)}{\%Solids: F}\right]}$$

$$\%SF: V = \frac{100}{\left[1 + 2.6 \times \dfrac{(100 - 29.17)}{29.17}\right]} = 13.67\%$$

Slurry Volumetric Rate = SVR

$$SVR = \left[\frac{Sand: F}{(S_g)_s} + \frac{(Sand: F)(100\% - \%Solids: F)}{\%Solids: F}\right]$$

$$SVR = \left[\frac{2239 \text{ kg/hr}}{2.60 \text{ kg/L}} + \frac{(2239 \text{ kg/hr})(100\% - 29.17\%)}{29.17\%}\right] = 6298 \text{ L/hr} = 6.30 \text{ m}^3/\text{hr} (27.7 \text{ gpm})$$

Nominal SVR per Spiral (Triple Start) = (3)(4.2 m³/hr) = 12.6 m³/hr (55.5 gpm)
Optimum Concentrate per Spiral (Triple Start) = (3)(300 kg/hr) = 900 kg/hr

Summary of Design Results

- Model: MD LG7, Triple Start
- Number of Spiral Concentrator Units Required = 1
- SVR (Triple Start) = (1 unit)(12.6 m³/hr) = 12.6 m³/hr (46 gpm)
- Concentrate (Triple Start) = (1 unit)(900 kg/hr) = 900 kg/hr
- The one spiral concentrator was acceptable for a soil feed rate of 4.64 t/hr (5.11 tons/hr) at 30% Solids.

J.10 DEWATERING SCREEN #3

Design Data

Design Balance Data: Table J.1

Description of Flow	kg/hr		% Distribution	
	Solid	Water	Solid	Water
Feed: Secondary Spiral Concentrate	751.3	751.3	50.0	50.0
Undersize: Water	0.0	618.7	0.0	100.0
Oversize: Concentrate	751.3	132.6	85.0	15.0
Totals Mass Input/Output	751.3	751.3		

Dewatering Screen Specifications

- Type: Rapped Sieve Bend Screen
- Mass Balance Sieve Bend Screen Feed = 751 g/hr (1657 lb/hr)
- Low Pressure Wash Water Requirement = 0.00 L/min per m^3 (0.00 gpm per yd^3)
- Mass Balance Solids Content of Feed Slurry = 50.0%
- Required Bar Slot (Screen) Opening =105 μm (150 mesh)
- Required Feed Velocity = 3.05 m/s (10 ft/s)
- Required Feed Fall Height = 102 cm (40 in.)

Dewatering Screen Design Calculations

Reference: Weiss, 1985, pp. 3E-20 to 3E-25

Non-rapped Sieve Bend Screen Capacity

Bulk Density of Soil = 1780 kg/m³ (111 lb/ft³)

Slot Opening (Bar Spacing) = 105 μm (150 mesh) = 0.105 mm

The diameter of separation (DOS) is the diameter of particles reporting 95% (or 50%) to the oversize fraction (Figure J.5):

Percentage Particles in Oversize at 95%: Diameter of Particles Oversize:

DOS at 95% = 824 × (Slot Opening, mm)$^{0.843}$ = 123 μm

Percentage Particles in Oversize at 50%: Diameter of Particles Oversize:

DOS at 50% = 521 × (Slot Opening, mm)$^{0.705}$ = 106 μm

Capacity (C) from Figure J.6:

Length of Sieve Bend Screen Surface, mm = L

L = 800 mm: C = (108)(Slot Opening, mm) = 11.4 m³/hr per m (15.3 gpm per ft)

L = 1600 mm: C = (138)(Slot Opening, mm) = 14.5 m³/hr per m (19.4 gpm per ft)

−105 μm (−150 mesh) Soil in Feed to Screen = 0.0 kg/hr (0.0 lb/hr)

Water in Feed Passing Through Screen = 619 kg/hr = 619 L/hr (2.72 gpm)

$$\text{Minimum Width} = \frac{\dfrac{619 \text{ L/hr}}{1000 \text{ L/m}^3}}{11.4 \text{ m}^3/\text{hr per m}} = 0.0543 \text{ m or } 54.3 \text{ mm } (0.178 \text{ ft})$$

Radius of Sieve Bend Screen Surface, mm = R

Length of Sieve Bend Screen Surface, mm = L

Capacity of Non-Rapped Sieve Bend Screen (Table J.9) with L = 760 mm, R = 915 mm, Width = 610 mm, and Slot Opening = 0.105 mm:

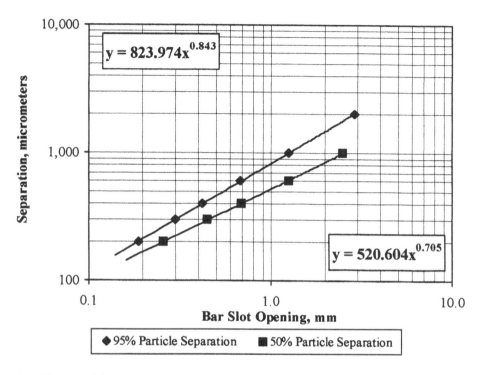

Figure J.5   Diameter of Separation as a Function of Bar Slot Opening.

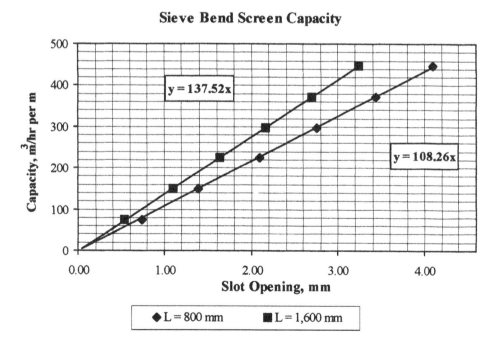

Figure J.6   Sieve Bend Screen Volumetric Capacity.

Table J.9 Description of Sieve Bend Screens

Type Sieve Bend Screen	Screen Surface		Available Widths mm	Slot Opening mm	Maximum Capacity (m³/hr)/m	Required Dimensions	
	R mm	L mm				Height m	Floor Area m²/m
Gravity Feed, Fixed Surface, 50°	915 (36 in.)	760 (30 in.)	610 1220 1830	0.35 mm to 3.50 mm	260	1.829	6.4 10.4 14.9
Gravity Feed, Fixed Surface, 45°	2030 (80 in.)	1530 (60 in.)	610 1220 1830	0.35 mm to 3.50 mm	260	2.438	7.3 12.2 17.1
Gravity Feed, Reversible Surface, 60°	760 (30 in.)	760 (30 in.)	1220	0.35 mm to 3.50 mm	150	2.438	14.6
Pressure Feed, Reversible, 270°	510 (20 in.)	2390 (94 in.)	460	0.20 mm to 0.60 mm	225	2.134	21.9
Gravity Feed, Fixed Surface plus Rapped, 45°	1625 (64 in.)	2390 (94 in.)	610	0.07 mm to 0.30 mm	110	2.743	8.2

R – Radius of Sieve-bend Screen Surface, mm

L – Length of Sieve-bend Screen Surface, mm

$$\text{Capacity} = \frac{(\text{Available Width, mm})}{(\text{Minimum Width, mm})} \times (\text{ScreenUndersize, L/hr})$$

$$\text{Capacity} = \frac{(610 \text{ mm})}{(54.3 \text{ mm})} \times (619 \text{ L/hr}) = 6954 \text{ L/hr} = 6.95 \text{ m}^3/\text{hr} \ (30.6 \text{ gpm})$$

Summary of Design Results

	Unit Size	Number of Units	Screen Capacity m³/hr (gpm)
Design Data	Non-Rapped Sieve Bend Screen R = 915 mm, L = 760 mm	1	7.0 (30)
Design Data	Rapped Sieve Bend Screen R = 1625 mm, L = 2390 mm	1	45 (195)

- Minimum Non-Rapped Sieve Bend Screen Size (Table J.9): Gravity Feed, Fixed Surface, 50°, 915 mm Surface Radius, 760 mm Surface Length, 610 mm Width
- The maximum capacity of the 915 mm by 760 mm sieve bend screen with a bar spacing of 3.50 mm should be about 260 m³/hr/m (350 gpm/ft). With a screen width of 610 mm, the maximum capacity would be about 160 m³/hr (~700 gpm). However, the capacity of the sieve-bend screen will be noticeably affected by the lower bar spacing of 0.105 mm.
- Design of Optimum Rapped Sieve bend Screen (Table J.10): Gravity Feed, Fixed Surface plus Rapped, 45°, 1625 mm Surface Radius, 2390 mm Surface Length, 610 mm Width. With a bar spacing of 0.300 mm, the maximum screen capacity should be about 68 m³/hr (300 gpm). With a bar spacing of 0.100 mm, the capacity should decrease to about 45 m³/hr (195 gpm).

Table J.10 Capacity of Rapped Sieve Bend Screen for Separations Finer than 300 μm

Slot Opening μm	Screen Bar Widths mm	Diameter of Separation μm	Screen Capacity m³/hr	gpm
100	1.5	95	44.3	195
150	1.5	135	52.2	230
200	1.5	190	60.2	265

Feed Falling Height of 102 cm (40 in.)

J.11 HUMATES SCREEN

Design Data

Design Balance Data: Table J.1

Description of Flow	kg/hr Solid	Water	% Distribution Solid	Water
Feed: Spiral Classifier Slime Fraction	1575.2	27,252.2	5.5	94.5
Feed: High Pressure Wash Water	0.0	875.6	0.0	100.0
Oversize: +2 mm Fraction (Humates)	31.5	31.5	50.0	50.0
Undersize: −2 mm Fraction	1543.7	28,096.3	5.2	94.8
Totals Mass Input/Output	1575.2	28,127.8		

Humates Screen Specifications

- Type – Vibrating Screen
- Screen Opening =2.00 mm (10 mesh)
- Mass Balance Wet Screen Feed = 1575 kg/hr (3473 lb/hr)
- High Pressure Wash Water Requirement =14.84 L/min per m³ (3.0 gpm per yd³)

Humates Screen Design Calculations

Reference: Weiss, 1985, pp. 3E-20 to 3E-25

Humates Screen Capacity

Bulk Density of Soil = 1780 kg/m³ (111 lb/ft³)
Capacity for Crushed Stone from Figure J.1:
Screen Opening =2.00 mm (10 mesh)
C (2.00 mm) = 4440 kg/hr per m²

2.0 mm (10 mesh) Screen Size

−2.00 mm (−10 mesh) Soil in Feed to Screen = 1575 kg/hr

$$B = \frac{1780 \text{ kg/m}^3}{1602 \text{ kg/m}^3} = 1.11$$

%O.A. = 33% (Table J.2)

$$O = \text{Open Area Factor} = \frac{33\%}{33\%} = 1.0$$

F = 1.00 (for −1.0 mm fines at ~65%, F = 1.60, Table J.3; used a conservative value of 1.0)
E = 1.00 (% Efficiency = 95%, Table J.3)
S = 1.00 (for square opening screens)

W = 1.00 (for 2.0 mm opening, W = ~3.0, Table J.4; used a conservative value of 1.00 assuming dry screening)

$$A = \frac{1575 \text{ kg/hr}}{4440 \text{ kg/hr/m}^2 \times 1.11 \times 1.00 \times 1.00 \times 1.00 \times 1.00 \times 1.00 \times 1.00}$$

A = 0.320 m² (3.44 ft²)

Summary of Design Results

- Vibrating Screen Size: Width = 0.91 m (3 ft); Length = 1.83 m (6 ft) with square screen opening of 2.00 mm (Table J.6)
- Single Deck Area = 1.44 m² (15.5 ft²)
- Bulk Density Factor = 1.11, Fines Factor F = 1.0, and Wet Screen Factor F = 1.0
- Maximum Solid Capacity = (1.44 m²)(4440 kg/hr-m²)(1.11)(1.0)(1.0) = 7097 kg/hr

J.12 CLARIFIER

Design Data

Design Balance Data: Table J.1

Description of Flow	kg/hr		% Distribution	
	Solid	Water	Solid	Water
Feed: −2.38 mm Fraction	34.5	1621.6	2.1	97.9
Feed: −2.00 mm Fraction	1543.7	28,096.3	5.2	94.8
Feed: −1.19 mm Fraction	9406.6	26,959.7	25.9	74.1
Feed: Polymer	0.0	0.0	0.0	100.0
Overflow: Clean Water	0.0	12,738.5	0.0	100.0
Underflow: −2.38 mm Fraction	10,984.8	43,939.0	20.0	80.0
Totals Mass Input/Output	10,984.8	56,677.5		

Clarifier Specifications

- Type – Lamella Clarifier
- Mass Balance Total Feed: C = 10,985 kg/hr (24,221 lb/hr) =10.99 t/hr
- Mass Balance Average Feed: %Solids = 16.23%

Clarifier Design Calculations

Reference: Svedala, 1996, pp. 6:5 to 6:10 and 6:25 to 6:36
Bulk Density of Soil = 1780 kg/m³ (111 lb/ft³)

$$\text{Feed: V} = \frac{(\text{Feed: C})}{(S_g)_s} + \frac{(\text{Feed: V})(100\% - \text{Feed: \%S})}{(\text{Feed: \%S})}$$

$$\text{Feed: V} = \frac{10,985 \text{ kg/hr}}{2.60 \text{ kg/hr}} + \frac{(10,985 \text{ kg/hr})(100\% - 16.23\%)}{(16.23\%)} = 60,923 \text{ L/hr } (60.9 \text{ m}^3\text{/hr})$$

$$\text{Feed V: \%S} = \frac{100}{1 + (S_g)_s \times \dfrac{(100\% - \text{Feed: \%S})}{(\text{Feed: \%S})}}$$

$$\text{Feed V: \%S} = \frac{100}{1 + (2.60 \text{ kg/L}) \times \dfrac{(100\% - 16.23\%)}{(16.23\%)}} = 6.93\%$$

Clarifier Design Data from Table J.11:
Use Flue Dust, Blast Furnace as Basis for Design

Surface Load = 1.5 to 3.7 m³ /hr per m²

Thickening Unit Area = 50 m² per (t/hr)

$$\text{Clarification Area} = \frac{\text{Feed: V}}{\text{Surface Load}} = \frac{60.9 \text{ m}^3/\text{hr}}{1.5 \text{ m}^3/\text{hr} \bullet \text{m}^2} = 40.6 \text{ m}^2$$

Thickening Area = (Feed: C, t/hr)(Unit Area, m²/t/hr) = (10.99 t/hr) × (50 m²/t/hr) = 550 m²

Table J.11 Surface Loads for Clarification

Material	Feed %Solids	Surface Load (m³/hr) / m²	Unit Area m²/ t/hr	Underflow %Solids	Comments
Sand Wash Water	1 to 5	1.0 to 4.0			With Flocculation
Ore Flotation Tailings	10 to 20	0.5 to 1.5			With Flocculation
Flue Dust, BOF	0.2 to 2.0	1.0 to 1.7	50	30 to 70	
Flue Dust, Blast Furnace	0.2 to 2.0	1.5 to 3.7	50	40 to 60	
Coal Refuse	0.5 to 6.0	0.7 to 1.7			

Summary of Design Results

- Clarifier Size: Lamella LTS, Model 100, Clarification Area at 80 m² and Thickening Area at 20 m² (Table J.12).
- The clarifier was acceptable for clarification but not for thickening of the settled slurry.
- The maximum clarifier capacity for clarification was 21.65 t/hr at 16.2% solids.

Table J.12 Lamella Clarifier Selection Data

Svedala Lamella Type	Model	Maximum Clarifying Area		Maximum Thickening Area	
		m²	ft²	m²	ft²
LT (S) (K)	15	12	129	3	32
	30	24	258	6	65
	50	40	430	10	108
	100	80	860	20	215
	200	160	1720	40	430
	350	280	3010	70	752
	500	400	4300	100	1075

J.13 CENTRIFUGE

Design Data

Design Balance Data: Table J.1	kg/hr		% Distribution	
Description of Flow	Solid	Water	Solid	Water
Feed: –2.38 mm Fraction	10,984.8	43,939.0	20.0	80.0
Overflow: Clean Water	0.0	36,315.8	0.0	100.0
Underflow: –2.38 Fraction	10,984.8	7293.3	60.0	40.0
Totals Mass Input/Output	10,984.8	43,939.0		

Centrifuge Specifications

- Type: Sharples, Super-D-Decanter
- Mass Balance Total Feed: C = 10,985 kg/hr (24,221 lb/hr) =10.99 tonnes per hour (t/hr)
- Mass Balance Average Feed: %Solids = 20.0%

Centrifuge Design Calculations

Reference: Svedala, 1996, pp. 4:11 to 4:13

Bulk Density of Soil = 1780 kg/m^3 (111 lb/ft^3)

Feed: C = 10,985 kg/hr (24,221 lb/hr)

$$\text{Feed: } V = \frac{(\text{Feed: C})}{(S_g)_s} + \frac{(\text{Feed: V})(100\% - \text{Feed: }\%S)}{(\text{Feed: }\%S)}$$

$$\text{Feed: } V = \frac{10,985 \text{ kg/hr}}{2.60 \text{ kg/hr}} + \frac{(10,985 \text{ kg/hr})(100\% - 20.0\%)}{(20.0\%)} = 48,165 \text{ L/hr (48.17 m}^3\text{/hr)}$$

$$\text{Feed V: } \%S = \frac{100}{1 + (S_g)_s \times \dfrac{(100\% - \text{Feed: }\%S)}{(\text{Feed: }\%S)}}$$

$$\text{Feed V: } \%S = \frac{100}{1 + (2.60 \text{ kg/L}) \times \dfrac{(100\% - 20.0\%)}{(20.0\%)}} = 8.77\%$$

Maximum Feed Flow (@ –25 mm Soil Feed of 18.1 t/hr) = 48.2 m^3/hr

Minimum Feed Flow (@ –25 mm Soil Feed of 9.1 t/hr) = 24.1 m^3/hr

Nominal Feed Flow (@ –25 mm Soil Feed of 13.6 t/hr) = 36.1 m^3/hr

Summary of Design Results

- Centrifuge Size: Sharples, Super-D-Canter (Table J.13)
  - Slurry Feed = 48.17 m^3/hr (212 gpm)
  - W:%Solids = 20.0%
  - V:%Solids = 8.8%
  - Model: P-3400
  - Bowl Diameter (D): 36 cm
  - Bowl Length (L): 125 cm
  - Rated Capacity = 4.5 to 34 m^3/hr (20 to 150 gpm)
  - Power Requirement = 37 kW (50 Hp) at 480 volts
- Number Centrifuges Required:

$$\text{Units} = \frac{\text{Maximum Feed Flow, m}^3/\text{hr}}{\text{Maximum Capacity, m}^3/\text{hr per unit}} = \frac{48.2 \text{ m}^3/\text{hr}}{34.0 \text{ m}^3/\text{hr per unit}} = 1.4$$

Number Required = 2 units
- Quoted Centrifuge: Sharples, Super-D-Canter, Model P-3400
  - Rated Capacity = 4.5 to 34.0 m^3/hr (20 to 150 gpm)
  - Power Requirement = 37 kW (50 Hp) at 480 volts

Table J.13 Sharples Super-D-Canter Centrifuge Data

Super-D-Decanter Model	Centrifugal Force		Bowl Size		Motor kW	Volumetric Capacity	
	Max.	RPM	D cm (in.)	L cm (in.)		minimum m³/hr (gpm)	maximum m³/hr (gpm)
P–2000	3180 × G	4000			to 19		
P-3000	3180 × G	4000	36 (14)	76 (30)	to 30	2.3 (10)	11.4 (50)
P-3400	3180 × G	4000	36 (14)	125 (49)	to 37	4.5 (20)	34.0 (150)
P-35000	2510 × G	3350		125 (49)	to 37		
P-36000	2510 × G	3350			to 37		

J.14 PUMPS

Design Data for Water Pumps

Design Balance Data: Table J.1		kg/hr		% Distribution	
Flow	Description of Flow	Solid	Water	Solid	Water
31	Screen Wash: High Pressure Water	0.0	33,608.4	0.0	100.0
42	Screen Wash: High Pressure Water	0.0	875.6	0.0	100.0
Total Flow High Pressure tWater		**0.0**	**34,484.0**	**0.0**	**100.0**
32	Jig Wash: Low Pressure Water	0.0	1413.2	0.0	100.0
33	Screen Wash: Low Pressure Water	0.0	383.0	0.0	100.0
34	Jig Wash: Low Pressure Water	0.0	14,688.2	0.0	100.0
39	Screen Wash: Low Pressure Water	0.0	4084.9	0.0	100.0
35	Spiral Dilution: Low Pressure Water	0.0	0.0	0.0	100.0
36	Spiral Dilution: Low Pressure Water	0.0	2124.0	0.0	100.0
37	Spiral Dilution: Low Pressure Water	0.0	0.0	0.0	100.0
38	Spiral Dilution: Low Pressure Water	0.0	697.0	0.0	100.0
Total Flow Low Pressure Water		**0.0**	**23,390.3**	**0.0**	**100.0**

Design Data for Slurry Pumps

Design Balance Data: Table J.1		kg/hr		% Distribution	
Flow	Description of Flow	Solid	Water	Solid	Water
9	Classifier: −150 mesh Slime Fraction	1575.2	27,252.2	5.5	94.5
12	Screen #1: Underflow Water	34.5	1621.6	2.1	97.9
16 & 35	Spiral Concentrator: Primary Feed Slurry and Dilution Water Flows	10,157.8	24,271.2	29.5	70.5
19	Spiral Concentrator: Primary Middlings	1487.6	2231.4	40.0	60.0
20 & 36	Spiral Concentrator: Primary Concentrate and Dilution Water Flows	1487.6	3611.6	29.2	70.8
22 & 38	Spiral Concentrator: Secondary Middlings and Dilution Water Flows	751.3	1824.0	29.2	70.8
23 & 37	Spiral Concentrator: Secondary Tailings and Dilution Water Flows	736.4	3557.3	17.1	82.9
18 & 44	Sands Pump: Primary Spiral Tailings and Screen #3 Undersize Flows	9406.6	26,959.7	25.9	74.1
26	Centrifuge: Slurry Feed	10,984.8	43,939.0	20.0	80.0
29	Clarifier Clean Water Overflow	0.0	12,738.5	0.0	100.0
30	Centrifuge Clean Water Overflow	0.0	36,615.8	0.0	100.0

Pump Design Calculations

Reference: Svedala, 1996, pp. 1:11 to 1:21

P1: High Pressure (HP) Water Pump for Double-Deck and Humates Screens

HP Screen Wash Flows 31 & 42

SG (Water) = 1.00 kg/L

Total HP Water Flow = (34,484.0 kg/hr)(1.0 L/kg) = 34,484.0 L/hr (152 gpm)

Line Loss Factor = 1.20; Additional 20% power required for pipeline loss
Design Flow = (Pump Flow, L/hr)(Line Loss Factor)
Design Flow = (34,484 L/hr)(1.20) = 41,381 L/hr (182 gpm)
@ 550 to 690 kPa (80 to 100 psi)

$$\text{Power, kW} = \frac{(\text{Design Factor})(\text{Design Flow, m}^3/\text{hr})(\text{Head, m})(\text{SG})}{(367.6)(\text{Pump Efficiency, }\%)}$$

Pump Design Factor = 1.20; Pump Efficiency (Assumed) = 55%

$$\text{Head,m} = \frac{\text{kPa}}{9.803 \text{ kPa/m}} = \frac{690 \text{ kPa}}{9.803 \text{ kPa/m}} = 70.4 \text{ m (230 ft)}$$

$$\text{Power,kW} = \frac{(1.20)(41.38 \text{ m}^3/\text{hr})(70.4 \text{ m})(1.00 \text{ kg/L})}{(367.6)(0.55)} = 17.29 \text{ kW (23.19 Hp)}$$

P2: Low Pressure (LP) Water Pump for Jigs, Screens, and Spiral Concentrator

LP Water Flows 32, 33, 34, 35, 36, 37, 38, and 39:
SG (Water) = 1.00 kg/L
Total LP Water Flow = (23,390 kg/hr)(1.0 L/kg) = 23,390 L/hr (103 gpm)
Line Loss Factor = 1.30; Additional 20% power required for pipeline loss
Design Flow = (23,390 L/hr)(1.30) = 30,407 L/hr (134 gpm) @ 209 kPa (30 psi)

$$\text{Head,m} = \frac{209 \text{ kPa}}{9.803 \text{ kPa/m}} = 21.3 \text{ m (70.0 ft)}$$

$$\text{Power,kW} = \frac{(1.20)(30.41 \text{ m}^3/\text{hr})(21.3 \text{ m})(1.00 \text{ kg/L})}{(367.6)(0.55)} = 3.85 \text{ kW (5.16 Hp)}$$

P3: Spiral Classifier Overflow Slurry Pump

Flow 9: Screw Classifier Overflow Solid Flow = 1575.2 kg/hr
%Solids = 5.5%

$$\text{Pump Flow} = \frac{\text{MassFlow, kg/hr}}{(S_g)_s} + \frac{(\text{MassFlow, kg/hr})(100 - \%\text{Solids}\%)}{(\%\text{Solids})}$$

$$\text{Pump Flow} = \frac{1575 \text{ kg/hr}}{(2.60 \text{ kg/L})} + \frac{(1575 \text{ kg/hr})(100 - 5.5\%)}{(5.5)} = 27,667 \text{ L/hr (122 gpm)}$$

Line Loss Factor = 1.20; Additional 20% power required for pipeline loss
Design Flow = (27,667 L/hr)(1.20) = 33,201 L/hr (146 gpm) @ 209 kPa (30 psi)

$$\text{SG} = \frac{\frac{(\text{Mass Flow, kg/hr})(100\%)}{(\%\text{Solids})}}{(\text{Pulp Volume, L/hr})} = \frac{\frac{(1575 \text{ kg/hr})(100\%)}{5.5\%}}{27,667 \text{ L/hr}} = 1.04 \text{ kg/L}$$

$$\text{Power, kW} = \frac{(1.20)(33.20 \text{ m}^3/\text{hr})(21.3 \text{ m})(1.04 \text{ kg/L})}{(367.6)(0.55)} = 4.37 \text{ kW (5.86 Hp)}$$

P4: Screen #1 Underflow Pump

Flow 12: Screen Underflow Water = (1622 kg/hr)(1.0 L/kg) =1622 L/hr (7.2 gpm)
Flow 12: Screen Underflow Solid = 34.5 kg/hr
%Solids = 2.1%
Line Loss Factor = 1.20; Additional 20% power required for pipeline loss
Design Flow = (1622 L/hr)(1.20) = 1946 L/hr (8.6 gpm) @ 209 kPa (30 psi)

$$SG = \frac{\dfrac{(34.5 \text{ kg/hr})(100\%)}{2.1\%}}{1622 \text{ L/hr}} = 1.01 \text{ kg/L}$$

$$\text{Power, kW} = \frac{(1.20)(1.95 \text{ m}^3/\text{hr})(21.3 \text{ m})(1.01 \text{ kg/L})}{(367.6)(0.55)} = 0.25 \text{ kW (0.33 Hp)}$$

P5: Primary Spiral Concentrator Feed Pump

Flows 16 and 35: Spiral Concentrator Feed Solid Flow = 10,157.8 kg/hr
%Solids = 29.5%

$$\text{Pump Flow} = \frac{10{,}158 \text{ kg/hr}}{(2.60 \text{ kg/L})} + \frac{(10{,}158 \text{ kg/hr})(100-29.5\%)}{(29.5)} = 28{,}183 \text{ L/hr (124 gpm)}$$

Line Loss Factor = 1.20; Additional 20% power required for pipeline loss
Design Flow = (28,183 L/hr)(1.20) = 33,820 L/hr (149 gpm) @ 209 kPa (30 psi)

$$SG = \frac{\dfrac{(10{,}158 \text{ kg/hr})(100\%)}{29.5\%}}{28{,}183 \text{ L/hr}} = 1.22 \text{ kg/L}$$

$$\text{Power, kW} = \frac{(1.20)(33.82 \text{ m}^3/\text{hr})(21.3 \text{ m})(1.22 \text{ kg/L})}{(367.6)(0.55)} = 5.22 \text{ kW (7.00 Hp)}$$

P6: Primary Spiral Concentrator Middlings Recycle Pump

Flow 19: Spiral Concentrator Middlings Recycle Solid Flow = 1488 kg/hr
%Solids = 40.0%

$$\text{Pump Flow} = \frac{1488 \text{ kg/hr}}{(2.60 \text{ kg/L})} + \frac{(1488 \text{ kg/hr})(100-40.0\%)}{(40.0)} = 2804 \text{ L/hr (12.3 gpm)}$$

Line Loss Factor = 1.20; Additional 20% power required for pipeline loss
Design Flow = (2804 L/hr)(1.20) = 3365 L/hr (14.8 gpm) @ 209 kPa (30 psi)

$$SG = \frac{\dfrac{(1488 \text{ kg/hr})(100\%)}{40.0\%}}{2804 \text{ L/hr}} = 1.33 \text{ kg/L}$$

$$\text{Power, kW} = \frac{(1.20)(3.36 \text{ m}^3/\text{hr})(21.3 \text{ m})(1.33 \text{ kg/L})}{(367.6)(0.55)} = 0.57 \text{ kW (0.77 Hp)}$$

P7: Secondary Spiral Concentrator (Primary Concentrate) Feed Pump

Flows 20 and 36: Spiral Concentrator Feed Solid Flow = 1488 kg/hr
%Solids = 29.2%

$$\text{Pump Flow} = \frac{1488 \text{ kg/hr}}{(2.60 \text{ kg/L})} + \frac{(1488 \text{ kg/hr})(100-29.9\%)}{(29.9)} = 4180 \text{ L/hr (18.4 gpm)}$$

Line Loss Factor = 1.20; Additional 20% power required for pipeline loss
Design Flow = (4180 L/hr)(1.20) = 5016 L/hr (22.0 gpm) @ 209 kPa (30 psi)

$$SG = \frac{\dfrac{(1488 \text{ kg/hr})(100\%)}{29.9\%}}{4180 \text{ L/hr}} = 1.22 \text{ kg/L}$$

$$\text{Power, kW} = \frac{(1.20)(5.02 \text{ m}^3/\text{hr})(21.3 \text{ m})(1.22 \text{ kg/L})}{(367.6)(0.55)} = 0.78 \text{ kW (1.05 Hp)}$$

P8: Secondary Spiral Concentrator Middlings Recycle Pump

Flows 22 and 38: Spiral Concentrator Middlings Recycle Solid Flow = 751 kg/hr
%Solids = 29.2%

$$\text{Pump Flow} = \frac{751 \text{ kg/hr}}{(2.60 \text{ kg/L})} + \frac{(751 \text{ kg/hr})(100 - 29.2\%)}{(29.2)} = 2110 \text{ L/hr (9.3 gpm)}$$

Line Loss Factor = 1.20; Additional 20% power required for pipeline loss
Design Flow = (2110 L/hr)(1.20) = 2532 L/hr (11.2 gpm) @ 209 kPa (30 psi)

$$SG = \frac{\dfrac{(751 \text{ kg/hr})(100\%)}{29.2\%}}{2110 \text{ L/hr}} = 1.22 \text{ kg/L}$$

$$\text{Power, kW} = \frac{(1.20)(2.53 \text{ m}^3/\text{hr})(21.3 \text{ m})(1.22 \text{ kg/L})}{(367.6)(0.55)} = 0.39 \text{ kW (0.52 Hp)}$$

P9: Secondary Spiral Concentrator Tailings Recycle Pump

Flows 23 and 37: Spiral Concentrator Tailings Recycle Solid Flow = 736 kg/hr
%Solids = 17.1%

$$\text{Pump Flow} = \frac{736 \text{ kg/hr}}{(2.60 \text{ kg/L})} + \frac{(736 \text{ kg/hr})(100 - 17.1\%)}{(17.1)} = 3851 \text{ L/hr (17.0 gpm)}$$

Line Loss Factor = 1.20; Additional 20% power required for pipeline loss
Design Flow = (3851 L/hr)(1.20) = 4621 L/hr (20.3 gpm) @ 209 kPa (30 psi)

$$SG = \frac{\dfrac{(736 \text{ kg/hr})(100\%)}{17.1\%}}{3851 \text{ L/hr}} = 1.12 \text{ kg/L}$$

$$\text{Power, kW} = \frac{(1.20)(4.62 \text{ m}^3/\text{hr})(21.3 \text{ m})(1.12 \text{ kg/L})}{(367.6)(0.55)} = 0.65 \text{ kW (0.88 Hp)}$$

P10: Sands (Primary Tailings) Pump

Flows 18 and 44: Spiral Tailings and Screen Underflow Solid Flows = 9407 kg/hr
%Solids = 25.9%

$$\text{Pump Flow} = \frac{9407 \text{ kg/hr}}{(2.60 \text{ kg/L})} + \frac{(9407 \text{ kg/hr})(100 - 25.9\%)}{(25.9)} = 30,532 \text{ L/hr (134.4 gpm)}$$

Line Loss Factor = 1.20; Additional 20% power required for pipeline loss
Design Flow = (30,532 L/hr)(1.20) = 36,638 L/hr (161 gpm) @ 209 kPa (30 psi)

$$SG = \frac{\dfrac{(9407 \text{ kg/hr})(100\%)}{25.9\%}}{30,532 \text{ L/hr}} = 1.19 \text{ kg/L}$$

$$\text{Power, kW} = \frac{(1.20)(36.64 \text{ m}^3/\text{hr})(21.3 \text{ m})(1.19 \text{ kg/L})}{(367.6)(0.55)} = 5.52 \text{ kW (7.40 Hp)}$$

P11: Centrifuge Feed Pump

Flow 26: Lamella Clarifier Underflow Solid Flow = 10,984 kg/hr
%Solids = 20.0%

$$\text{Pump Flow} = \frac{10,984 \text{ kg/hr}}{(2.60 \text{ kg/L})} + \frac{(10,984 \text{ kg/hr})(100 - 20.0\%)}{(20.0)} = 48,161 \text{ L/hr (212 gpm)}$$

Line Loss Factor = 1.20; Additional 20% power required for pipeline loss

Design Flow = (48,161 L/hr)(1.20) = 57,792 L/hr (255 gpm) @ 209 kPa (30 psi)

$$SG = \frac{\dfrac{(10,984 \text{ kg/hr})(100\%)}{20.0\%}}{48,161 \text{ L/hr}} = 1.14 \text{ kg/L}$$

$$\text{Power, kW} = \frac{(1.20)(57.92 \text{ m}^3/\text{hr})(21.3 \text{ m})(1.14 \text{ kg/L})}{(367.6)(0.55)} = 8.35 \text{ kW (11.19 Hp)}$$

P12: Clarifier Overflow Pump

Flow 29: Clarifier Clean Water Overflow = 12,738 kg/hr

SG (Water) = 1.00 kg/L

Overflow Water = (12,738 kg/hr)(1.0 L/kg) = 12,738 L/hr (56.1 gpm)

%Solids = 0.0%

Line Loss Factor = 1.20; Additional 20% power required for pipeline loss

Clarifier Pump: (12,738 L/hr)(1.20) = 15,286 L/hr (67.3 gpm) @ 209 kPa (30 psi)

$$\text{Power, kW} = \frac{(1.20)(15.29 \text{ m}^3/\text{hr})(21.3 \text{ m})(1.00 \text{ kg/L})}{(367.6)(0.55)} = 1.93 \text{ kW (2.59 Hp)}$$

Pump Selection

P1: High Pressure (HP) Water Pump for Double-Deck and Humates Screens

Q = 690 L/min (182 gpm)

Head = 550 to 690 kPa (184 ft to 230 ft)

Required Power = 17.29 kW (23.19 Hp)

18.5 kW; 3 × 4 × 6-3/4 in.; 3500 rpm; Goulds Model 3656, 16 AI/BF

P2: Low Pressure (LP) Water Pump

Q = 507 L/min (134 gpm)

Head = 209 kPa (21.3 m or 70 ft)

Required Power = 3.85 kW (5.16 Hp)

3.75 kW; 2-1/2 × 3 × 8-5/8 in.; 1750 rpm; Goulds Model 3656, 11 AI/BF

P3: Spiral Classifier Overflow Slurry Pump

Q = 554 L/min (146 gpm)

%Solids = 5.5%

Head = 209 kPa (70 ft)

Required Power = 4.37 kW (5.86 Hp)

7.50 kW; 2-1/2 × 2 × 10 in.; 1600 rpm; Denver Model SRL Pump

P4: Screen #1 Underflow Pump

Q = 33 L/min (8.6 gpm)

%Solids = 2.1%

Head = 209 kPa (70 ft)
Required Power = 0.25 kW (0.33 Hp)
1.50 kW; 1-1/2 × 1-1/4 × 10 in.; 1800 rpm; Denver Model SRL Pump

P5: Primary Spiral Concentrator Feed Pump

Q = 564 L/min (149 gpm)
%Solids = 29.5%
Head = 209 kPa (70 ft)
Required Power = 5.22 kW (7.00 Hp)
7.50 kW; 2-1/2 × 2 × 10 in.; 1800 rpm; Denver Model SRL Pump

P6: Primary Spiral Concentrator Middlings Recycle Pump

Q = 56 L/min (14.8 gpm)
%Solids = 40.0%
Head = 209 kPa (70 ft)
Required Power = 0.57 kW (0.77 Hp)
1.50 kW; 1-1/2 × 1-1/4 × 10 in.; 1600 rpm; Denver Model SRL Pump

P7: Secondary Spiral Concentrator (Primary Concentrate) Feed Pump

Q = 83 L/min (22.0 gpm)
%Solids = 29.2%
Head = 209 kPa (70 ft)
Required Power = 0.78 kW (1.05 Hp)
1.50 kW; 1-1/2 × 1-1/4 × 10 in.; 1800 rpm; Denver Model SRL Pump

P8: Secondary Spiral Concentrator Middlings Recycle Pump

Q = 42 L/min (11.2 gpm)
%Solids = 29.2%
Head = 209 kPa (70 ft)
Required Power = 0.39 kW (0.52 Hp)
1.50 kW; 1-1/2 × 1-1/4 × 10 in.; 1800 rpm; Denver Model SRL Pump

P9: Secondary Spiral Concentrator Tailings Recycle Pump

Q = 77 L/min (20.3 gpm)
%Solids = 17.1%
Head = 209 kPa (70 ft)
Required Power = 0.65 kW (0.88 Hp)
1.50 kW; 1-1/2 × 1-1/4 × 10 in.; 1800 rpm; Denver Model SRL Pump

P10: Sands (Primary Tailings) Pump

Q = 611 L/min (161 gpm)
%Solids = 25.9%
Head = 209 kPa (70 ft)
Required Power = 5.52 kW (7.39 Hp)
7.50 kW; 2-1/2 × 2 × 10 in.; 1800 rpm; Denver Model SRL Pump

P11: Centrifuge Feed Pump

Q = 963 L/min (255 gpm)
%Solids = 20.0%
Head = 206 kPa (70 ft)
Required Power = 8.35 kW (11.19 Hp)
Denver SLR Pump: 7.50 kW; 2-1/2 × 2 ×10 in.; 1800 rpm; Denver Model SRL Pump
Proposed Pump: AllWeiler Progressive Cavity Pump; Model AE1E 550 – ID/13
75 to 375 L/min; 3.8 kW; 2 to 440 rpm;

P12: Clarifier Overflow Pump

Q = 255 L/min (68 gpm)
%Solids = 0.0%
Head = 206 kPa (70 ft)
Required Power = 1.93 kW (2.59 Hp)
3.75 kW; 1-1/2 × 2-10 in.; 1750 rpm; Goulds Model 3656, 11 AI/BF

J.15 WATER AND SLURRY LINES

Line-Size Design Calculations

Water Lines	Flow Rate L/min (gpm)	Pipe Diameter cm (in.)	Velocity m/sec (ft/sec)
P1: High Pressure Feed	575 (152)	7.6 (3.0)	2.0 (6.6)
P2: Low Pressure Feed	390 (103)	7.6 (2.5)	2.1 (6.9)
P4: Screen #1 underflow	27 (7)	2.5 (1.0)	0.8 (2.6)
Primary Jig Wash	24 (7)	2.5 (1.0)	0.7 (2.3)
Secondary Jig Wash	245 (65)	3.8 (2.0)	1.9 (6.2)
Spiral Concentrator Wash	47 (13)	2.5 (1.0)	1.4 (4.6)
Screen Wash	7 (2)	2.5 (1.0)	0.2 (0.6)
Screen Wash	68 (18)	2.5 (1.0)	2.0 (6.7)
Clarifier Return Water	212 (56)	7.6 (2.0)	1.6 (5.4)

Slurry Lines	Flow Rate L/min (gpm)	Pipe Diameter cm (in.)	Velocity[1] m/sec (ft/sec)
P3: Classifier OverFlow	461 (122)	6.4 (2.5)	2.5 (8.2)
P5: Reichert Spiral Feed	470 (124)	6.4 (2.5)	2.5 (8.3)
P7: Reichert Spiral Feed	70 (18)	2.5 (1.0)	2.1 (6.8)
P6 & P8: Reichert Spiral Recycle	47 (12)	2.5 (1.0)	1.4 (4.6)
P9: Reichert Spiral Recycle	64 (17)	2.5 (1.0)	1.9 (6.3)
P10: Reichert Spiral Tailings	509 (135)	6.4 (2.5)	2.7 (9.0)
P11: Centrifuge Feed	803 (212)	7.6 (3.0)	2.8 (9.2)
P12: Clarifier OverFlow	212 (56)	3.8 (1.5)	2.7 (8.8)

1. Minimum Slurry Velocity at 1.52 m/sec (5 ft/sec)

J.16 ELECTRICAL POWER REQUIREMENTS

Soil Feed System

Apron Feeder	3.75 kW
Conveyor 1	3.75 kW
Subtotal Feed System	**7.50 kW**

High-pressure Water Pump Skid

P1: High-pressure Water Pump	18.50 kW
Subtotal HP Pump Skid	**18.50 kW**

Low-pressure Water Pump Skid

P2: Low-Pressure Water Pump	3.75 kW
Subtotal LP Pump Skid	**3.75 kW**

Skid 1

Double-Deck Wash Screen	7.50 kW
Primary Jig	2.25 kW
Dewatering Screen #1	3.75 kW
P4: Screen #1 Underflow Pump	1.50 kW
Conveyor 2	3.75 kW
Subtotal Skid 1	**18.75 kW**

Skid 2

Secondary Jig	3.75 kW
Dewatering Screen #2	3.75 kW
P5: Primary Spiral Feed Pump	7.50 kW
P6: Primary Middlings Recycle Pump	1.50 kW
P7: Secondary (Concentrate) Feed Pump	1.50 kW
P8: Secondary Middlings Recycle Pump	1.50 kW
P9: Secondary Tailings Recycle Pump	1.50 kW
P10: Sands (Primary Tailings) Pump	7.50 kW
Dewatering Screen #3	3.75 kW
Conveyor 3	3.75 kW
Subtotal Skid 2	**36.00 kW**

Skid 3

Spiral Classifier	6.00 kW
P3: Spiral Overflow Pump	7.50 kW
Subtotal Skid 3	**13.50 kW**

Dewatering Trailer (DW)

Polymer Metering Pump	0.75 kW
Clarifier	0.00 kW
Two Centrifuges (@37.5 kW)	75.00 kW
Humates Screen	3.75 kW
P11: Centrifuge Feed (Moyno) Pump	3.75 kW
P12: Clarifier Overflow Pump	3.75 kW
Conveyor 4	3.75 kW
Subtotal DW Trailer	**90.75 kW**
Total System kW	**188.75 kW**

Power Consumption

Power Requirement   **189 kW (253 Hp)**

Power Factor = 0.85
Power Line Efficiency = 95%
Line Voltage = 460 volts

$$\text{Amps} = \frac{(189{,}000 \text{ Watts})}{(1.732)(460 \text{ volts})(95\%/100\%)(0.85)} = 294 \text{ amps (Full Load)}$$

$$\text{KVA} = \frac{(1.732)(460 \text{ volts})(294 \text{ amps})}{(1000 \text{ VA/KVA})} = 234 \text{ KVA}$$

Generator

250 KVA (200 kW) Motor-Generator Set

Electric Power Available

Cost = $0.14/kw-hr
Power = 189 kW (253 Hp)

Design Figures for Full-Scale Plant

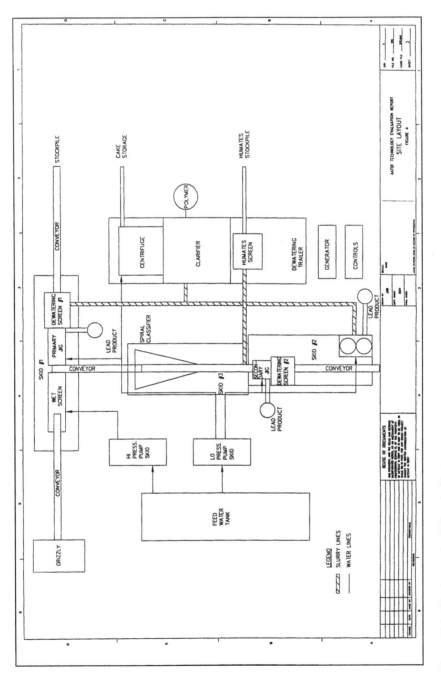

Figure K.1 Site Layout for Full-scale Plant.

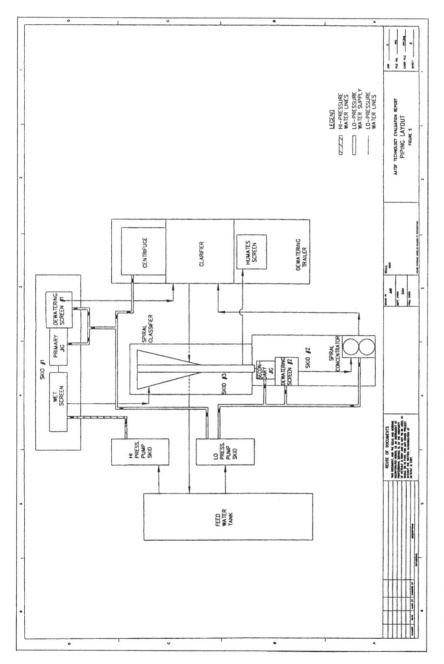

Figure K.2 Piping Layout for Full-scale Plant.

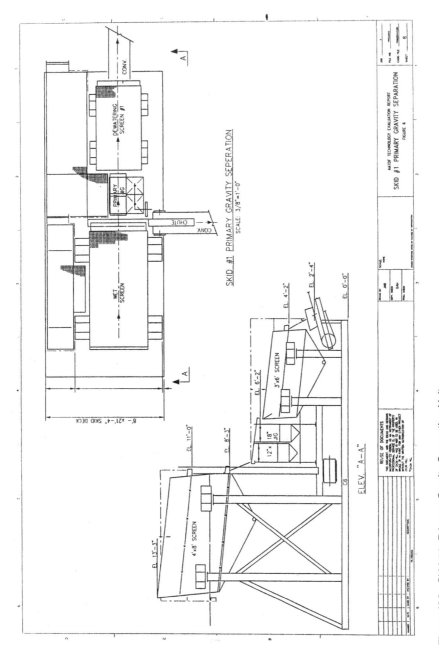

Figure K.3 Skid #1, Primary Gravity Separation Units.

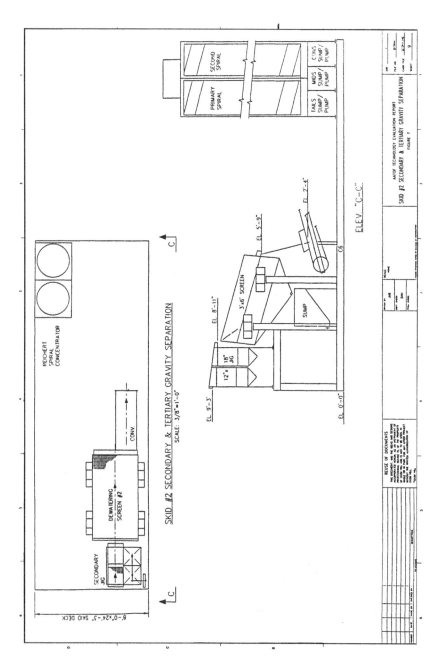

Figure K.4 Skid #2, Secondary and Tertiary Gravity Separation Units.

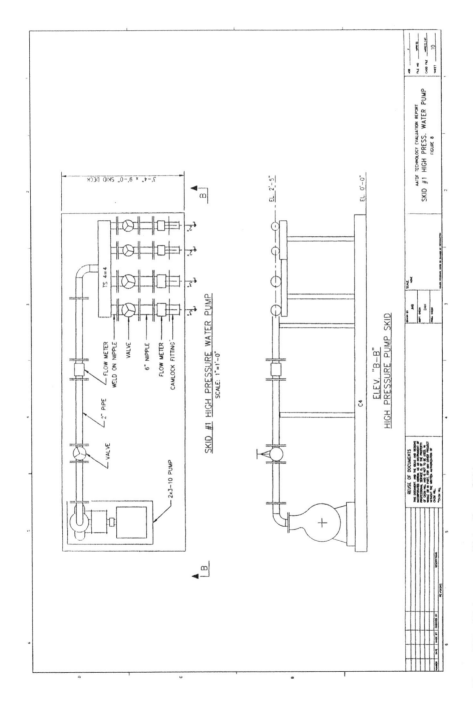

Figure K.5 High-pressure Water Pump Skid.

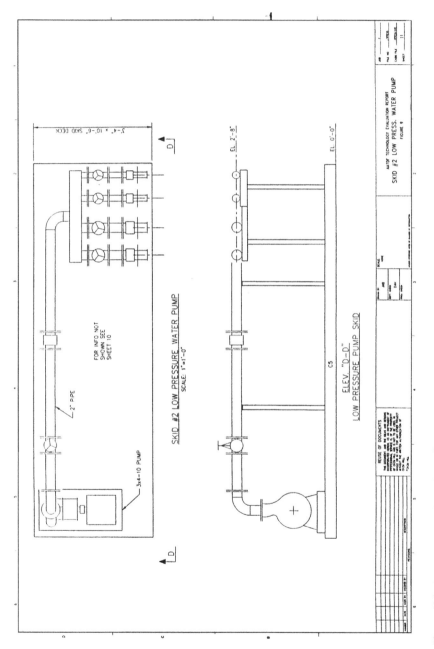

Figure K.6 Low-pressure Water Pump Skid.

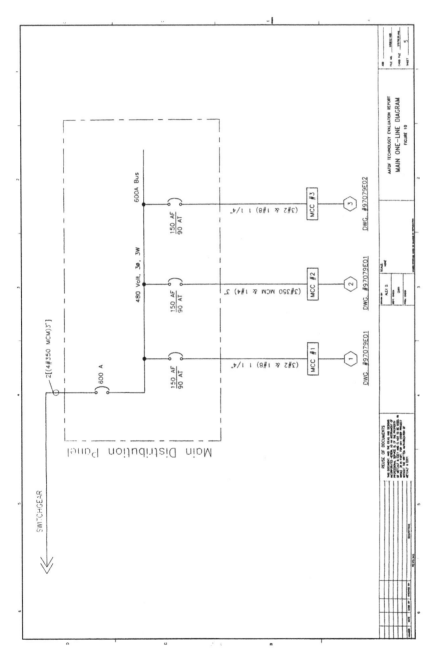

Figure K.7 Main Electrical Diagram.

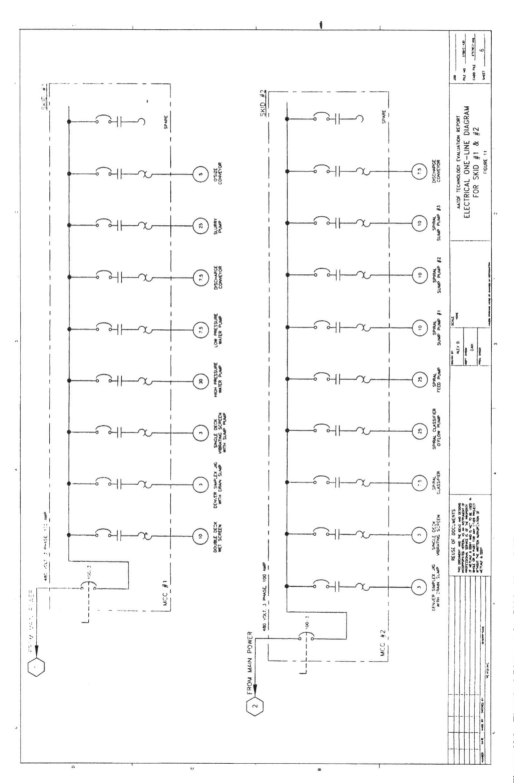

Figure K.8 Electrical Diagram for Skids #1 and #2.

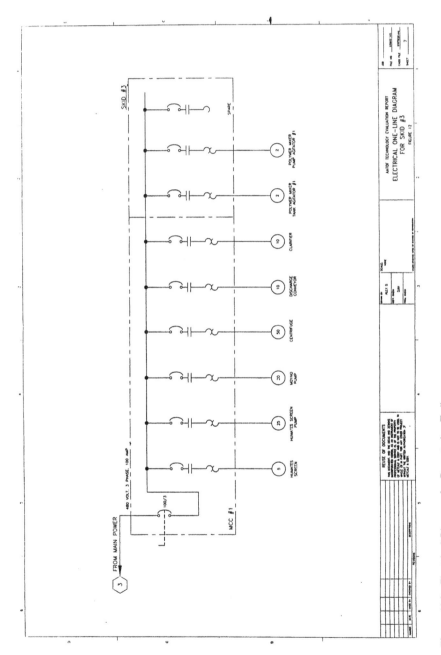

Figure K.9 Electrical Diagram for Dewatering Trailer.

Cost Analysis Backup Data

CONTENTS

L.1 PROJECT DURATIONS AND LABOR RATES

Durations

3.4.1	Mobilization and Construction	1	week, 5 days/week @ 10 hr/day	50	hr	5	days
	Site Preparation	2	weeks, 5 days/week @ 8 hr/day	80	hr	10	days
3.4.2	Treatment	10	weeks, 5 days/week @ 10 hr/day	500	hr	50	days
3.4.3	Demobilization	1	week, 5 days/week @ 10 hr/day	50	hr	5	days
		14	weeks	680	hr	70	days

Labor Rates

Office	Rate/Hr
Project Manager	$99.50
Project Engineer	$77.00
Clerical	$33.00
Field	
Site Manager	$75.00
Lead Process Operator	$62.00
Process Operator (local hire)	$54.00
Rental Equipment Operator I (local hire)	$28.00
Rental Equipment Operator II (local hire)	$27.00
Truck Driver (local hire)	$22.00

Laborer (local hire)	$56.00	
Technician	$66.00	

Other Rates

Air Fare Costs	$500.00	per round trip
Per Diem Rate	$150.00	per day

L.2 WBS 3.4.1: MOBILIZATION AND PREPARATORY WORK

L.2.1   WBS 3.4.1.1: Preconstruction Submittals and Implementation Plan

Labor		Salary		Totals
Project Manager	Regular Time	30 hours @ $99.50	$2985.00	
Project Engineer	Regular Time	40 hours @ $77.00	$3080.00	
Clerical	Regular Time	10 hours @ $33.00	$330.00	
Certified Industrial Hygenist	Regular Time	8 hours @ $99.50	$796.00	
			$7200.00	$7200.00

Treatability Study	Item		
Sampling			
Technician	8 hours @ $66.00/hr	$530.00	
Air Fare	1 @ $500.00	$500.00	
Auto Rental	1 @ $65.00	$65.00	
Sampling Equipment	1 @ $150.00	$150.00	
PPR	1 @ $600.00	$600.00	
Backhoe Rental	1 day @ $600.00	$600.00	
Laboratory Analysis	1 @ $1500.00	$1500.00	
		$3495.00	~$3500.00
Treatability Study	1 @ $12,000.00	$12,000.00	
Data Validation			
Technician	30 hours @ $66.00/hr	$1980.00	
		$13,980.00	~$14,000.00

Data Evaluation and Plant Design			
	Salary		
Project Manager	30 hours @ $99.00	$2985.00	
Project Engineer	80 hours @ $77.00	$6160.00	
Clerical	10 hours @ $33.00	$330.00	
		$9475.00	~ $9500.00

L.2.2   WBS 3.4.1.2: Mobilization of Construction Equipment and Facilities

Duration of Mobilization = 5 days (1 week)

Fuel carried under treatment

Transportation

Transport Distance = 1000 miles

Transport Cost = $1.65 per mile per truck

Number of Trucks to Transport Treatment Plant = 5

Transportation Cost = (5 trucks) (1000 miles/truck)($1.65/mile) = $8250.00

Equipment

25 ton crane, rental with operator and rigger:

2 days @ $1100/day = $2200.00

Fork lift, prorated from monthly rental @ $1150/month for 20 operating days/month:

5 days @ $57.50/day = ~$300/week

Total Equipment Cost

Crane	$2200.00
Fork Lift	$300.00
Total	$2500.00

Labor

5 days (1 week), 10 hours per day, 40 hours Regular Time, 10 hours Overtime

		Salary		Travel		Per Diem		Total
Lead Process Operator	Regular Time	40 hrs @ $62.00	$2480.00	1 @ $500.00	$500.00	5 @ $150.00	$750.00	$3730.00
	Overtime	10 hrs @ $93.00	$930.00					$930.00
Two Process Operators	Regular Time	80 hrs @ $54.00	$4320.00					$4320.00
	Overtime	20 hrs @ $81.00	$1620.00					$1620.00
Laborer	Regular Time	40 hrs @ $56.00	$2240.00					$2240.00
	Overtime	10 hrs @ $84.00	$840.00					$840.00
Total Labor Cost			**$12,430.00**		**$500.00**		**$750.00**	**$13,680.00**

Note: Process Operator & Laborer - local hires. No travel or per diem.

Supplies

Item	Description		Total
Flocculent	Assume 0.5 lbs/ton; (5800 tons)(0.5 lbs/ton) = 2900 lbs @ $1.50/lb		$4350.00
Startup Sand	30 tons @ $7.50/tons		$230.00
10 mil HDPE Liner	(100 ft)(100 ft) = 10,000 ft^2; 15 % for laps: (10,000 ft^2)(1.15) = 11,000 ft^2 @ $0.50/ ft^2		$5750.00
Hay bales	400 Lf per 3 ft coverage per Bale; ~150 bales @ $3.50/bale		$530.00
Misc. Lumber, etc.	Allow		$2500.00
Health & Safety	Tyvek Clothing	(6 person)(2 changes/day)(40 days) = 480; ~500/ (25/case) = 20 cases @ $70.00/case	$1400.00
	Rubber Boots, Steel Toe	10 pair @ 14.00	$140.00
	Gloves, Nitrile	3 boxes @ $22.00	$70.00
	Gloves, Work	1 case (144) @ $150.00	$150.00
	Ear Protection (hardhat)	8 @ $30.00	$240.00
	Earplugs, Foam	1 Box	$40.00
	Safety Glasses	3 boxes (30) @ $40.00	$120.00
	Emergency Eye Wash	1 @ $340.00	$340.00
	First Aid Kit	2 @ $40.00	$80.00
	Fire Extinguisher	4 @ $35.00	$140.00
	Miscellaneous Items	Allow	$500.00
Total Supplies			**$16,870.00**

L.2.3   WBS 3.4.1.3: Site Preparation

Duration of Site Preparation = 10 days (2 weeks)
Total amount of soil to be processed = 5800 tons
Bulk Density of Excavated Soil = 111 lbs/ft^3 (1780 kg/m^3)

EQUIPMENT

50 minutes per hour; 83% efficiency
All equipment to operate over concurrent time frame.

Excavator

Capacity Excavator = 1.5 yd^3 (1.15 m^3)
Excavator Rate = 72 tons/hr
(72 tons/hr)(8 hr/day)(5 days/week) = 2880 tons/week
(5800 tons)/(2880 tons/week) = ~2 weeks

Truck

Capacity Truck = 12.0 yd^3 (9.2 m^3)
Time per load = 15 minutes
(4 loads/hr)(18 tons/load)(8 hr/day)(5 days/week) = 2880 tons/week
(5800 tons)/(2880 tons/week) = ~2 weeks

Loader

Capacity Loader Bucket = 3.0 yd^3 (2.29 m^3), 2 weeks to trim stockpile

Fuel

Excavator	3 gallons/hour
Loader	5 gallons/hour
Truck	7 gallons/hour
Total	**15 gallons/hour**

Costs

Excavator	2 weeks @ $2100/week	$4200.00
Truck	2 weeks @ $775/week	$1550.00
Loader	2 weeks @ $1025/week	$2050.00
	(prorated from a monthly rate)	
Fuel	(15 gallons/hour)(2 weeks)(40 hr/wk)	
	= 1200 gal @ $1.25/gal	$1500.00
Total Equipment Cost		**$9300.00**

Labor

Note: All labor - local hires
2 weeks @ 5 days per week, 8 hours per day = 80 hours regular time

Equipment Operator	80 hours @ $28.00	$2240.00
Truck Driver	80 hours @ $22.00	$1760.00
Total Labor Cost		**$4000.00**

L.2.4   Recap of WBS 3.4.1: Mobilization and Preparatory Work

Recap of WBS 3.4.1.1: Preconstruction Submittals and Implementation Plan

Labor	$7200.00
Sampling & Analysis	$3500.00
Treatabilty Study & Validation	$14,000.00
Data Evaluation & Design	$9500.00
Total	**$34,200.00**

Recap of WBS 3.4.1.2: Mobilization of Construction Equipment and Facilities

Transportation	$8250.00
Equipment	$2500.00
Labor	$13,680.00
Supplies	$16,870.00
Total	**$41,300.00**

Recap of WBS 3.4.1.3: Site Preparation

Equipment	$9300.00
Labor	$4000.00
Total	**$13,300.00**

L.3 WBS 3.4.2: PHYSICAL TREATMENT

L.3.1 WBS 3.4.2.1: Equipment Costs

Soil Processed = 5.261 kt (5800 tons)
Processing Rate of Treatment Plant = 13.60 t/hr (15 tons/hr)
Processing Time = 50 hours/week (5 days/week at 10 hours/day)
Processing Efficiency = 90%

$$\text{Processing Time} = \frac{(5.261 \text{ t})(100\%)}{(13.60 \text{ tons/hour})(50 \text{ hours/week})(90\%)} = 8.6 \text{ weeks} \approx 9 \text{ weeks (450 hours)}$$

Plant Setup Time = 1 week
Plant Startup Time = 1 week
Demobilization Time = 1 week
Total Plant On-site Time = 12 weeks

Capital Equipment

$$\text{Annualized Equipment Cost} = \frac{(V - V_s)(i)(1 + i)^n}{(1 + i)^n - 1}$$

where　i = Annual Interest = 7%
　　　　n = Time of Depreciation = 3 years
　　　　V = Equipment and Fabrication Cost = $1,723,300
　　　　V_S = Salvage Cost = 10% of V = $172,300

$$\text{Annualized Equipment Cost} = \frac{(\$1,723,300 - \$172,300)(0.07)(1 + 0.07)^3}{(1 + 0.07)^3 - 1} = \$591,000 \text{ per year}$$

Equipment Cost per Year	$591,000
Contingency 10% of Equipment Cost	$59,100
Insurance 7.5% of V	$129,200
Total Annual Cost	$779,300

$$\text{Monthly Equipment Costs} = \frac{\$779,300/\text{year}}{12 \text{ months/year}} = \$64,942/\text{month}$$

Total Treatment Time = 3.0 months
Total Equipment Cost = (3.0 months)($64,950/month) = $194,825

Rental Equipment

Loader	3 months @ $4100/mo.	$12,300.00
Forklift	3 months @ $1150/mo.	$3450.00
20,000 gallon Frac tank	3 months @ $1110/mo.	$3330.00
Total Rental Equipment Cost		**$19,100.00**

Equipment Labor carried under labor.

Maintenance

Plant Startup Time = 1 week
Processing Time = 9 weeks
Total Plant Operating Time = 10 weeks (2.30 months)
Annual Maintenance Cost = 10% of Purchased Equipment Cost
Annual Maintenance Cost = (0.10)($1,006,300) = $100,600/year

$$\text{Project Maintenance} = \frac{(\$100,600/\text{year})}{(12 \text{ months/year})} (2.3 \text{ months}) = \$19,300$$

Repair Materials

Repair Material Cost = 60% of Annual Maintenance Cost
Repair Material Cost = (0.60)($100,600/year) = $60,400/year

$$\text{Project Repair Material Cost} = \frac{\$60,400/\text{year}}{12 \text{ months/year}} \times (2.30 \text{ months}) = \$11,600$$

L.3.2 WBS 3.4.2.2: Labor Costs

Labor		Salary		Travel		Per Diem		Total
Site Manager	- regular time	400 hrs @ $75.00	$30,000.00	5 @ $500.00	$2500.00	50 @ $150	$6750.00	$39,250.00
	- overtime	100 hrs @ $112.50	$11,250.00					$11,250.00
Lead Process Operator	- regular time	400 hrs @ $62.00	$24,800.00	5 @ $500.00	$2500.00	50 @ $150	$6750.00	$34,050.00
	- overtime	100 hrs @ $93.00	$9300.00					$9300.00
Two Process Operators	- regular time	800 hrs @ $54.00	$43,200.00					$43,200.00
	- overtime	200 hrs @ $81.00	$16,200.00					$16,200.00
Laborer	- regular time	400 hrs @ $56.00	$22,400.00					$22,400.00
	- overtime	100 hrs @ $84.00	$8400.00					$8400.00
Technician	- regular time	360 hrs @ $66.00	$23,760.00	4 @ $500.00	$2000.00	45 @ $150	$6000.00	$31,760.00
	- overtime	90 hrs @ $99.00	$8910.00					$8910.00
Equipment Operator II	- regular time	360 hrs @ $73.00	$26,280.00					$26,280.00
	- overtime	90 hrs @ $109.50	$9855.00					$9855.00
			$234,355.00		**$7000.00**		**$19,500.00**	**$260,855.00**

Administration		Salary		Travel	Per Diem	Ttotal
Project Manager	- regular time	20 hrs @ $99.00	$1980.00	$0.00	$0.00	$1990.00
Clerical	- regular time	10 hrs @ $33.00	$330.00	$0.00	$0.00	$330.00
			$2310.00	**$0.00**	**$0.00**	**$2310.00**

Notes: Travel: 1 round trip every 2 weeks for site manager, lead process operator, and technician

Per Diem: Only site manager, lead process operator, and technician

L.3.3   WBS 3.4.2.3: Consumables

Office Trailer Rental	3 months @ $400/month	$1200.00
Pickup Truck Rental (2)	3 months @ $800/month each	$4800.00
Phone	3 months @ $400/month	$1200.00
Fax Machine Rental	3 months @ $250/month	$750.00
Copy Machine Rental	3 months @ $250/month	$750.00
Computer Rental	3 months @ $350/month	$1050.00
Office Furniture Rental	3 months @ $400/month	$1200.00
Water – Drinking	(7 men)(5 gal/day)(84 days) = 2940 gallons Cost = $1.10/gallon	$3234.00
Noise Meter	10 weeks @ $165/week	$1650.00
Shipping/Mail	100 pieces @ $7.00	$700.00
Sample Bottles, etc.	Allow	$1000.00
Decon Equipment	Allow	$500.00
Misc. Office Supplies, etc.	3 months @ $500/month	$1500.00
Fuel		
Loader	5 gal/hr	
Forklift	1 gal/hr	
	6 gal/hr	
	(6 gal/hr)(10 hr/day)(5 day/wk)(10 wk processing) = 3000 gallon @ $1.25/gal	$3750.00
Total Consumables		$23,280.00

L.3.4   WBS 3.4.2.4: Utilities

Equipment

Water Truck	3 months @ $3600/month	$10,800.00
Generator	3 months@ $2700/month	$8100.00
		$18,900.00

Consumables

Generator Fuel	13 gal/hours × 10 hr/day × 5 day/work × 10 weeks = 6500 gallons @ $1.25 / gallon	$8100.00
		$8100.00

Process Make-up Water

Process Water	653,000 gallons @ $0.010 per gallon	$6500.00
		$6500.00

Electric Power

None		$0.00
		$0.00
Total Utilities		**$33,500.00**

L.3.5   Recap of WBS 3.4.2: Physical Treatment

3.4.2.1 Equipment Costs	$194,825.00
3.4.2.1 Rental Equipment	$19,100.00
3.4.2.1 Maintenance	$19,300.00
3.4.2.1 Repair Materials	$11,600.00
3.4.2.2 Labor	$260,900.00
3.4.2.2 Administration	$2300.00
3.4.2.3 Consumables	$23,300.00
3.4.2.4 Utilities	$33,500.00
	$564,825.00

L.4 WBS 3.4.3: DEMOBILIZATION

Duration of Demobilization = 5 days (1 week)

Transportation

Transport Distance = 1000 miles
Transportation Cost = $1.65 per mile per truck
Number of Trucks to Transport Treatment Plant = 5
Transportation Cost = (1000)(5)($1.65) = $8250.00

Labor Costs

Labor		Salary	Travel	Per diem	Total
Lead Process Operator	Regular Time	40 hrs @ $62.00	$2480.00	1 @ $500.00	5 @ $150.00	$3730.00
	Overtime	10 hrs @ $93.00	$930.00	—	—	$930.00
Process Operator	Regular Time	40 hrs @ $54.00	$2160.00	—	—	$2160.00
	Overtime	10 hrs @ $81.00	$810.00	—	—	$810.00
Laborer (2)	Regular Time	80 hrs @ $56.00	$4480.00	—	—	$4480.00
	Overtime	20 hrs @ $84.00	$1680.00	—	—	$1680.00
			$12,540.00	$500.00	$750.00	$13,790.00

Note: Process operator and laborers - local hires

Reports

Labor		Salary	Travel	Per diem	Total
Project Manager	Regular Time	30 hrs @ $99.50	$2985.00	$0.00	$0.00	$2985.00
Project Engineer	Regular Time	60 hrs @ $77.00	$4620.00	$0.00	$0.00	$4620.00
Clerical	Regular Time	20 hrs @ $33.00	$8265.00	$0.00	$0.00	$8265.00
			$15,870.00	$0.00	$0.00	$15,870.00

Equipment

25 ton crane, with operator, and Rigger	(2 days)($1100 per day)	$2200
All-Terrain forklift	($1150/month)(1 month/30 days) × 5 days	$200
		$2400

Recap of WBS 3.4.3: Demobilization Costs

Transportation	$8250.00
Labor	$13,790.00
Equipment	$2400.00
Reports	$15,870.00
Total	**$40,310.00**

L.5 VENDOR QUOTES

Rental Equipment

Item Description	Vendor	Quote
250 KVA Generator	Hertz Equipment	$2700/mo
20,000 gallon Fractionation Tank	Hertz Equipment	$1360/mo
	J.D. Cox	$1110/mo
25 ton Crane	Phillips Cranes & Rigging	$89/hr, min 4 hr
Rigger	Phillips Cranes & Rigging	$45/hr, min 4 hr
Cat 950 Wheel Loader	Hertz Equipment	$4100/mo
5000 lb all Terrain Fork Lift	Hertz Equipment	$1150/mo
4000 gallons Water Truck	Hertz Equipment	$3570/mo

Purchased Equipment

Item Description	Vendor	Design Number	Quoted Cost Each	Size Scale Factors		Design Cost Formula[1]	Design Cost Each
				(B) Design	(A) Quote		
420 Feed Hopper	Wright Equipment Company	1	$37,000			None	$37,000
530 Conveyor	Wright Equipment Company	1	$19,500			None	$19,500
TD 4 ft x 8 ft Wash Deck	Wright Equipment Company	1	$24,500	(3ft)(6 ft)	(4 ft)(8 ft)	Cost = ($24,500)(B/A)$^{0.4256}$	$19,200
18 in. x 30 ft Conveyors	M&E Quote	3	$12,000			None	$12,000
12 in. x 18 in. Denver Simplex Jig	Pumps +	1	$22,200	(16 in.)(24 in.)	(12 in.)(18 in.)	Cost = ($22,200)(B/A)$^{0.4603}$	$28,700
12 in. x 18 in. Denver Duplex Jig	No Vendor Quote	1	$0	(24 in.)(36 in.)	(12 in.)(18 in.)	Cost = ($31,700)(B/A)$^{0.4589}$ (Duplex = 1.44*Simplex)	$59,900
2 ft x 3 ft Dewatering Screens	M&E Quote	2	$5000	(3 ft)(6 ft) (3 ft)(8 ft)	(2 ft)(3 ft) (2 ft)(3 ft)	Cost = ($5000)(B/A)$^{0.4589}$ Cost = ($5000)(B/A)$^{0.4589}$	$7800 $8800
Denver Simplex Spiral Classifier	Pumps +	1	$80,900	Spiral D (60 in.)	Spiral D (48 in.)	Cost = ($80,900)(B/A)$^{1.986}$	$126,000
48 in. spiral, full flare, SP, 7.5 Hp Reichert Spirals Model MD LG7	M&E Quote	4	$10,000			None	$10,000
Derrick Sieve-bend Screen Model IS48-30	Derrick Equipment Corp.	1	$5100			None	$5100
2 ft x 3 ft Humates Screen	M&E Quote	1	$5000	(3 ft)(6 ft)	(2 ft)(3 ft)	Cost = ($5000)(B/A)$^{0.4069}$	$7800
Lamella LTS Model 100 Clarifier	Pumps +	1	$134,400	Settling Area (80 ft²)	Settling Area (80 ft²)	Cost = ($134,400)(B/A)$^{0.71}$	$134,000

				Bowl DxLn (14 in.)(49 in.)	Bowl DxLn (14 in.)(49 in.)		
Sharples P3400 Centrifuge With Centrifuge Stand and Pump	Derrick Equipment Corp.	2	$155,000			Cost = ($155,000)(B/A)$^{0.4589}$	$155,000
Polymer Injection System	M&E Quote	1	$5500			None	$5500
(P1) Goulds Water Pump (1) Model 3656, Size 2.5 x 3 - 8 835 L/min	Eagle Pump & Equipment	1	$2200	690 L/min	835 L/min	Cost = ($2200)(B/A)$^{0.5478}$	$2000
(P2) Goulds Water Pump (2) Model 3656, Size 1.5 x 3 - 10 340 L/min	Eagle Pump & Equipment	1	$1800	510 L/min	340 L/min	Cost = ($1800)(B/A)$^{0.5478}$	$2200
(P3) Goulds Slurry Pump (4) Model 5800, Size 2 x 2 - 8 680 L/min	Eagle Pump & Equipment	1	$7300	560 L/min	680 L/min	Cost = ($7300)(B/A)$^{0.277}$	$6900
(P4) Goulds Water Pump (2) Model 3656, Size 1.5 x 3 - 10 340 L/min	Eagle Pump & Equipment	1	$1800	35 L/min	340 L/min	Cost = ($1800)(B/A)$^{0.5478}$	$500
(P5) Goulds Slurry Pump (4) Model 5800, Size 2 x 2 - 8 680 L/min	Eagle Pump & Equipment	1	$7300	575 L/min	680 L/min	Cost = ($7300)(B/A)$^{0.277}$	$7000
(P6) Goulds Slurry Pump (5) Model 5800, Size 1.5 x 2 - 8 325 L/min	Eagle Pump & Equipment	1	$6000	60 L/min	325 L/min	Cost = ($6000)(B/A)$^{0.277}$	$3800
(P7) Goulds Slurry Pump (5) Model 5800, Size 1.5 x 2 - 8 325 L/min	Eagle Pump & Equipment	1	$6000	90 L/min	325 L/min	Cost = ($6000)(B/A)$^{0.277}$	$4200
(P8) Goulds Slurry Pump (5) Model 5800, Size 1.5 x 2 - 8 325 L/min	Eagle Pump & Equipment	1	$6000	45 L/min	325 L/min	Cost = ($6000)(B/A)$^{0.277}$	$3500
(P9) Goulds Slurry Pump (5) Model 5800, Size 1.5 x 2 - 8 325 L/min	Eagle Pump & Equipment	1	$6000	80 L/min	325 L/min	Cost = ($6000)(B/A)$^{0.277}$	$4100
(P10) Goulds Slurry Pump (4) Model 5800, Size 2 x 2 - 8 680 L/min	Eagle Pump & Equipment	1	$7300	615 L/min	680 L/min	Cost = ($7300)(B/A)$^{0.277}$	$7100
(P11) AllWeiler Progressive Cavity Pump Model AE1E 550 - ID/131 3.8 kW, 75 - 375 L/min 2 - 440 rpm	Shanley Pump & Equipment	1	$5600	975 L/min	305 L/min	Cost = ($5600)(B/A)$^{0.277}$	$7700

| (P12) Goulds Water Pump (2) Model 3656, Size 1.5 x 3 - 10 340 L/min | Eagle Pump & Equipment | 1 | $1800 | 235 L/min | 340 L/min | Cost = ($1800)(B/A)$^{0.5478}$ | $1100 |

Item Description	Vendor	Number	Cost Each
GOULDS WATER PUMP QUOTES			
(1) Goulds End Suction Pump Model 3656 Size 2.5 x 3 - 8, 835 L/min 690 kPa head, 18.5 kW, 3500 rpm	Eagle Pump & Equipment	1	$2200
(2) Goulds End Suction Pump Model 3656 Size 1.5 x 2 - 10, 340 L/min 209 kPa head, 3.73 kW, 1750 rpm	Eagle Pump & Equipment	1	$1800
(3) Goulds End Suction Pump Model 3656 Size 2.5 x 3 - 10, 850 L/min 209 kPa head, 5.60 kW, 1750 rpm	Eagle Pump & Equipment	1	$2100
GOULDS SLURRY PUMP QUOTES			
(4) Goulds Rubber-Lined Slurry Pump Model 5800 Size 2 x 2 - 8, 680 L/min 209 kPa head, 14.9 kW, 1800 rpm	Eagle Pump & Equipment	1	$7300
(5) Goulds Rubber-Lined Slurry Pump Model 5800 Size 1.5 x 2 - 8, 325 L/min 209 kPa head, 11.2 kW, 1800 rpm	Eagle Pump & Equipment	1	$6000

Notes: These cost equations related purchased equipment cost to equipment size and capacity.

1. The "Design Cost Formula" used exponents from cost equations found in Mular (1982) and Ulrich (1984).

References

7.1 MONOGRAPH REFERENCES

Burkin, A. R. 1966. *The Chemistry of Hydrometallurgical Processes*, D. Van Nostrand, Princeton, New Jersey.

Butler, J. N. 1964. *Ionic Equilibria: A Mathematical Approach*, Addison Wesley, Reading, Massachusetts.

Carpco. 1992a. *Three-inch Mozley Hydrocyclone Performance Brochure*, Carpco, Jacksonville, Florida.

Carpco. 1992b. *Two-inch Mozley Hydrocyclone Performance Brochure*, Carpco, Jacksonville, Florida.

Caterpillar. Undated. "Caterpillar Performance Handbook," Edition 26, Caterpillar, Inc.

Chemineer. 1985. "Liquid Agitation," Preprint, *Chemical Engineering*, McGraw-Hill, New York.

Dataquest. 19AED Greenbook: Rental Rates & Specifications for Construction Equipment, 1997," Dataquest, Inc., San Jose, California.

Denver. 1954. Denver Equipment Handbook, Denver Equipment Company (Svedala Pump & Process), Colorado Springs, Colorado.

Evans, G. E. 1997. "Estimating Innovative Technology Costs for the SITE Program," *Journal of Air Waste Management Association*, Volume 40, No. 7.

Federal Remediation Technologies Roundtable (FRTR). 1995. "Guide to Documenting Cost and Performance for Remedial Technologies," prepared by member agencies of the FRTR, U.S. Environmental Protection Agency (USEPA), Technology Innovation Office (TIO), Washington, D.C.

Freiser, H. and Fernando, Q. 1963. *Ionic Equilibria in Analytical Chemistry*, John Wiley & Sons, New York.

Gates, L. E., Morton, J. R., and Fondy, P. L. 1976. "Selecting Agitator Systems to Suspend Solids in Liquids," *Chemical Engineering*, Liquid Agitation, CE Refresher, May 24, 1976, McGraw-Hill, New York.

Green, D. W., Perry, R. H., and Maloney, J. O. 1984. *Perry's Chemical Engineering Handbook*, Sixth Edition, McGraw-Hill, New York.

Kunin, R. 1982. "Special Index and Literature Review," No. 172, Winter, 1982–83, Rohm & Haas Company, Fluid Process Chemicals Department, Philadelphia, Pennsylvania.

Litz, J. E. 1996. *Metcalf & Eddy, Inc. Firing Range Treatability Studies*, Project No. 96017, November 12, 1996, Colorado Minerals Research Institute, Golden, Colorado.

Means. 1997. "R. S. Means Heavy Construction Cost Data, 1997," 11th Edition, Construction Publishers and Consultants, Kingston, Massachusetts.

Metcalf & Eddy, Inc. (M & E). 1998. Remediation of Lead Contaminated Soils at Small-arms Firing Ranges, Using Mining and Beneficiation Technologies, AATDF Technology Evaluation Report, prepared for DOD/AATDF at Rice University, Houston, Texas.

Mular, A. L. 1982. *Mining and Mineral Processing Equipment Costs and Preliminary Capital Cost Estimation*, Special Volume 25, The Canadian Institute of Mining and Metallurgy, Montreal, Quebec.

NEESA. 1984. Initial Assessment Study of Naval Air Station Miramar, San Diego, California, Naval Energy & Environmental Support Activity (NEESA) 13-058, September 1984.

Portec. 1992. *Facts and Figures*, Portec Construction Equipment Division, Pioneer Products, Yankton, South Dakota.

Sillen, L. G. and Martell, A. E. 1964. *Stability Constants of Metal-ion Ligands*, Section 1: Inorganic Ligands and Section 2: Organic Ligands, Publication No. 17, The Chemical Society, London.

Stumm, W. and Morgan, J. J. 1970. *Aquatic Chemistry: An Introduction Emphasizing Chemical Equilibria in Natural Waters*, Wiley-Interscience, John Wiley & Sons, New York.

Svedala. 1996. *Basic – Selection Guide for Process Equipment*, Edition 4, Svedala Pump & Process (Denver Equipment Company), Colorado Springs, Colorado.

Ulrich, G. D. 1984. *A Guide to Chemical Engineering Process Design and Economics*, John Wiley & Sons, New York.

Warminsky, M.F. and Shekher, C. Unpublished. "Soil Washing - A Viable Cleanup Approach for Contaminated Firing Ranges."

Weiss, N.L. 1985. *SME Mineral Processing Handbook*, American Institute of Mining, Metallurgical, and Petroleum Engineers, New York.

7.2 RELATED REFERENCES

U.S. Environmental Protection Agency (USEPA). 1993. "Biogenesis Soil Washing Technology," Innovative Technology Evaluation Report, EPA/540/R-93/510, September 1993.

U.S. Environmental Protection Agency (USEPA). 1995a. "BESCORP Soil Washing System for Lead Battery Site Treatment," Applications Analysis Report, EPA/540/AR- 93/503, January 1995.

U.S. Environmental Protection Agency (USEPA). 1995b. "Bergman USA Soil Sediment Washing Technology," Applications Analysis Report, EPA/540/AR-92/075, September 1995.

Index

Printed and bound by CPI Group (UK) Ltd, Croydon, CR0 4YY

23/10/2024

01778249-0014